THE ELEMENTS

	1B	2B	3A	4A	5A	6A	7A	8A
							1 **H** 1.0079	2 **He** 4.00260
			5 **B** 10.81	6 **C** 12.011	7 **N** 14.0067	8 **O** 15.9994	9 **F** 18.99840	10 **Ne** 20.179
			13 **Al** 26.98154	14 **Si** 28.086	15 **P** 30.97376	16 **S** 32.06	17 **Cl** 35.453	18 **Ar** 39.948
28 **Ni** 58.70	29 **Cu** 63.546	30 **Zn** 65.38	31 **Ga** 69.72	32 **Ge** 72.59	33 **As** 74.9216	34 **Se** 78.96	35 **Br** 79.904	36 **Kr** 83.80
46 **Pd** 106.4	47 **Ag** 107.868	48 **Cd** 112.40	49 **In** 114.82	50 **Sn** 118.69	51 **Sb** 121.75	52 **Te** 127.60	53 **I** 126.9045	54 **Xe** 131.30
78 **Pt** 195.09	79 **Au** 196.9665	80 **Hg** 200.59	81 **Tl** 204.37	82 **Pb** 207.2	83 **Bi** 208.9804	84 **Po** (210)	85 **At** (210)	86 **Rn** (222)

63 **Eu** 151.96	64 **Gd** 157.25	65 **Tb** 158.9254	66 **Dy** 162.50	67 **Ho** 164.9304	68 **Er** 167.26	69 **Tm** 168.9342	70 **Yb** 173.04	71 **Lu** 174.97	2 8 18 32 9 2

95 **Am** (243)	96 **Cm** (247)	97 **Bk** (247)	98 **Cf** (251)	99 **Es** (254)	100 **Fm** (257)	101 **Md** (258)	102 **No** (255)	103 **Lr** (256)	2 8 18 32 32 9 2

Second Edition

CHEMISTRY
An Introduction

STANLEY M. CHERIM
Delaware County Community College
Media, Pennsylvania

LEO E. KALLAN
El Camino College
Torrance, California

 SAUNDERS GOLDEN SUNBURST SERIES

SAUNDERS COLLEGE

Philadelphia

Saunders College
West Washington Square
Philadelphia, PA. 19105

Front cover photograph is Tower Falls and Creek in Yellowstone National Park, by David Sumner.

CHEMISTRY: An Introduction   ISBN 0–03–056762–9

0 1 2 3   090   9 8 7 6 5 4 3

PREFACE

Students who aspire to become chemists, physicists, biologists, engineers, physicians, or professionals in the health sciences are required to take comprehensive and rigorous chemistry courses. The first-year general chemistry course in colleges and universities presumes a certain amount of advance preparation, including recent exposure to a high-level high school course in chemistry. Qualified students have an understanding of the fundamentals of writing formulas, balancing equations, using the metric system, and solving simple stoichiometry problems.

But what about the capable and highly motivated student who has not taken high school chemistry? Or who took it years ago? Or who found his previous chemistry class a dismal encounter? We wrote *Chemistry: An Introduction* for these students.

The format and content are designed to encourage a mastery of basic skills. Difficult concepts are described in simplified terms. Emphasis is placed on systematic methods of interpreting problems, organizing relevant data, and understanding the importance of dimensional analysis. The use of scientific notation and the handling of dimensional units are stressed in numerous examples and practice exercises.

We try to explain in clear and readable language such relatively difficult concepts as atomic structure, chemical bonding, acids and bases, and chemical equilibrium. We develop specific skills in dealing with chemical formulas, nomenclature, gases, types of reactions, solutions, and the techniques for balancing equations.

A rationale for the introduction of topics is included deliberately. Rather than imply, ''Learn this because it is a traditional part of a chemistry course,'' we suggest valid reasons why it is to the student's advantage to study a topic.

Many devices are used to convince the student that the mole concept is indispensable in chemistry. We also explain that a minimum amount of physics is appropriate in a chemistry text, since many theories and concepts are fundamental to both disciplines.

Too often do otherwise educated people shudder at the mention of a chemistry course. We hope to show our readers that they need not feel intimidated by this exciting and dynamic field. The relevance of chemistry to modern concerns such as ecology, energy sources, pollution, and drugs elevates this subject above a mere intellectual exercise. We believe that the mastery of problem solving techniques, the understanding of basic physical concepts, and an appreciation for the rich historical aspects of chemistry can generate a desire to inquire more deeply into our physical world. And the student who acquires these abilities will close the book at the end of the course with much satisfaction and a feeling of personal accomplishment.

Some special features should aid the student. *Chemistry: An Introduction* employs **boldface** type to call attention to technical terms, and *italics* are used for emphasis. Numerous sample problems are solved in annotated steps. Each chapter is prefaced with a list of learning objectives. The answers to numerical problems are provided in the Appendix.

We are indebted to a number of outstanding educators for their reviews of the text and for their many helpful suggestions, constructive criticism, and encouragement during the writing of the manuscript. Our gratitude specifically is extended to Donald G. Slavin of Philadelphia Community College, Sheldon I. Clare of the University of Pittsburgh, Alan Cunningham of Monterey Peninsula College, Richard B. Moreau of Alpena Community College, Fred Redmore of Highland Community College, Edwin L. DeYoung of The Loope College, Dorothy M. Goldish of California State University, and Ralph Burns of East Central Junior College.

We were fortunate to have been able to work closely with Joan Garbutt, Kay Dowgun, Suzanne Rommel, Lloyd Black, John Hackmaster, George Laurie, Amy Shapiro, and other staff members of W. B. Saunders Co.

We thank Cheryl Elliot for her excellent job of typing the manuscript. Finally, a note of thanks and love to our respective families, whose encouragement and patience were so important.

Stanley M. Cherim
Leo E. Kallan

PREFACE

TO THE STUDENT

It seems to us that an important question a student should try to answer as he or she begins the study of chemistry is, "Why bother?" There must be some convincing rationale to make the sweat and toil of studying chemistry a worthwhile and enjoyable experience rather than an anticipation of dread. No one should cling to the illusion that a rigorous chemistry course is easy. We all know that insistence on quality demands a high price. What, then, can we offer to justify the effort?

Besides the obvious and rather prosaic reason that chemistry is a requirement in the curriculum or, perhaps, an essential for admission to a professional or graduate school, we believe that the study of chemistry can improve our scientific literacy and gratify our intellectual curiosity about the nature of the physical world.

Getting the Most out of the Course

Your successful mastery of the content of a chemistry course will probably depend on more than sheer talent and desire. If those sparks of interest, curiosity, or even enthusiasm are to be fanned into a fire, a disciplined approach to your study is essential. To this end we have some suggestions regarding study habits for chemistry:

1. Study the learning objectives that precede each chapter. These will tell you something of what you should be able to understand, define, and do when you have mastered the content. We believe that it is helpful to know what is expected of you before you begin.

2. Don't hesitate to make marginal notes and underline those sections where you have questions to be raised or where your instructor has clearly indicated emphasis.

3. Keep complete and legible notes during the class lecture. It is very likely that your instructor will supplement the text with useful analogies, additional information, and sample problems.

4. Do the exercises that appear throughout the chapters and the additional questions and problems at the end of the chapters *even if they aren't assigned as homework*. There is often a large gap to be bridged between the understanding of how a problem is solved systematically and the actual solving of problems with pencil in hand. You will find the answers to problems in Appendix I so that you can keep a constant check on your progress.

5. Get into the habit of studying with paper and pencil. First read and then list key terms, outlining briefly those concepts having a sequential character, and attacking mathematical problems systematically.

6. Now let's deal with the mathematical problems—especially for those among you who feel overwhelmed in your lifelong battle with numerical problems. We suggest that the observance of the following steps may inspire you to confront "word problems" with a renewed determination to establish just who is the boss. The steps are as follows:

(a) Read the entire problem and determine exactly what it is that you are looking for.

(b) Organize the data in tabular form. Make sure that the units, or dimensions, are compatible.

(c) When the problem is of the type that permits you to visualize actual quantities, to make simple sketches, or to construct mental models—do it! You can justify almost any device that helps you to move from the abstract to the tangible.

(d) Most importantly, use dimensional analyses. Make sure that all units factor-out except for those required in the answer. Rarely will the answer to a problem be a unitless number. For example, the number 2.34 can mean anything or nothing, but 2.34 *grams* very clearly states a definite mass.

(e) Check your calculations in an effort to catch careless mistakes in computation. Estimate your answers roughly as a preliminary check before you use a slide rule or calculator. Ask yourself whether the answer seems reasonable. If it seems good, allow yourself a few moments' warm glow of satisfaction or a thoughtfully modulated shout of triumph.

7. Make it a habit to come to class prepared to ask questions. Studying the homework assignment will help you to understand the lecture and to get the most out of it.

Stanley M. Cherim
Leo E. Kallan

CONTENTS

CONTENTS

CHEMISTRY
An Introduction

CHAPTER ONE

LEARNING OBJECTIVES

At the completion of this chapter, you should be able to:

1. Distinguish among the metric units of length, volume, and mass.
2. Explain the difference between mass and weight.
3. List and define the metric system prefixes.
4. Convert measurement units within the metric system.
5. Apply dimensional analysis in problem solving.
6. Choose appropriate conversion factors from available tables.
7. Develop conversion factors when data are available.
8. Write numbers in the form of scientific notation.
9. List four types of volumetric glassware.
10. Distinguish between temperature and heat.
11. Convert temperatures among Fahrenheit, Celsius, and Kelvin scales.
12. Define the terms *density* and *specific gravity*.
13. Calculate densities, volumes, and masses from measurement data.

MEASUREMENT

When the great English scientist James Clerk Maxwell was a small child he began to demonstrate the powerful driving force of human curiosity. For Maxwell, life in its most meaningful and exuberant sense meant asking questions about the mysteries of our environment. As young Maxwell examined a mechanical device and asked his father, "What is the 'go' of that," he was raising the kind of question that characterizes the scientific mentality regardless of age. What is it made of? How does it work? What is it used for?

Maxwell typified the scientific approach by going first to the wellspring of recorded human experience. In other words, by learning something about the laws of nature as they were understood in the nineteenth century, he was better able to distinguish between the physical principles that might provide a sound basis for progress and those that needed to be questioned. It takes a trained mind to see clearly and objectively the cause-and-effect relationships present in natural phenomena, and it requires that peculiarly scientific mind to evaluate subjectively the significance of the observations. To make an orderly fabric (scientific discipline) out of a chaotic tangle of threads (isolated facts) can be a beautiful and creative human endeavor.

We should be glad that the aesthetic aspects of science persist. It is true that we have moved through the centuries from the art and magic of the alchemists to the very different objectives and methods of modern chemistry, but there is a debt to be acknowledged. Despite the fact that the energies of our ancestors were fruitlessly devoted to the creation of the "philosopher's stone" with which common metals might be changed into gold and to the discovery of the "elixir of life" that would assure eternal youth, science had its raw beginnings in their arcane laboratories.

When the age of charlatans passed and with them their magical incantations of occult symbols, some light began to creep into the darkness of the Dark Ages. Soon men dared to question "sacred" authority, and they suffered for it. (Was something really true because Aristotle said so?) Nevertheless, it was during these ages of antiscience that glassblowers first produced retorts, beakers, funnels, and crude volumetric glassware. The potters fashioned earthenware crocks, crucibles, mortars, pestles, and ovens. And the blacksmiths gave us support stands, forceps, tongs, and spatulas. And so we moved from authoritarianism and the black arts to the modern arts, and from there to science—from magic and absolutes to tentative assumptions and systematic experimental testing. This is the scientific method. To paraphrase an old anonymous saying: in science we have moved from cocksure ignorance to thoughtful uncertainty.

Unfortunately, science has not always been a pure and beautiful monument to man's nobler instincts. Chemistry, the science concerned with the structure and behavior of matter, has provided more than an adventure in living. It has been and still is the basic tool for death and destruction in war. Problems of environmental pollution provide us with unique and critical challenges and teach us about moral responsibility. The poisoning of our waters and air bring the relationship between science and life into a new and very special focus. The young Maxwells of our age are likely to ask, "What is the 'nice' of it?" in addition to the "go" of it. The old and rather meaningless adage, "science for science's sake" is hopefully giving way to "science for mankind's sake."

1.2 THE AREAS OF CHEMISTRY

The broad physical science of chemistry is commonly subdivided into specialty areas. **Inorganic** chemistry deals with all the known elements and combinations of elements in terms of their structures and changes. **Organic** chemistry is specially concerned with compounds of carbon, which are most familiar to us in forms of living things and in synthetic fabrics and plastics. **Analytical** chemistry is involved with finding out what a substance is composed of and how much is there. **Physical** chemistry is most directly concerned with the complexity and mechanism of chemical change and its relationship to energy. **Biochemistry** is tied in with the nature of the chemical changes that sustain life. **Nuclear** chemistry, which deals with the unique changes related to atomic nuclei, has special relevance to current ecological problems. In another context, nuclear chemistry is one aspect of a larger study of the topics of pollution, drugs, and energy conservation—all of which are placed under the general heading of **environmental** chemistry. It is difficult to find any aspect of life in which chemistry does not predominate. Think of ways in which the beauty of chemistry is apparent in economics, philosophy, politics, theology, business, and art. Furthermore, and very importantly, we shall see how chemistry serves as an interface between biology and physics.

1.3 MEASUREMENT

Any serious attempt to learn about the forces and energy effects that relate to the structure and behavior of matter requires an understanding of the concept of measurement. How else can the observations

FUNKY WINKERBEAN   Tom Batiuk

CHEMISTRY 1 -

In Chemistry 1, you'll learn how to use the periodic table, how to combine elements, and how to copy experiment results from someone else's lab book. Chemistry is the foundation of the hard sciences, and when you flunk your first quiz, you'll realize how hard it really is.

FUNKY WINKERBEAN by Tom Batiuk.
© Field Enterprises, Inc., 1978.
Courtesy of Field Newspaper Syndicate.

of "how much," "how many," "how far," and "how fast" be described? Measurement is a fundamental effort needed to increase the orderly expansion of our knowledge. Dimensional units must be used if sense is to be made out of linear distance, mass, time, and temperature. And an understanding of how these basic units are interrelated is necessary so that the more sophisticated phenomena of energy, density, force, and volume can be understood. The international standards for measurement today constitute what is known as the **metric system.** Throughout the book the fact that all measurements include both a *number* and a *dimension* will become obvious.

Now let us consider how a scientist handles numbers conveniently (scientific notation), sensibly (significant figures), and with an ability to discriminate between correctness and reproducibility (accuracy and precision). Then we can see how the dimensions (unit labels) are applied to the numbers in such a way that we can carefully reason out and check the problem solving process (dimensional analysis).

1.4 SCIENTIFIC NOTATION

Scientific notation is a method of increasing efficiency in calculations by expressing cumbersome, many-digit numbers in a compact form. It consists of writing with exponents. The exponent indicates the number of times a value is multiplied by itself. Some examples are as follows:

$$2^2 = 2 \cdot 2 = 4$$

$$2^3 = 2 \cdot 2 \cdot 2 = 8$$

$$2^{-2} = \frac{1}{2^2} = \frac{1}{2 \cdot 2} = \frac{1}{4}$$

$$2^{-3} = \frac{1}{2^3} = \frac{1}{2 \cdot 2 \cdot 2} = \frac{1}{8}$$

$$N^4 = N \cdot N \cdot N \cdot N$$

$$N^{-4} = \frac{1}{N^4} = \frac{1}{N \cdot N \cdot N \cdot N}$$

The last example provides a useful focal point, because the base number 10 is the heart of the international system (metric system) of weights and measures. Note the relationship between the exponent and the number of digits (zeros in this case):

$$10^{③} = 1000 \qquad \text{3 digits}$$

The following examples support the observation that the exponent of the base 10 is exactly the same as the number of places that the decimal point is moved in the coefficient.

$$1.0 \times 10^{⓪} = 1$$

$$1.0 \times 10^{①} = 10$$

$$1.0 \times 10^{②} = 100$$

$$2.0 \times 10^{⑤} = 200,000$$

$$1.0 \times 10^{⑥} = 1,000,000$$

$$2.5 \times 10^{③} = 2500$$

This direct relationship between the exponent and the number of digits in the coefficient greatly simplifies the process of writing a large number in exponential form:

$$100,000 = 10^5 \text{ or } 1.0 \times 10^5$$

$$3220 = 3.22 \times 10^3$$

$$96,500 = 9.65 \times 10^4$$

A single exponential value of the base 10 is known also as an **order of magnitude** (a term that comes from astronomy). For example, 10^6 is four orders of magnitude larger than 10^2. When a chemist desires to decrease a quantity by two orders of magnitude, he would use $1/100$ of the original amount. This last point raises the question of how some fraction, or any number less than *one*, is written in the form of scientific notation. The method is to shift the decimal point in such a way as to express the number as a value between *one and ten*, and indicate this operation by use of a *negative* exponent. For example

$$0.1 = 1 \times 10^{-1}$$

$$0.003 = 3 \times 10^{-3}$$

$$0.000251 = 2.51 \times 10^{-4}$$

Although the last example could be written correctly,

$$0.000251 = 25.1 \times 10^{-5}$$

or

$$0.000251 = 251 \times 10^{-6}$$

the conventional method is to shift the decimal far enough to make the number between one and ten, multiplied by the appropriate exponential value (i.e., so that there is only one digit to the *left* of the decimal point).

A useful rule that can be derived from the previous examples is that *the process of shifting a decimal point corresponds to a change in the exponent*. Every digit included in a decimal shift is equivalent to a change of one order of magnitude (or one decimal place). Some examples are as follows:

$$100 \times 10^2 = 10 \times 10^3 = 1 \times 10^4 = 0.1 \times 10^5 = 0.01 \times 10^6, \text{etc.}$$

$$230 \times 10^2 = 23.0 \times 10^3 = 2.30 \times 10^4$$

$$0.05 \times 10^{-3} = 0.5 \times 10^{-4} = 5 \times 10^{-5}$$

$$2170 \times 10^{-2} = 2.17 \times 10^1 = 0.217 \times 10^2$$

Exercise 1.1

1. Use scientific notation to express the following numerical values as numbers between one and ten, times the correct power of 10. For example, $7283 = 7.283 \times 10^3$.
   (a) 762
   (b) 538,000
   (c) 42,600,000
   (d) 0.038
   (e) 0.0000112
2. Increase the following values by two orders of magnitude:
   (a) 3.35×10^3
   (b) 7.6×10^5
   (c) 2.8×10^{-7}
   (d) 2.8×10^{-2}
3. Decrease the following values by four orders of magnitude:
   (a) 6.2×10^4
   (b) 7.18×10^{11}
   (c) 3.08×10^{-3}
   (d) 9.2×10^{-8}

Changing the Sign of an Exponent

It is often convenient in the process of performing a calculation to change the sign of an exponent. This may be done by writing the **reciprocal** value of a number. The reciprocal is really the result of inverting the numbers in a fraction. For example

$$\frac{100}{1} = \frac{1}{\frac{1}{100}} = \frac{1}{0.01}$$

In other words

$$\frac{10^2}{1} = \frac{1}{10^{-2}}$$

Other examples are

(a) $\dfrac{1}{10^{-4}} = 10^4$

(b) $2 \times 10^{-3} = \dfrac{2}{10^3}$

(c) $\dfrac{6}{2 \times 10^{-2}} = \dfrac{6 \times 10^2}{2} = 3 \times 10^2 = 300$

Rules for Addition and Subtraction of Exponents

In a list of numbers having different exponents, the coefficients must be adjusted so that their *exponents are identical.* For example if we wish to add the following numbers,

$$\begin{aligned}
& 1.25 \times 10^4 \\
+ & 30.60 \times 10^3 \\
+ & 602.0 \times 10^2
\end{aligned}$$

we must change their appearance so that they are all of the same *order of magnitude* (i.e., the same power of 10).

$$\begin{aligned}
& 1.25 \times 10^4 \\
+ & 3.06 \times 10^4 \\
+ & 6.02 \times 10^4 \\
\hline
& 10.33 \times 10^4, \text{ or } 1.033 \times 10^5
\end{aligned}$$

Notice that the exponential values are *not* added. The "$\times 10^4$" is a unit, a tag, a designation, like apples or grams or meters. The error of adding the exponents in an addition operation may be illustrated by the following example:

$$1 \times 10^3 = 1000 \qquad\qquad 1 \times 10^3 = 1000$$
$$+2 \times 10^3 = 2000 \qquad +2 \times 10^3 = 2000$$
$$\underline{+3 \times 10^3 = 3000} \qquad \underline{+3 \times 10^3 = 3000}$$
$$6 \times \mathbf{10^3} = 6000 \qquad 6 \times \mathbf{10^9} = 6{,}000{,}000{,}000$$

$$\textbf{correct} \qquad\qquad\qquad \textbf{wrong}$$

If the exponents were added, the answer would be 6×10^9, or 6,000,-000,000. This is obviously a gross error.

The same rule applies to subtraction. For example

$$2.76 \times 10^{-5} = \quad 2.76 \times 10^{-5}$$
$$\underline{-4.3 \times 10^{-6}} = \underline{-0.43 \times 10^{-5}}$$
$$2.33 \times 10^{-5}$$

Exercise 1.2

1. Write the reciprocals of the following:

2. (a) x^{-a} $\quad x^A$
   $\quad 10^7$

   (b) $\dfrac{1}{10^{-7}}$

   (c) $3^{-1/2}$ $\quad 3^{1/2}$
   $\quad 2^{-5}$

   (d) 2^5

2. Add the following terms:
   $(36.2 \times 10^4) + (1.12 \times 10^3) + (5.22 \times 10^5)$
   36.2×10^4
   1.12×10^4
   52.2×10^4
   88.5×10^4

3. Calculate the difference between
   581.3×10^4 and 4.14×10^6
   5.813×10^6
   $\underline{4.14 \times 10^6}$
   1.67×10^6

Rules for the Multiplication and Division of Exponents

The rules governing the multiplication and division of exponential values may be deduced from examples. Consider $10^2 \times 10^4$, substituting a parenthesis for the "times" sign ($\times$):

$$(10^2)(10^4) = (10 \cdot 10)(10 \cdot 10 \cdot 10 \cdot 10)$$

Observe that the multiplication in its expanded form shows the number 10 multiplying itself six times. Therefore, the answer would be

$$(10^2)(10^4) = 1 \times 10^6$$

This amounts to *adding* the exponents rather than multiplying them. If it were assumed that $(10^2)(10^4) = 10^8$, the answer would be incorrect. The derived rule is as follows: *When multiplying exponential values, add the exponents.* It must be understood that any coefficients of the exponential values are handled traditionally, and the exponential values themselves must have the *same* base, such as 10. Some examples are

$$(10^2)(10^5) = 1 \times 10^7$$
$$(10^3)(10^4)(10^{-2}) = 1 \times 10^5$$
$$(2 \times 10^3)(3 \times 10^7) = 6 \times 10^{10}$$

Consider the example 10^5 *divided* by 10^2:

$$\frac{1 \times 10^5}{1 \times 10^2} = \frac{10 \cdot 10 \cdot 10 \cdot 10 \cdot 10}{10 \cdot 10} = 1 \times 10^3$$

In this case the exponents are *subtracted.* The rule is as follows: *When dividing exponential values, subtract the exponents.* The coefficients, once again, are handled normally. Some examples are

$$\frac{1 \times 10^7}{1 \times 10^5} = 1 \times 10^2$$

$$\frac{10^3}{10^8} = 10^{-5}$$

$$\frac{8 \times 10^5}{2 \times 10^2} = 4 \times 10^3$$

An alternative method is to use the reciprocals, thereby converting division operations to multiplication. For example

$$\frac{10^7}{10^5} = \frac{(10^7)(10^{-5})}{1} = 10^2$$

$$\frac{6 \times 10^3}{4 \times 10^8} = \frac{(1.5 \times 10^3)(10^{-8})}{1} = 1.5 \times 10^{-5}$$

Rules for Raising an Exponent to a Power or Finding a Root of an Exponent

Analysis of the expression $(2^3)^2$ suggests a simplified form:

$$(2^3)(2^3) = (2 \cdot 2 \cdot 2)(2 \cdot 2 \cdot 2) = 2^6$$

The rule that may be quickly extracted from this observation is as follows: *When raising an exponent to a power, multiply the exponents.* Consider the following examples:

$$(a^4)^3 = a^4 \cdot a^4 \cdot a^4 = a^{12}$$

$$(2 \times 10^4)^2 = 4 \times 10^8$$

$$(180)^2 = (1.8 \times 10^2)^2 = 3.2 \times 10^4$$

$$(2 \times 10^{-3})^2 = 4 \times 10^{-6}$$

$$(3 \times 10^{-2})^{-3} = 3^{-3} \times 10^6 = \frac{1 \times 10^6}{3^3} = \frac{1 \times 10^6}{27}$$

$$= 0.037 \times 10^6 = 3.7 \times 10^4$$

Examine the figure $(90{,}000)^{1/2}$. This value may be expressed as $(9 \times 10^4)^{1/2}$. The fact to be remembered is that any value to the $1/2$ power may be described as the *square root* of that value. Therefore, $(9 \times 10^4)^{1/2} = \sqrt{9 \times 10^4}$. However, if $(10^4)^{1/2}$ is considered by itself, it can be observed that the previous rule indicates an answer of 10^2, since that is the result of the multiplication of the exponents. The rule deduced from this example is *when taking the square root of an exponential value, multiply the exponent by one-half.* This rule may be extrapolated to cube roots, in which case the exponential value would be multiplied by one-third. The answer to the example is

$$\sqrt{9 \times 10^4} = 3 \times 10^2$$

Some further examples are as follows:

$$\sqrt{8.1 \times 10^5} = \sqrt{81 \times 10^4} = 9 \times 10^2$$

$$\sqrt{640 \times 10^{-7}} = \sqrt{64.0 \times 10^{-6}} = 8 \times 10^{-3}$$

$$(2.5 \times 10^9)^{1/2} = \sqrt{25 \times 10^8} = 5 \times 10^4$$

$$(2.7 \times 10^{-5})^{1/3} = \sqrt[3]{27 \times 10^{-6}} = 3 \times 10^{-2}$$

1.5 SIGNIFICANT FIGURES

Sci. notation first?

The term "significant figures" in a measurement refers to the number of digits that has been obtained with a generally acceptable level of accuracy. The accuracy of the numerical value and the number of digits representing the order of magnitude to which that accuracy extends depend on the *nature of the object being measured, the technical skill of the measurer, and the* **precision** *allowed by the instrument used to perform the measurement.* It should be made clear that *precision* and *accuracy* are related, but they are not synonymous. Precision refers to a degree of *reproducibility* of measurements, whereas accuracy is a matter of exactness as opposed to approximation. The distinction between these two terms is illustrated by the analogy of shooting at a target (Fig. 1–1).

Good accuracy,
poor precision

Good precision,
poor accuracy

Good accuracy,
good precision

Bad news

Figure 1–1   The difference between accuracy and precision.

Because no measurement is exact, the uncertainty of a measurement or calculations involving measurements is communicated by the proper use of significant figures. A measurement of 22.63 cm has four significant figures; the last digit recorded is in doubt. The measurements 1.0675 mm and 0.0025 g have five and two significant figures, respectively. In numbers less than one, the zeros are significant if measured but not significant if used only to indicate the magnitude of a measurement. For example, in 0.005740 mL the two zeros following the decimal point are there to show magnitude and are not significant digits. But the final zero following the 4 was measured and is a significant figure. Thus, 0.005740 mL has four significant figures, and 0.00574 mL has three, as does 0.574. The position of the decimal point has no effect on the number of significant figures. For example notice the number of significant figures in the following measurements:

	number of significant figures
16.3 g	3
16.0 mg	3
1280.05 mL	6
0.0020 L	2
0.00202 km	3
0.02100 m	4

When using significant figures in *multiplication* and *division*, the answer must not contain a greater number of significant figures than that necessary measurement having the least number of significant figures. For example, 12.62 cm $\times$ 1.2 cm is numerically equal to 15.144 cm^2, but only two significant figures may be retained (because 1.2 cm has only two), and the answer is 15 cm^2. Similarly, 1.0056 g divided by 0.0343 cm is 29.317784 g/cm, but only three significant figures should be retained, because 0.0343 cm has three significant figures. The answer is 29.3 g/cm.

2
not addition + subtraction
see p. 14

MEASUREMENT

Exercise 1.3

1. Multiply the following and express the coefficients as one-digit numbers before the decimal point:
   (a) $(10^3)(10^5) =$ $10^8 = 1 \times 10^8$
   (b) $(4 \times 10^2)(6 \times 10^6) =$ $24 \times 10^8 = 2.4 \times 10^9$
   (c) $(2 \times 10^2)(1 \times 10^{-3})(3 \times 10^4) =$ 6×10^3
   (d) $(216.5 \times 10^2)(0.018 \times 10^2) =$ 3.897×10^4

2. Divide the following and express the coefficients as one-digit numbers before a decimal point:

   (a) $\dfrac{10^4}{10^2} =$ 10^2 (c) $\dfrac{15 \times 10^7}{30 \times 10^{11}} =$ $.5 \times 10^{-4} = 5 \times 10^{-5}$

   (b) $\dfrac{1 \times 10^{-3}}{2 \times 10^2} =$ (d) $\dfrac{617 \times 10^3}{0.012 \times 10^8} =$ $51416 \times 10^{-5} =$

   $1/2 \times 10^{-5} = .5 \times 10^{-5} = 5 \times 10^{-6}$ 5.1×10^{-1}

3. Raise the following values to the indicated power:
   (a) $(x^2)^5 =$ x^{10}
   (b) $(3 \times 10^3)^3 =$ $27 \times 10^9 = 2.7 \times 10^{10}$
   (c) $(4 \times 10^2)^{-2} =$ $4^{-2} \times 10^{-4} = 1 \times 10^{-4} = .0625 \times 10^{-4} = 6.25 \times 10^{-6}$
   (d) $(20000)^3$ $(2 \times 10^4)^3$ 8×10^{12} 4^2

4. Perform the indicated operation on the following values:
   (a) $\sqrt{360 \times 10^3} =$ $\sqrt{36.0 \times 10^4} = 6 \times 10^2$
   (b) $\sqrt{0.049 \times 10^{-3}} =$ $\sqrt{49 \times 10^{-6}} = 7 \times 10^{-3}$
   (c) $(3.6 \times 10^5)^{1/2} =$ $\sqrt{3.6 \times 10^5} = \sqrt{36 \times 10^4} = 6 \times 10^2$
   (d) $(80 \times 10^8)^{1/3} =$ $\sqrt[3]{8.0 \times 10^9} = 2 \times 10^3$

Example 1.1

A student calculates the area of a circle having a diameter of 6.30 centimeters.
1. The equation is, area of circle $= \pi r^2$.
2. If the diameter is 6.30 centimeters, the radius is 3.15 centimeters.
3. If π is 3.14

$$\text{area} = 3.14(3.15 \text{ cm})^2$$

4. Using a calculator

$$\text{area} = 31.15665 \text{ cm}^2$$

This is absurd! The accuracy of the answer is limited to *three* significant figures because the information given in the beginning (the 6.30 cm) has a certainty of only three figures. The answer cannot be more accurate than the measurement. The answer should read, area = 31.2 centimeters squared. A common misconception is that the greater the number of digits in the answer, the greater the accuracy. This is false unless the digits are *significant*.

Other common errors appear when values of many significant figures are mysteriously obtained with the use of relatively crude, low-precision instruments. To suggest that a meter stick is an inappropriate instrument to measure the length of a flea would hardly be described as being unreasonably fussy. It is simply a matter of common sense. This common-sense approach to handling numbers and ''rounding off'' is the process of recognizing and using significant figures.

Rounding off numbers is the usual method of avoiding an implication toward accuracy that is unjustified. The convention for rounding off is as follows: *When the final digit (the point of questionable accuracy) is 4 or less, the digit immediately before it remains unchanged, and the last digit is dropped. If the final digit is greater than 5, the digit immediately before it is increased by 1, and the final digit is dropped.* For example:

five-digit number	rounded off to 4 sig. figs.
23.716	23.72
23.713	23.71

If the final digit is 5, the value of the digit immediately before it is increased by one if the digit is odd and left unchanged if it is even. Thus 23.715 rounds off to 23.72, but 23.725 rounds off to 23.72. An application of the rule for rounding off in addition or subtraction is illustrated in Example 1.2, which is based on the following principle: *The accuracy indicated by the answer to an addition or subtraction problem is limited to the least number of decimal places found in any single value.*

Example 1.2

Calculate the average of the following masses:
12.22 g
11.50 g
11.6 g
12.272 g

1. Note the third number. It is 11.6, not 11.60. Therefore, the number could be anywhere between 10.55 and 11.64.
2. Since the number 11.6 has only one decimal place, the answer to the problem is restricted to the same number of decimal places.
3. Round off all the numbers to three digits and calculate the answer:

$$\begin{array}{l}12.2 \text{ g}\\11.5 \text{ g}\\11.6 \text{ g}\\\underline{12.3 \text{ g}}\\47.6 \text{ g}\end{array}\qquad \frac{47.6 \text{ g}}{4.00} = 11.9 \text{ g}$$

average = 11.9 g

Exercise 1.4

1. How many significant figures are contained in each of the following:
   (a) 163.5   (c) 0.0003
   (b) 0.015   (d) 5.17 × 10³
2. Perform the following calculations within the limits of significant figures and decimal places.
   (a) 2.02 × 3.5 =

   (b) $\dfrac{4.15}{2.077}$ =

   (c) (8.1 × 10²)(1.23 × 10³) =
   (d) 4.22 + 0.308 + 65.4 =
   (e) 1.13 × 10² − 0.18 × 10³ =
3. Round off the following values to three significant figures:
   (a) 3.081   (c) 20809
   (b) 0.07635   (d) 14.22

Dimensional analysis is a highly recommended technique for problem solving. By starting with information that is provided, we can proceed toward the solution by a series of steps using fundamental equalities (conversion factors). Dimensional analysis provides a continuous check on the correctness of our work by allowing us to "cancel out" the original dimensions as they are converted to the desired dimensions. The most common method for handling the basic equalities (for example: 1 ft = 12 in, 1 yd = 3 ft, 10 dimes = 1 dollar) is to write them as fractions so that the numerators and denominators are arranged as they *must* be arranged in order to produce the desired answer. Consider the following examples:

1. Express a measured distance of 0.52 ft in inches.

$$0.52 \, \text{ft} \times \frac{12 \text{ in}}{1 \text{ ft}} = \boxed{6.2 \text{ in}}$$

plug in formula

Notice how the fundamental equality, 1 ft = 12 in, is arranged as a fraction so that the "feet" dimension cancels out and the answer appears in inches.

2. Calculate the number of dimes in 0.5 dollar.

$$0.5 \, \text{dollar} \times \frac{10 \text{ dimes}}{1 \text{ dollar}} = \boxed{5 \text{ dimes}}$$

3. Express a 90-mile distance betwen two cities in meters, given the following conversion factors:

$$1 \text{ m} = 39.37 \text{ in}$$
$$1 \text{ ft} = 12 \text{ in}$$
$$1 \text{ mi} = 5280 \text{ ft}$$

$$90 \, \text{mi} \times \frac{5280 \text{ ft}}{1 \text{ mi}} \times \frac{12 \text{ in}}{1 \text{ ft}} \times \frac{1 \text{ m}}{39.37 \text{ in}} = \boxed{1.4 \times 10^5 \text{ m}}$$

Although the dimensions used in chemistry have little to do with inches, feet, dimes, or dollars, the technique of dimensional analysis is applicable to any dimensions—including those of the metric system.

1.7 THE METRIC SYSTEM

The metric system, which is an international system of weights and measures, is the foundation of today's modern measuring system used throughout most of the world in everyday living and by practically all chemists. Regardless of what is being measured, be it length, volume, mass (or weight as a loosely applied synonym), or temperature, the system has the advantage of being directly related to the commonly used base of ten. Conversion among different units amounts to a simple change in the order of magnitude (a factor of ten). Our familiar English system of weights and measures does not enjoy this advantage. One cannot proceed from inches to feet to yards to miles by using a factor or a multiple of ten each time.

A marked increase in efficiency is gained when the metric system is used exclusively. Conversions of metric units to inches, quarts, and pounds is time consuming and unproductive. Such a process is analogous to communication in a foreign language; until it is possible to think in the other tongue instead of mentally constructing an English sentence and translating word for word, conversation is slow and frustrating. Similarly, chemistry students should reach the point at which they think quickly, confidently, and directly in metric units.

It takes time and effort to master the various units and prefixes that describe the "what," "how much," and "how many" of our observations. But the effort is worthwhile. It's helpful to remember that our American monetary system is basically a metric one.

1.8 METRIC UNITS OF LENGTH AND AREA

The basic unit of linear measurement is the meter (m). Think of this as being roughly the length of a large stride (Fig. 1–2).

One hundredth of a meter is a **centimeter (cm).** The prefix *centi-* means a hundredth part of something, just as a cent is one hundredth of a dollar.

$$1 \text{ cm} = 0.01 \text{ m} = \frac{1}{100} \text{ m} = 10^{-2} \text{ m}$$

Figure 1–2   A large stride is a rough approximation of one meter.

1m

Figure 1–3   The width of your little finger helps you visualize a centimeter.

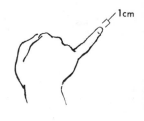

One centimeter is about equal to the diameter of a stick of chalk or the width of the tip of your little finger (Fig. 1–3).

The large practical unit of length is equivalent to 1000 meters and is called a **kilometer (km)**. The prefix *kilo-* indicates a multiple of 1000. For example, a "kiloglop" would mean 1000 "glops." A 60-mile distance between two towns may be visualized as being about 100 km.

Examining smaller units of linear measure brings us to the **millimeter (mm)**. This is one thousandth of a meter, or one tenth of a centimeter. The millimeter is approximately equal to the thickness of a dime (Fig. 1–4). The prefix *milli-* means a thousandth part of the fundamental unit of comparison.

$$1 \text{ mm} = 0.001 \text{ m} = \frac{1}{1000} \text{ m} = 10^{-3} \text{ m}$$

$$1 \text{ mm} = 0.1 \text{ cm} = \frac{1}{10} \text{ cm} = 10^{-1} \text{ cm}$$

The word "millimeter" is familiar because photographic film is commonly described in this way. We take 8 mm home movies or use 35 mm film for color slides.

Further subdivisions are more difficult to visualize. The **micrometer (μm),** in which the "micro" is represented by the Greek letter *mu,* is one thousandth of a millimeter and one millionth of a meter. The dimensions of bacteria are measured in micrometers. The prefix micro- means a millionth part of a basic unit:

$$1 \ \mu\text{m} = 0.000001 \text{ m} = \frac{1}{1,000,000} \text{ m} = 10^{-6} \text{ m}$$

Figure 1–4   The thickness of a dime is about one millimeter.

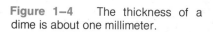

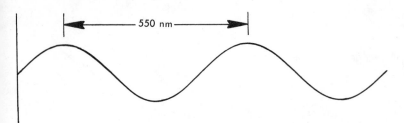

Figure 1–5   A wavelength of green light, 550 nanometers.

Subdividing the micrometer into a thousand parts produces the **nanometer (nm).** This tiny unit of measure is most appropriate for measuring the dimensions of large molecules and lengths of light waves. The wavelength of green light is illustrated in Figure 1–5.

The nanometer will be observed to have special application to the solving of problems in the chapter covering light and the quantum model (Chap. 2).

$$1 \text{ nm} = 0.000000001 \text{ m} = \frac{1}{1,000,000,000} \text{ m} = 10^{-9} \text{ m}$$

The smallest useful unit of linear measure is the angstrom (Å), which is one tenth of a nanometer. Angstroms are used to measure the dimensions of small molecules, individual atoms, and very short wavelengths of light. A typical atom has a diameter of one or two angstroms.

$$1 \text{ Å} = 0.1 \text{ nm} = \frac{1}{10} \text{ nm} = 10^{-1} \text{ nm} = 10^{-8} \text{ cm} = 10^{-10} \text{ m}$$

Three incredibly small units that should be mentioned briefly are the **picometer (pm), femtometer (fm),** and **attometer (am).** These represent 10^{-12} m, 10^{-15} m, and 10^{-18} m, respectively. Although they will not be included in further discussions because their practical application in chemistry is limited, students should be aware of the existence of such units. Before continuing with other metric units let us examine linear measurement and see how to perform conversions.

[handwritten margin notes:]

$1 nm = 10^{-9} m$

$A° = 10^{-10} m$

$pm = 10^{-12} m$

$fm = 10^{-15} m$

$am = 10^{-18} m$

Reprinted by permission of the Chicago Tribune—New York News Syndicate, Inc.

A practically foolproof way of interconverting metric units is found in the use of **dimensional analysis.** A conversion factor is obtained by the following steps:

1. Take a fundamental equivalence such as 1 cm = 0.01 m.

$$\frac{1}{100} = \frac{1}{10^2} = 10^{-2}$$

2. Express the equivalence by scientific notation:

$$1 \text{ cm} = 10^{-2} \text{ m}$$

3. For uniformity and convenience we can restate the equivalence in terms of positive exponents:

$$10^2 \text{ cm} = 1 \text{ m}$$

4. If 100 centimeters equals 1 meter, the relationship can be stated as follows:

$$10^2 \text{ cm } \textbf{per } \text{m}$$

5. The term **per** signifies a fraction:

$$10^2 \text{ cm/m or } \frac{10^2 \text{ cm}}{1 \text{ m}}$$

Example 1.3

How many meters are there in 12,500 centimeters?
1. In scientific notation, 12,500 cm = 1.25×10^4 cm (assuming 3 significant figures). The conversion factor is 1 meter = 10^2 cm. Conversion factors such as this one are exact, by definition.
2. Visualizing the change, it would seem that many centimeters (small units) are equal to few meters (comparatively large units). This kind of reasoning serves as an additional check on your answer.

$$\frac{1.25 \times 10^4 \text{ cm}}{1} \times \frac{1 \text{ m}}{10^2 \text{ cm}} = \boxed{1.25 \times 10^2 \text{ m}}$$

3. This simple calculation shows dimensional analysis in action. The centimeter units cancel out and the remaining unit is meters—exactly the unit required. The final value for the order of magnitude is found according to the rules for division or multiplication of exponents.

$$\frac{10^4}{10^2} = 10^2; \quad \text{or} \quad (10^4)(10^{-2}) = 10^2$$

subtract 2 from 4   *add* 4 and −2

Example 1.4

Convert 5.50×10^{-5} cm to nm.
1. Obtain an equivalence between centimeters and nanometers from Table 1-1:

$$1 \text{ cm} = 10^7 \text{ nm.}$$

2. Using dimensional analysis:

$$\frac{5.50 \times 10^{-5} \text{ cm}}{1} \times \frac{10^7 \text{ nm}}{1 \text{ cm}} = \boxed{5.50 \times 10^2 \text{ nm}}$$

TABLE 1-1 Conversion Factors (Linear Measurement)

Unit	Symbol	Conversion Factor
kilometer	km	$1 \text{ km} = 10^3 \text{ m}$
meter	m	basic unit
centimeter	cm	$1 \text{ m} = 10^2 \text{ cm}$
		$1 \text{ km} = 10^5 \text{ cm}$
millimeter	.mm	$1 \text{ cm} = 10^1 \text{ mm}$
		$1 \text{ m} = 10^3 \text{ mm}$
micrometer	μm	$1 \text{ m} = 10^6 \text{ } \mu\text{m}$
		$1 \text{ cm} = 10^4 \text{ } \mu\text{m}$
		$1 \text{ mm} = 10^3 \text{ } \mu\text{m}$
nanometer	nm	$1 \text{ m} = 10^9 \text{ nm}$
		$1 \text{ cm} = 10^7 \text{ nm}$
		$1 \text{ mm} = 10^6 \text{ nm}$
		$1 \text{ } \mu\text{m} = 10^3 \text{ nm}$
angstrom	Å	$1 \text{ m} = 10^{10} \text{ Å}$
		$1 \text{ cm} = 10^8 \text{ Å}$
		$1 \text{ mm} = 10^7 \text{ Å}$
		$1 \text{ } \mu\text{m} = 10^4 \text{ Å}$
		$1 \text{ nm} = 10^1 \text{ Å}$

Exercise 1.5

Perform the following linear measure conversions:
(a) 1.53 cm = _1.53×10⁻²_ m
(b) 0.027 km = _2.7×10¹_ m
(c) 153 nm = _1.53×10⁻⁵_ cm
(d) 2.41×10^3 Å = _2.41×10⁻⁵_ cm
(e) 1.2 μm = _1.2×10⁻⁴_ m

(f) 8.1×10^{-5} m = _8.1×10⁴_ nm
(g) 2.3×10^7 nm = _2.3×10⁻⁵_ km
(h) 1.5×10^{-6} cm = _1.5×10¹_ nm
(i) 0.44 Å = _4.4×10⁻⁵_ μm
(j) 3.5×10^{-3} m = _3.5mm_ mm

1.10 METRIC UNITS OF VOLUME

The basic unit of volume is the **liter, L,** which is just slightly larger than a quart. By international agreement, the liter is equal to 1000 cubic centimeters (cm^3 or cc). Thus cm^3 and milliliter (one thousandth of a liter) can be used interchangeably. The double system of volume measure is similar to the English system choice of cubic inches,

Figure 1-6 1 mL = 0.001 liter = $\frac{1}{1000}$ liter = 10^{-3} liter.

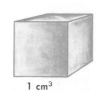

1 cm³

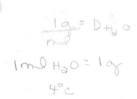

1 mL   eye level

cubic feet, pecks, and bushels; or ounces, pints, quarts, and gallons. In short, the liter and its subdivisions are commonly applied to liquid and gas volumes, while solids are likely to be measured in cm³.

Because 1000 cm³ equals one liter, a single cm³ equals one **milliliter (mL)** (Fig. 1–6). It is helpful to note that 1 mL of cold water weighs about a gram since the volume occupied by a sample of water varies with temperature. The units of mass are based on the weight of a milliliter of water at 4°C.

There is one other useful unit of volume: the **microliter,** symbolized **μL,** or by the Greek letter *lambda* (λ). The microliter is extremely small. It is one millionth of a liter (one thousandth of a milliliter). Microliter volumes are used in the laboratory, and there are small (and expensive) types of **pipets** and syringes available for very small measurements (Fig. 1–7).

A variety of devices, including pipets, cylinders, burets, and flasks, are used to measure volumes. These items vary in capacity, purpose, and precision. The **volumetric flasks** and **graduated cylinders,** for example, are designed to measure more accurately what they contain than what they deliver when poured. The amount of liquid adhering to the walls of such vessels prevents accurate measurement of poured volumes. A piece of glassware designed for contained volume measurement is usually marked *TC* (to contain). See Figure 1–8 for illustrations of volumetric flasks and graduated cylinders.

A **buret** or **pipet** (the latter is really a simplified buret) is more likely to be designed to accurately measure the volume delivered, and they are appropriately marked *TD* (to deliver). Pipets are designed to be used much like a straw in a soft drink. When poisonous or otherwise

volume mark

(transfer)

Figure 1–7   Micropipets.

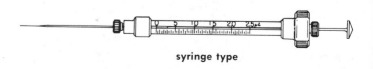

syringe type

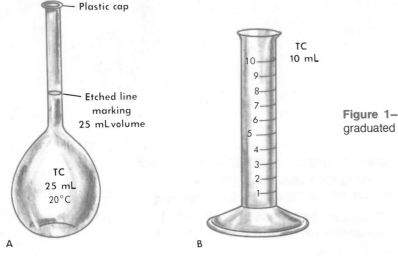

Figure 1–8   Volumetric flask (*A*) and graduated cylinder (*B*).

dangerous solutions are to be drawn, a rubber bulb or a more sophisticated controlled-volume plunger attachment should be used rather than sucking by mouth. Many people prefer this more cautious approach in every case. Some examples of burets are illustrated in Figure 1–9.

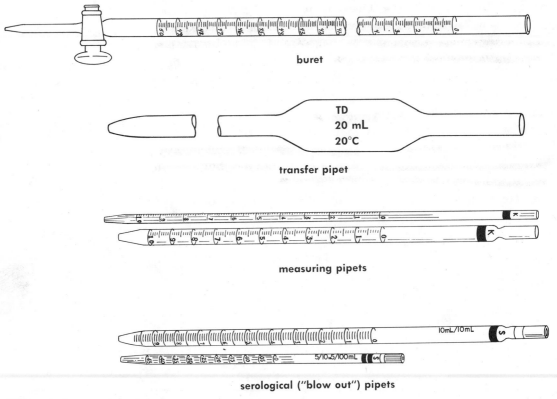

Figure 1–9   Variety of pipets.

MEASUREMENT

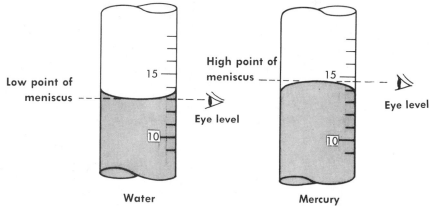

Figure 1–10 Determining the lowest or highest point of a meniscus in volume reading.

In reading the volume of liquid contained in a graduated piece of glassware, one must take into account the tendency of liquids to adhere to the walls of vessels. The standard procedure is to hold the vessel at eye level and read the lowest point of the liquid when it sags or the highest point when it bulges upward. The name given to the surface of the liquid is the **meniscus** (Fig. 1–10). Certain plastic measuring vessels eliminate this bulge, because water or water solutions do not adhere to such material.

TABLE 1–2 Conversion Factors (Volume)

Unit	Symbol	Conversion Factor
liter	L	basic unit
milliliter	mL	$1\ L = 10^3\ mL$
microliter	μL	$1\ mL = 10^3\ \mu L$
(lambda)	(λ)	$1\ L = 10^6\ \mu L$
cubic meter	m^3	basic unit
cubic centimeter	cm^3	$1\ m^3 = 10^6\ cm^3$
	(cc)	

(handwritten notes in right margin:)
$\lambda = \mu l$
lambda

(handwritten notes below table:)
$(1\ m)^3 = (00\ cm)^3$
$1\ m^3 = 10^6\ cm^3$

Example 1.5

Convert 300 mL to liters.
1. Change 300 mL to scientific notation form:
   300 mL = 3.00×10^2 mL (assuming 3 sig. figs.)
2. The conversion factor is $1\ L = 10^3$ mL
3. Considering that the liter is a much larger unit of volume than the milliliter, it would seem correct to predict that many mL equal few liters.

$$\frac{3.00 \times 10^2\ \text{mL}}{1} \times \frac{1}{10^3\ \text{mL}} = \boxed{0.300L}$$

(handwritten notes in right margin:)
$300\ ml = ?\ l$
$300\ ml \times \dfrac{1\ l}{10^3\ ml} = 300 \times 10^{-3}\ l$ or
$.300\ l$

Example 1.6

Express 0.4 μL of water in mL units.
1. Change 0.4 μL to scientific notation form:
   0.4 μL = 4 × 10⁻¹ μL
2. The conversion factor is 10³ μL = 1 mL
3. Because one microliter is such a small fraction of a milliliter, a very small numerical value is expected as an answer.

$$\frac{4 \times 10^{-1}\ \mu L}{1} \times \frac{1\ mL}{10^3\ \mu L} = \boxed{4 \times 10^{-4}\ mL}$$

Exercise 1.6

Perform the following volume conversions:

(a) 27.2 mL = _____ L   (f) 0.002 L = _____ λ or μℓ
(b) 0.084L = _____ mL   (g) 5.4 cm³ = _____ mL
(c) 176 μL = _____ mL   (h) 260 cc = _____ L
(d) 20 mL = _____ μL   (i) 3.1 × 10⁻⁵ m³ = _____ cm³
(e) 5.6 × 10⁷ μL = _____ L   (j) 0.049 mL = _____ μL

1.11   METRIC UNITS OF MASS (WEIGHT)

The terms "mass" and "weight" are loosely equated in practice. However, there is a difference that should be pointed out for the sake of accuracy. Mass is a description of the quantity of matter in an object; weight describes the attraction of gravity on the object. Consider a block of copper, for example. Regardless of where this block of copper is measured or what the conditions of temperature of pressure may be, the *amount* of copper is constant. Under the sea, in space, or on the moon, this block of copper represents a precise amount of matter measured in units of *mass*. Its *weight* is something else. In space, the block may be weightless. On the moon it weighs only 1/6 of its weight on earth. The reasons for these changes are (1) in space the effects of gravity may be canceled by the movement of the block away from earth, and (2) on the moon the gravity is only about 1/6 that of earth.

Why, then, are mass and weight commonly used as synonymous terms? Very simply, because the standards for mass were established on the surface of the earth and because they were originally derived from weight measurements in surface-of-the-earth laboratories, the distinction was unnecessary. Until scientists find themselves in extraterrestrial laboratories, this distinction need not be a critical one.

MEASUREMENT

The most frequently used unit of mass is the **gram (g)**. A gram is equal to about half the mass of a dime. The units of mass are definitely preferred to accurately describe the quantity of a solid.

The unit of mass that is a multiple of a gram is the **kilogram (kg).**

$$1 \text{ kg} = 1000 \text{ g} = 10^3 \text{ g}$$

An example will help you to visualize the magnitude of a kilogram of mass: a five-pound bag of potatoes is roughly equivalent to two kilograms.

High-precision work often involves the use of the **milligram (mg),** which is three orders of magnitude less in mass than the gram.

$$1 \text{ mg} = 0.001 \text{ g} = \frac{1}{1000} \text{ g} = 10^{-3} \text{ g}$$

A small "pinch" of salt would probably have a mass (or "weight") in the milligram range. The weight of ink used in the writing of a person's signature is about three to four milligrams.

When extremely dangerous or expensive materials are required in an experiment, it may be necessary to use a unit of mass that is even smaller than those described thus far. The **microgram (μg)** is one millionth of a gram and is sometimes symbolized by the Greek letter *gamma* (γ).

$$1 \gamma = 0.000001 \text{ g} = \frac{1}{1,000,000} \text{ g} = 10^{-6} \text{ g}$$

TABLE 1–3 **Conversion Factors (Mass)**

Unit	Symbol	Conversion Factor
kilogram	kg	basic unit
gram	g	1 kg = 10^3 g
milligram	mg	1 g = 10^3 mg
microgram	μg	1 mg = 10^3 μg
	(γ)	1 g = 10^6 μg

$\gamma = \mu g$

gamma

In the laboratory the device used to weigh a substance is called a **balance.** The actual type of balance that should be used to determine mass depends on the amount of material to be weighed and the desired or required precision. It would not make sense to attempt to weigh several hundred grams of a substance of an analytical balance that is designed to weigh small samples in the milligram range. It would be equally poor procedure to use a coarse triple-beam or double-pan balance when directions call for weighing out 0.055 g of a solid. The guidelines are based on common sense. When a precision (repro-

Figure 1-11   Typical coarse balances.

ducibility) of 0.1 g is called for, a relatively coarse triple-beam or double-pan balance is unsatisfactory. Examples of such balances are illustrated in Figure 1-11.

If the directions call for a precision of 0.01 g, 0.001 g, or 0.0001 g, analytical balances of higher precision are used. Some examples of modern analytical balances are illustrated in Figure 1-12.

Example 1.7

Convert 4.2 mg to grams.
1. Select the conversion factor:

$$1 \text{ g} = 10^3 \text{ mg}$$

2. Since milligrams are small compared to grams, many mg will equal few g, and the answer will be less than 10^3 in magnitude.

$$\frac{4.2 \text{ mg}}{1} \times \frac{1 \text{ g}}{10^3 \text{ mg}} = \boxed{4.2 \times 10^{-3} \text{ g}}$$

Figure 1-12   Analytical balances.

MEASUREMENT

Example 1.8

Change 0.0067 mg to micrograms.

1. Express the value in scientific notation:

$$0.0067 \text{ mg} = 6.7 \times 10^{-3} \text{ mg}$$

2. Find the conversion factor:

$$1 \text{ mg} = 10^3 \text{ } \mu g$$

3. Reasoning that micrograms are very small units, a larger answer is expected.

$$\frac{6.7 \times 10^{-3} \text{ mg}}{1} \times \frac{10^3 \text{ } \mu g}{1 \text{ mg}} = \boxed{6.7 \text{ } \mu g}$$

Exercise 1.7

Perform the following conversions of mass:

(a) 325 mg = _____ g

(b) 1.4 g = _____ kg

(c) 0.0033 kg = _____ mg

(d) 3×10^{-2} mg = _____ μg

(e) 1.5×10^{-7} g = _____ μg

(f) 4280 γ = _____ g

(g) 0.09 mg = _____ g

(h) 6×10^{-4} kg = _____ g

(i) 0.023 γ = _____ mg

(j) 72 mg = _____ μg

1.12 TEMPERATURE

Temperature is a measure of the degree of random motion of particles as they move rapidly in straight lines. Temperature and heat are not the same. Heat is a form of thermal energy that tends to flow from regions of higher temperature to other regions where the tempera-

THE WIZARD OF ID   by Brant parker and Johnny hart

ture is lower. The topic of heat will be considered in the next chapter; we now turn our attention to the measurement of temperature.

Fahrenheit and Celsius Scales

A feature common to both the Fahrenheit and Celsius (formerly called Centigrade) temperature scales is the arbitrary zero point. An arbitrary zero is a starting point that has been selected for convenience. All temperatures above the starting point are given positive values, while those below are negative. The difference between the zero points on the Fahrenheit and Celsius scales is 32 Fahrenheit units, because 0°F is set 32 Fahrenheit degrees lower than the freezing point of water (which is the starting point for the Celsius scale). See Figure 1–13.

From Figure 1–13 we see that the conversion factor relating Celsius and Fahrenheit is as follows:

$$1.8 \text{ Fahrenheit degrees} = 1 \text{ Celsius degree}$$

Celsius is less than fahrenheit

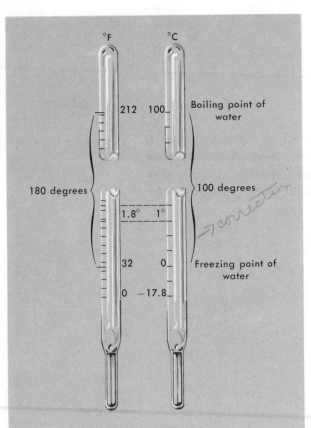

Figure 1–13   Comparison between the Fahrenheit and the Celsius temperature scales.

Also, the 32 Fahrenheit degree difference between the zero points must be included as a correction. The equation that emerges is as follows:

$$°C = \frac{°F - 32}{1.8}$$

Although most of us would argue that we have no use for Fahrenheit degrees in chemistry, there are industrial applications that justify an ability to interconvert between the scales. The equation for finding °F can be developed algebraically.

$$1.8 \, °C = \frac{°F - 32}{1.8} \qquad (1)$$

Now, multiply both sides of the equation by 1.8:

$$1.8 \, °C = 1.8 \left(\frac{°F - 32}{1.8} \right)$$

$$1.8°C = °F - 32 \qquad (2)$$

$$°F = 1.8°C + 32 \qquad (3)$$

Now let us consider some examples.

Example 1.9

If the temperature in the laboratory is 72°F, what is the value in degrees Celsius?

1. Select the appropriate form for the equation:

$$°C = \frac{°F - 32°}{1.8}$$

2. Substitute the data:

$$°C = \frac{72° - 32°}{1.8}$$

3. Simplify the numerator and divide:

$$°C = \frac{40°}{1.8} = \boxed{22°C}$$

Example 1.10

The temperature of a mixture of dry ice and alcohol is about $-80°C$. Convert this to degrees Fahrenheit.

1. Select the appropriate form of the equation:

$$°F = 1.8 \ °C + 32°$$

2. Substitute the data:

$$°F = 1.8 \ (-80°) + 32°$$

3. Estimating the answer to be about -100, multiply and add 32:

$$°F = -144° + 32° = \boxed{-112°F}$$

(handwritten in left margin:)
$0 F = 1.8 °C + 32$
$°F = (1.8)(-80°C) + 32$
$°F = -144 + 32$
$°f = 112 °F$
$sig. fig. 2$

The Kelvin (Absolute) Temperature Scale

The striking difference between the Kelvin scale and other temperature scales is the fact that the Kelvin zero point is not an arbitrary starting point. It is the point that indicates theoretical absence of temperature; that is, the point at which molecular motion theoretically stops. Basing the definition on the behavior of a "perfect" gas (see

(handwritten:) $0°K = -273°C$

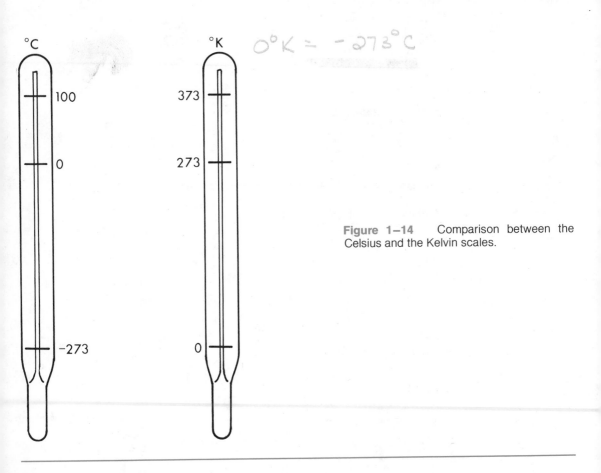

Figure 1–14   Comparison between the Celsius and the Kelvin scales.

Chap. 6), it is the point at which the volume of a perfect gas would be reduced to zero. It should be mentioned that a gas volume reduction to zero is a mathematical concept and not a reality. There is no known substance that exists in the gaseous state below 4°K. The relationship between gas volume and temperature will be discussed in the chapter on gases (Chap. 6).

Using the Celsius scale units to measure absolute zero, it can be shown from the behavior of gases that 273.15° (usually rounded off to 3 significant figures) is the difference between 0°C and 0°K (Fig. 1–14). The equation, easily extracted from a comparison of the two scales, incorporates a correction term of 273°. It is expressed as follows:

$$°K = °C + 273°$$

Example 1.11

The temperature of a water bath is 27°C. Convert this value to Kelvin degrees.
1. Write the appropriate equation:

$$°K = °C + 273°$$

2. Substitute to obtain the answer:

$$°K = 27° + 273°$$

$$\boxed{= 300°K}$$

Example 1.12

Liquid oxygen boils at 90°K. What is this in degrees Celsius?
1. Write the appropriate equation:

$$°C = °K - 273°$$

2. Substitute to obtain the answer:

$$°C = 90° - 273°$$

$$\boxed{= -183°C}$$

Exercise 1.8

Perform the following temperature conversions:
(a) 17°F = ___ °C   (e) 10°C = ___ °K
(b) 200°C = ___ °F   (f) 115°K = ___ °C
(c) 6.82°F = ___ °C   (g) −30.5°C = ___ °K
(d) −220°C = ___ °F   (h) 40°F = ___ °K

A useful physical property of matter is **density.** Densities of materials are used as a fundamental basis for comparison, and they have extensive application to the solution of many practical problems in chemistry. Density is a precise type of description that takes into account both mass and volume.

If a person were asked to compare a roomful of cork to a small cube of gold, there might be a nagging suspicion that this is a meaningless basis for comparison. Such a suspicion would be well founded. If any comparison is to be meaningful, and perhaps useful, pieces of cork and gold having equal size or equal mass are required. Density is the label applied when we measure the mass of a definite volume of some specific substance. The definite volume is usually taken to be one milliliter (or one cubic centimeter) for liquids and solids, and one liter for gases. These values are known as *unit volumes.* Therefore, by definition, the density of a substance is its mass per unit of volume.

$$d = \frac{m}{V}$$

Example 1.13

Find the density of a plastic substance if a 57 cm³ block weighs 52 g.

1. Organize the data:

$$d = ?$$
$$m = 52 \text{ g}$$
$$V = 57 \text{ cm}^3$$

2. Solve for density:

$$d = 52 \text{ g} \times \frac{1}{57 \text{ cm}^3}$$

$$\boxed{d = 0.91 \text{ g/cm}^3}$$

Example 1.14

Find the density of liquid benzene in a 214 mL sample having a mass of 184 g at 16°C.*

---

*It is important to note the temperature at which the density is measured, because the volume of an object varies with temperature.

MEASUREMENT

1. Organize the data:

$$d = ?$$
$$m = 184 \text{ g}$$
$$V = 214 \text{ mL}$$

2. Solve for density:

$$d = 184 \text{ g} \times \frac{1}{214 \text{ mL}}$$

$$\boxed{d = 0.860 \text{ g/mL}}$$

Example 1.15

Find the mass of 30.0 mL of a liquid having a density of $1.84 \dfrac{g}{mL}$.

1. Organize the data:

$$d = 1.84 \frac{g}{mL}$$

$$m = ?$$

$$V = 30.0 \text{ mL}$$

$V \cdot D = m \quad \checkmark \qquad A \quad m = V \cdot D$

2. Using dimensional analysis:

$$\text{mass} = 30.0 \text{ mL} \times \frac{1.84 \text{ g}}{1 \text{ mL}} = \boxed{55.2 \text{ g}}$$

Example 1.16

If a metal has a density of $9.71 \dfrac{g}{cm^3}$ and a mass of 205.6 g, what volume will it occupy?

1. Organize the data:

$$d = 9.71 \frac{g}{cm^3}$$

$$m = 205.6 \text{ g}$$

$$V = ?$$

$V \cdot D = m \quad \checkmark \quad A$

$\dfrac{m}{D} = \dfrac{V \cdot V}{V}$

$V = \dfrac{m}{D}$

2. Using dimensional analysis:

$$V = 205.6 \text{ g} \times \frac{1 \text{ cm}^3}{9.17 \text{ g}}$$

$$\boxed{V = 21.2 \text{ cm}^3}$$

The fact that density has the dimensions of mass and volume is what distinguishes it as a physical property of matter from **specific gravity (sp. gr.)**. Specific gravity is a comparison in the form of a fraction, or ratio, between the density of some substance and the density of a standard. Water has a density of exactly 1.00 g per milliliter at 4°C $\left(d_{water} = \dfrac{1.00 \text{ g}}{\text{mL}} \right)$ and is the standard for solids and liquids at specified temperatures. At this temperature, specific gravity is numerically equal to density. Assume this condition for all problems unless exceptions are noted specifically.

The equation for specific gravity is:

$$\text{sp. gr.} = \frac{d_{substance}}{d_{standard}}$$

Example 1.17

What is the specific gravity of sulfuric acid if its density is $1.84 \dfrac{\text{g}}{\text{mL}}$?

$$\text{sp. gr.} = \frac{d_{sulfuric\ acid}}{d_{water}} \qquad \text{sp. gr.} = \frac{1.84 \ \dfrac{\cancel{g}}{\cancel{mL}}}{1.00 \ \dfrac{\cancel{g}}{\cancel{mL}}} = \boxed{1.84}$$

With this answer we can see that the numerical values of density and specific gravity for solids and liquids are the same; only the units of mass and volume are missing from the specific gravity. It is a dimensionless number.

Exercise 1.9

1. What is the density of a liquid if a 35.0 mL sample has a mass of 22.5 g?
2. If the density of mercury is $13.6 \dfrac{\text{g}}{\text{mL}}$, what volume of it must be added to 3.2 g of tin to give a total mass of 5.8 g?
3. What volume of brass $\left(d = 8.0 \dfrac{\text{g}}{\text{cm}^3} \right)$ is equal in mass to a 2 cm³ block of gold $\left(d = 19.3 \dfrac{\text{g}}{\text{cm}^3} \right)$?
4. What is the specific gravity of sulfuric acid if 0.100 L has a mass of 184.0 g?
5. What is the density of oxygen if 22.4 L has a mass of 32.0 g?

QUESTIONS & PROBLEMS

1.1   Define an order of magnitude. Change 2.1×10^{-3} by four orders of magnitude larger and three orders of magnitude smaller.

1.2   What determines the number of digits retained to the right of the decimal in addition and/or subtraction?

1.3   Explain the rule followed when retaining significant figures in multiplying and/or dividing.

1.4   What is the difference between accuracy and precision?

1.5   Round off the following measurements to two significant figures.
(a) 2.571 cm
(b) 0.03819 g
(c) 8325 m
(d) 7.02 mL

1.6   How many significant figures are in each measurement?
(a) 0.2676 cm
(b) 12.0220 g
(c) 1.81×10^{-3} mL
(d) 1027 m

1.7   Use scientific notation to write the following as numbers having one digit before the decimal point.
(a) 8400
(b) 0.025
(c) 176000000
(d) 0.00087

1.8   Express in scientific notation:
(a) 0.267617
(b) 0.00015
(c) 21.67
(d) 801.61

1.9   Give answers in scientific notation:
(a) $39.0 + 42.6 + 1.39 =$
(b) $1.008 + 32.06 + 149.1 =$
(c) $14.1 \times 3.84 =$
(d) $55.841 \div 4.10 =$

1.10   Round off the following numbers to three significant figures:
(a) 9241
(b) 7.3932
(c) 2.516×10^2
(d) 9.1196×10^{-3}
(e) 0.001137
(f) 87.994

1.11   Perform the indicated operations:
(a) $(2 \times 10^4)^2 =$
(b) $(8 \times 10^{-2})^{-3}$
(c) $\sqrt{0.00064 \times 10^{-5}}$
(d) $(270 \times 10^7)^{1/3}$

1.12   Round off to four significant figures and express answers in scientific notation:
(a) 0.020815
(b) 120.55
(c) 11.00274
(d) 2627.05

1.13   Perform the indicated operations:
(a) $(2.3 \times 10^2)(1.1 \times 10^{-5}) =$
(b) $\dfrac{2.3 \times 10^2}{1.1 \times 10^{-5}} =$
(c) $(8.2 \times 10^1) + (2.4 \times 10^2) =$
(d) $\dfrac{1}{10^7}(100 \times 10^5) =$

1.14   Perform the indicated operations:
(a) $\dfrac{26.5 \times 2.816 \times 34}{19.8} =$
(b) $\dfrac{(18.50)(5.12 \times 10^{-4})(4.61 \times 10^2)}{0.516} =$
(c) $\dfrac{67.9 - 12.73}{0.5152} =$
(d) $12.91 - 3.7 + 91.312 =$

1.15   Express in grams:
(a) 7527 kg
(b) 212.0 mg
(c) 0.7121 cg
(d) 1.0×10^{-7} dg

1.16   Perform the following metric system mass conversions:
(a) 0.4 kg = _____ g
(b) 512 mg = _____ g
(c) 7×10^4 µg = _____ g
(d) 4.5 mg = _____ µg
(e) 2×10^{-3} g = _____ mg
(f) 6.7×10^5 γ = _____ mg

1.17   Express in millimeters:
(a) 23.2 cm
(b) 0.5851 km
(c) 1.676 Å
(d) 96.11 µm

1.18   Calculate the volume of these solids:
(a) a cube 16.2 mm on a side
(b) a rectangular prism having dimensions 6.2 cm $\times$ 1.7 cm $\times$ 17 mm
(c) a rectangular prism having dimensions 11.6 mm $\times$ 1.8 mm $\times$ 0.416 cm
(d) a cylinder $\left(\text{vol.} = \dfrac{\pi d^2 L}{4} \right)$ 3.61 cm long and 1.8 cm in diameter

1.19   Perform the following metric system conversions for linear measurement:
(a) 351 m = _____ km
(b) 14 mm = _____ cm
(c) 650 nm = _____ cm
(d) 52 nm = _____ Å
(e) 0.02 mm = _____ µm
(f) 1.3×10^{-6} cm = _____ nm

1.20   Perform the following metric system volume conversions:
(a) 0.12 L = _____ mL
(b) 3.7×10^{-5} mL = _____ µL
(c) 185 mL = _____ L
(d) 28 mL = _____ cm³
(e) 0.08 µL = _____ mL
(f) 5×10^3 λ = _____ mL

1.21   When a metallic object weighing 16.6 g is placed in a graduated cylinder containing 8.10 mL of water, the final water level rises to the 11.10 mL mark. Calculate the density of the metallic object.

1.22   An empty graduated cylinder weighs 26.22 g. When 25.00 mL of liquid is added, the mass of cylinder plus liquid is 52.10 g. What is the density of the liquid?

1.23   A completely filled flask holds 17.58 g of benzene or 20.0 g of water. Calculate the specific gravity of benzene.

1.24   A graduated cylinder contains 110.0 mL of a liquid whose density equals 0.98 g/mL. How many grams of liquid is this? What volume of alcohol (d = 0.78 g/mL) would have an identical mass?

1.25   If one calorie equals 4.2×10^7 ergs, write a conversion factor for converting ergs to calories.

1.26   What volume is occupied by 454 g of mercury having a density of 13.6 g/mL?

1.27   What is the density of a cylindrical rod that is 15.0 cm long and 2.0 cm in diameter, and has a mass of 45.0 g?

1.28   What is the mass of 600 mL of sulfuric acid that has a specific gravity of 1.54?

1.29   What total volume results when 160.0 g of silver is added to 35.0 mL of mercury? The density of silver is 10.5 g/mL.

1.30   Find the density of a ring that weighs 7.3 g in air and 6.9 g when suspended in water. (Hint: the weight of water displaced is equal to the weight reduction of the ring.)

1.31   Compare heat and temperature.

1.32   Perform the following temperature scale conversions:
(a) 5°C = _____ °F
(b) −70°C = _____ °F
(c) 160°F = _____ °C
(d) −80°C = _____ °K
(e) 230°K = _____ °C
(f) 15°F = _____ °K

1.33   What is the mass of 787 mL of an alcohol having a density of 0.79 g/mL?

1.34   What is the mass of a cube of metal 2.2 cm on a side if the metal's density is 19.6 g/mL?

1.35   The sp. gr. of marble is 2.70. Some marble chips were dropped into a graduated cylinder containing 32.1 mL of water, raising the water level to 39.5 mL. How many grams of chips were used?

1.36   The densities of aluminum and molybdenum are 2.70 g/mL and 10.2 g/mL, respectively. Which occupies the greater volume, 82 g of aluminum or 26 g of molybdenum?

1.37   A graduated cylinder contains 50.0 mL of water. Pellets weighing 6.22 grams and having a density of 2.71 g/mL are dropped into the cylinder. What is the minimum number of pellets required to raise the water level by 125 mL?

1.38   Copper has a density of 8.96 g/mL. If 25.0 g of copper is placed in a dry 100 mL graduated cylinder, how many mL of water are needed to fill the cylinder to the 100 mL graduation level?

1.39   A piece of gold has the dimensions 1.23 mm $\times$ 0.00220 cm $\times$ 1.21 $\times$ 10^{-4} m. What is its mass if gold has a specific gravity of 19.3?

1.40   A flask has a mass of 9.75 g when empty and 35.51 g when completely filled with water. The same flask completely filled with an organic solvent has a mass of 29.66 g. What is the density of the solvent?

CHAPTER TWO

LEARNING OBJECTIVES

At the completion of this chapter, you should be able to:

1. Define matter.
2. Classify and give examples of the types of matter.
3. List the states of matter.
4. Define *element* and *compound.*
5. List four fundamental forces of attraction.
6. Define *entropy* in terms of a driving force in nature.
7. Distinguish between physical and chemical change.
8. List four observations that serve as evidence for a chemical change.
9. List four conditions *that may be needed* to permit a chemical change to occur.
10. Define the term *catalyst.*
11. Explain the laws of conservation of matter and energy.
12. Distinguish among the following terms: *energy, force,* and *work.*
13. Define *kinetic energy* in words and by a mathematical equation.
14. Distinguish between *kinetic* and *potential* energy.
15. Define the term *specific heat.*
16. Calculate heat energy values in calories from given data.
17. Use energy conversion factors to change energy units.

MATTER, ENERGY, AND CHANGE

Chemistry is often defined as the systematic investigation of the properties, structure, and behavior of matter and of the changes that matter undergoes. Such a generalized definition raises a number of questions. What is matter? What is meant by change? What are the driving forces behind changes? What is it about the structure of matter that lends itself to respond to these "driving forces"? What are the results of change? These questions and many others of a similar fundamental nature are all involved in the study of chemistry.

The development of new alloys, plastics, and medications comes about as a result of the practical application of the whole network of assumptions, hypotheses, theories, and laws that chemists have formulated from their research into the nature of matter, energy, and change.

2.1 MATTER

We usually describe matter as anything having mass and occupying space (volume). In other words, matter has a measurable density. Minerals, air, milk, plants, and animals are a few of the countless examples of matter. While the examples are obvious, the definition is limited to that aspect of the word familiar to us. For example, "things" such as electrons and photons of light do not fit the definition and distinctions that scientists have constructed for convenience. Familiar types of matter may be subdivided into pure substances and mixtures.

Mixtures may be either **homogeneous** or **heterogeneous.** In many cases it is easy to see the components of a mixture, such as the glittering quartz and mica in a piece of granite and the yellow sulfur in a sulfur-sugar mixture. More often, however, the separate ingredients of a mixture cannot be seen; it appears to be homogeneous. It is only by experimentation, for instance, that we can establish that air is a mixture of gases or that sugar in water constitutes a mixture. Both mixtures appear to be uniform throughout, but they can be resolved into simpler pure substances by nonchemical methods. Such *homogeneous mixtures* form a special category called **solutions,** which will be studied in detail in Chapter 9.

experiment
3

Familiar types of matter are observed to have variable states. A **solid** can usually be changed to a **liquid** by increasing the temperature to a value described as the **melting point** of the substance. A further increase in temperature brings the liquid to its **boiling point,** where the liquid is changed to the **vapor** (gas) state. This changing of state is linked necessarily to energy because it is energy that enables molecules to move more vigorously in the process of overcoming attractive forces. Change in state is a story of forces in conflict. There is a tendency for matter to achieve maximum randomness in arrangement. By "maximum randomness" we mean the tendency toward disorder or disarray of particles. (Entropy)

The forces of attraction between unit particles of matter (atoms, tions, or molecules) are universal phenomena. Without fully understanding the nature of these forces, they are classified, nevertheless, as **nuclear, electrical, magnetic,** and **gravitational.** The nuclear and gravitational forces provide an interesting contrast. Although the nuclear force is extremely powerful, it operates effectively over a distance of less than an angstrom. On the other hand, the gravitational force is effective over distances of millions of miles, yet it is comparatively weak. Regardless of the way in which the forces of attraction are classified, they all tend to pull together particles of matter in an orderly geometric arrangement that characterizes stability (Fig. 2–1). There are forces of repulsion that must be reckoned with also. Such forces will be investigated in the chapters on atomic structure and chemical bonding (Chaps. 7 and 8).

The tendency of matter to sacrifice energy as it strives for stability is one of the fundamental driving forces involved in the changes of matter. There are many common examples to illustrate this observation. A ball on a hill rolls down and finally stops. A machine wears out

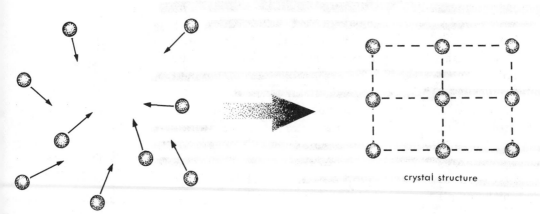

crystal structure

Figure 2–1   Attractive forces among unit particles, resulting in a stable crystal-like structure.

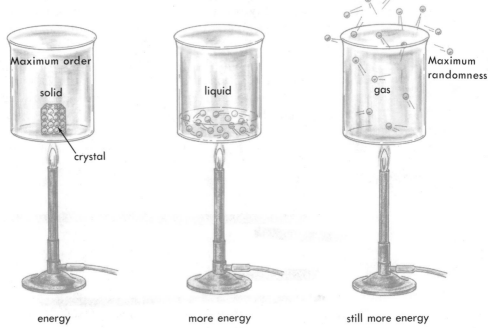

Figure 2–2 Increasing the entropy of a system as the temperature of a crystal changes from the melting point to the boiling point.

and eventually ends on a scrap pile. Living organisms grow old, break down, and die. However, nature seems to provide a counterforce for every one-directional tendency, just as each skeletal muscle in a body has another muscle operating in opposition to it.

When the unit particles have some external source of energy that can be absorbed, there is a movement toward disorder. With the absorption of energy, the unit particles in the solid and liquid state can overcome the forces of attraction by their increased ability to shift to the gaseous state, in which there is a maximum disorder. This tendency toward randomness is another fundamental driving force behind changes in matter. The degree of randomness is often associated with the measure of the **entropy** of a system (Fig. 2–2). *unorganization*

Figure 2–3 The drive toward randomness, resulting in the absorption of energy.

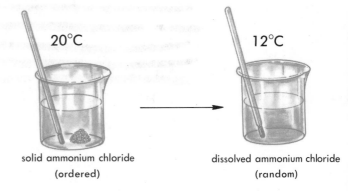

Occasionally the drive toward an increase in entropy is so powerful that a highly ordered solid may spontaneously absorb energy in order to achieve greater randomness. An example of this phenomenon is the considerable drop in the temperature of water that occurs when solid crystalline ammonium chloride is dissolved in it (Fig. 2–3). As a result of the absorption of heat energy from the water, the dissolved particles of ammonium chloride are much more random in their arrangement than was the original solid that they composed.

2.3 CHANGES OF MATTER

The changes of matter are conveniently divided into *physical* and *chemical* changes.

Physical Change

Changes of state are examples of physical change. Water, ice, and steam are chemically the same substance, all being composed of exactly the same unit particles (water molecules). Whereas the arrangement of the water molecules may vary from ordered (ice) to random (steam), it is only when the molecules are changed into new types of particles (hydrogen and oxygen molecules) that a chemical change results.

If a pane of glass is ground into powder, it remains glass. If black iron powder is blended with yellow sulfur, a gray mixture is formed. The mixture can be separated into its original components by one of two simple physical processes. The iron can be removed by a magnet, or the sulfur can be dissolved in liquid carbon disulfide and the iron then removed by filtration. Grinding, melting, dissolving, and blending are all physical changes. If no new substances are formed, the change is said to be physical.

Chemical Change

A chemical change is one which begins with **reactants** having a set of unique characteristics (properties) quite different from those of the **products** that result from the change. When a **reaction** (a chemical change) goes to completion—that is, when essentially all of the reactants are converted to products at the end—the reaction is said to be **stoichiometric** (Greek *stoikheion* = element). Stoichiometry is concerned with very specific masses or volumes of reactants and products. The subject of stoichiometry will be discussed in Chapter 5.

stoichiometric reaction

$$A + B \xrightarrow{\text{produce}} C + D$$

reactants   products

Observations suggesting that a chemical reaction has occurred include the following:
1. A **gas** is produced.
2. A **precipitate** (visible solid particles) is formed and separates from the bulk of the mixture.
3. An **energy change** is observed in the form of light or heat or both.
4. A distinct color change occurs.

Conditions necessary for a chemical change to take place may be:
1. Simple contact between two reacting substances at room temperature:
   (a) colorless ammonia gas plus colorless hydrogen chloride gas → white ammonium chloride solid
   (b) solid yellow phosphorus plus solid iodine crystals → solid red phosphorus triiodide
2. Supplying energy via Bunsen burner, for example, to jostle sluggish reactants into an active condition, which allows breakage of the original bonds of attraction and the formation of new ones.
3. Getting the reactants into water **solution** often is necessary. Figure 2–4 illustrates the absence of noticeable change when white crystals of silver nitrate are mixed with white crystals of table salt. The addition of water provides a medium through which the interacting particles can move and form the white precipitate silver chloride.
4. The use of a **catalyst** frequently is necessary if a reaction is to occur within a reasonable length of time (Fig. 2–4). A catalyst is anything that alters the speed of a chemical reaction without becoming permanently altered itself. Large complex substances called **enzymes** often function as catalysts in living organisms. For example, the breakdown of our food into simple substances (the process of diges-

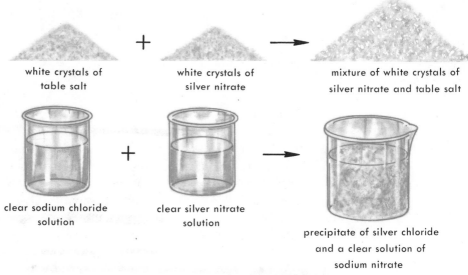

Figure 2–4 Reactants in aqueous solution leading to chemical change.

tion) involves complicated chemical changes that must be catalyzed by enzyme action. If it were not for enzymes, the first meal ever eaten by a person would probably not by fully digested even after a period of many years. A catalyst works by acting as a gathering point for reactants that normally are spread out and randomly distributed, or it may actually be formed in a reaction as a temporary intermediate product.

2.4 THE LAW OF CONSERVATION

Another consideration of matter that is central to the process of stoichiometry is the fact that *under ordinary conditions,* matter cannot be created or destroyed. When a reaction begins with 10 grams of reactants, it must end with 10 grams of products. Only in extraordinary changes such as a nuclear reaction is the "annihilation" of matter a measurable quantity. The term "annihilation" is slightly misleading: what happens in this case is the conversion of matter into an equivalent amount of energy. The popularized Einsteinian equation states that in reality (although not obviously), matter and energy are two aspects of the same thing.

$$E = mc^2$$

where

E = energy (in ergs)
m = mass (in grams)
c^2 = the proportionality constant squared, which is the speed of light squared, or $(3 \times 10^{10} \text{ cm/sec})^2$.

2.5 ENERGY

In the previous discussion of matter, we often referred to the concept of energy. Indeed, the law of conservation of matter and energy most directly underscores the fundamental unity of matter and energy. This raises a number of important questions regarding the definition, uses, types, and transformations of energy.

Energy is difficult to define adequately in a single concise statement. The usual attempt—"Energy is the ability to do work"—is at best a starting point. It remains for examples and analogies to give substance to a poor skeleton. **Work,** rigidly defined as a physical action, means the acceleration or deceleration of an object. When a push or

MATTER, ENERGY, AND CHANGE

pull is applied to an object, changing its state of rest or motion, a force is said to be exerted. Forces can come about only if there is energy.

Energy has many familiar faces, including heat, light, magnetic, mechanical, electrical, and atomic energy, to mention a few. These forms are often easily interconverted. For example, the mechanical energy of falling water at a hydroelectric plant is converted to electrical energy, which can be converted again to heat and light energy in the home or factory. The scientist's need to deal quantitatively with energy leads to a classification of types of energy, the units of measurement, and the methods of conversion between units. All forms of energy can be classified as either potential or kinetic energy.

2.6 TYPES OF ENERGY

Potential Energy

A sharpshooting basketball player is described as a scorer. When actually playing in a game he is scoring. Before a game, the coach plans his strategy on the basis of his star's *potential* scoring ability. Energy, the *ability* to do work, is described in much the same way. An iron, for example, perched on a table edge has the potential to fall and move a hapless insect closer to the ground (Fig. 2–5).

Potential energy is the *stored energy* an object has by virtue of position with respect to other objects. A stretched rubber band, a tightly wound watch spring, a magnet, a charged condenser, and a storage battery all possess potential energy. The chemical energy in gasoline and other fuels is a form of potential energy.

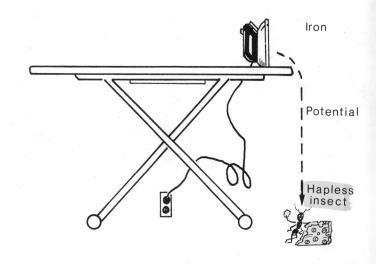

Figure 2–5   Potential energy.

Kinetic Energy

When the book actually does fall or burn, it is effectively converting *potential* energy into **kinetic** energy. The symbols we will use for these forms of energy are **PE** for potential and **KE** for kinetic. The interrelationship of the two can be illustrated by a boy on a swing (Fig. 2–6). In just the same way, if the book were allowed to burn, it would convert its chemical potential energy into heat.

Kinetic energy is a property of matter that is due to its motion. The energy of a moving car, a stream of falling water, and a moving projectile are examples of kinetic energy. A moving car has the capacity to do work (on a wall, or another car, for example) merely because the car is in motion. Moving particles of all matter possess kinetic energy. The amount of KE of a body is proportional to its mass (m) and the square of its velocity (v) in the relationship

$$KE = 1/2 \, mv^2$$

Therefore, a speeding bullet of comparatively low mass could have the same amount of kinetic energy as a slowly moving car of high mass. The bullet makes up in velocity what it lacks in mass. One of the criteria for classifying anything as a particle of matter is its ability to

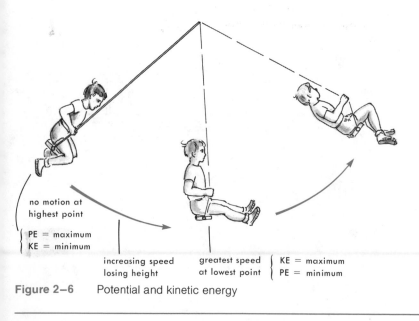

no motion at
highest point

PE = maximum
KE = minimum

increasing speed   greatest speed   KE = maximum
losing height   at lowest point   PE = minimum

Figure 2–6   Potential and kinetic energy

MATTER, ENERGY, AND CHANGE

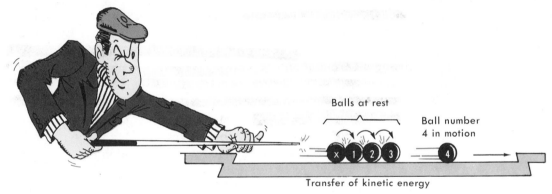

Balls at rest

Ball number
4 in motion

Transfer of kinetic energy

Figure 2–7   The transfer of kinetic energy to ball number 4, which is set in motion.

transfer kinetic energy. This concept can be illustrated by a line of billiard balls that rapidly perform such a transfer (Fig. 2–7).

In every energy conversion, some energy cannot be recovered for human use. It is *unavailable* energy, but it is not destroyed energy. For example, in the change of electrical energy into radiant energy in an incandescent lamp, some of the electrical energy is converted to heat and hence is unavailable for illumination. Similarly, the chemical energy of gasoline is only partially converted to energy of motion of an automobile; most of it is lost to the atmosphere as heat from the engine, muffler, and brakes.

Exothermic and Endothermic Change

We have suggested that every physical and chemical process is accompanied by a change of energy between reactants and products. If the change is such that energy is given to the atmosphere and is taken from the process, the change is said to be **exoenergetic** or **exoergic**. In cases where energy is absorbed during a reaction, the process is called

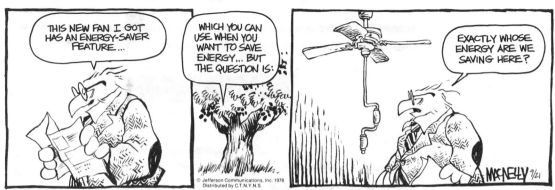

Reprinted by permission of the Chicago Tribune–
New York News Syndicate, Inc.

endoenergetic or **endoergic.** These terms include electrical, light, heat, and all other forms of energy that may be involved in a chemical or physical process. In chemistry, however, we are principally interested in the heat energy involved in a chemical reaction; hence we use the term **exothermic** to denote a process in which heat energy is given off to the surroundings and **endothermic** to describe a process that absorbs heat from its surroundings. The terms "exothermic" and "endothermic" are now loosely used to refer to changes in which any type of energy is involved, whether it be heat, light, or electrical in nature.

2.7 ENERGY UNITS AND CONVERSIONS

We described kinetic energy mathematically as being equivalent to one half of the product of the mass of the particle and its velocity squared. If the units used are mass in grams and velocity in centimeters per second, the product of these units is gram centimeters squared per second squared, or

$$\frac{g \ cm^2}{sec^2} = \text{energy in ergs}$$

This cumbersome group of units is more commonly known as an **erg.** One erg is roughly the amount of energy required by a mosquito to do a push-up.

Just as a distance can be measured by different kinds of units (centimeters, inches, miles, rods), energy can be treated similarly. There are times when it is not convenient to use ergs. The energy required to lift a load of ship's cargo is so great that the use of ergs would be as awkward as measuring the distance between New York and Paris in centimeters. Energy values can be expressed by using equivalent units. If a conversion factor is available, mechanical energy can be expressed as an *equivalent* amount of heat or electrical energy.

A frequently used energy unit is the **joule,** symbolized **J,** named after the British physicist James P. Joule. As a measure of the kinetic energy of particles, a joule is equal to the product of a kilogram weight moving at a velocity of one meter squared per second squared.

$$\frac{kg \ m^2}{sec^2} = \text{energy in joules}$$

The great utility of the joule lies in its direct relationship to the measurement of electrical energy. When the gram and centimeter units of the erg are converted to the kilogram and meter units of the joule, the resulting relationship is:

$$10^7 \ \text{erg} = 1 \ \text{joule}$$

Example 2.1

Electrical measurements on a motor indicate that 2.5×10^4 joules of energy are being used. What is the mechanical equivalent in ergs?

$$2.5 \times 10^4 \text{ joule} \times \frac{10^7 \text{ erg}}{1 \text{ joule}} = \boxed{2.5 \times 10^{11} \text{ erg}}$$

2.8 HEAT ENERGY

The unit commonly used to describe the magnitude of heat energy in the metric system is the **calorie** (cal). Today, the calorie is in the process of being displaced by the joule. We will use both units, however. While the calorie may be defined as the amount of heat needed to raise 1 gram of water from $14.5°C$ to $15.5°C$ (i.e., a 1 degree change), it is more properly defined as being equal to 4.184 J.

$$1 \text{ cal} = 4.184 \text{ J}$$

cal = amt of heat to raise 1 g H₂O 1°C

Example 2.2

The kinetic energy of a moving ball is 5×10^7 erg. What is this energy value in joules?

$$5 \times 10^7 \text{ erg} \times \frac{1 \text{ J}}{10^7 \text{ erg}} = \boxed{5 \text{ J}}$$

Chemical reactions, while fundamentally a matter of electrical interactions among atoms, are most often described in terms of the amount of heat absorbed (endothermic reactions) or heat liberated (exothermic reactions). On the other hand, there are numerous reactions, especially in the investigations of biological chemistry, in which electrical measurements are more readily obtained than direct heat measurements. There is no question about the practicality of being able to interconvert the heat and electrical units of energy measurement, and this is usually done in terms of calories.

The specific heat capacity (**specific heat**) of a substance is the number of calories required to raise the temperature of one gram of that substance by $1°C$:

$$\text{sp. ht} = \frac{\text{cal}}{\text{g} °C}$$

Water, by definition, has a specific heat of one cal/(g °C). The specific heat of a substance varies slightly with temperature, but the variations are so small that they are normally disregarded. The variation of the specific heat of water is minimal between 14.5°C and 15.5°C, and we assume that it remains constant regardless of temperature.

Table 2–1 lists specific heats for several common substances. Note that specific heat values of most elements and compounds listed are much smaller than that for water. Water has one of the highest specific heats known. We will discuss reasons for this in subsequent chapters.

TABLE 2–1 **Specific Heats of Common Substances (at 25°C)**

Substance	Specific Heat $\dfrac{cal}{g\ °C}$
aluminum	0.215
gold	0.030
sulfur	0.177
mercury	0.033
hydrogen	3.42
water (liquid)	1.0
water (ice)	0.50
water (steam)	0.48
alcohol	0.581
beach sand	0.188
glass	0.199
air	0.291

Calories cannot be measured directly; they must be calculated. The amount of heat associated with a reaction or process is determined by measuring the change in temperature of a substance with a known mass and a known specific heat. A calorimeter is a device used to obtain data that enable us to calculate the heat liberated or absorbed when a given amount of material reacts. It consists of an insulated vessel containing a weighed amount of water, a thermometer, and a stirring device. The heat produced by the reaction changes the temperature of the water, which is stirred in order to maintain a uniform temperature. The energy absorbed by the water is equal to the energy produced by the reaction. Suitable corrections are made for small heat losses to the reaction chamber, thermometer, and stirrer. The amount of rise in water temperature is obtained by reading the thermometer before and after the reaction. Since we know the specific heat of water, the total number of calories liberated can be calculated. To calculate the heat lost or gained, a useful relationship is:

$$cal = mass \times specific\ heat \times temperature\ change$$

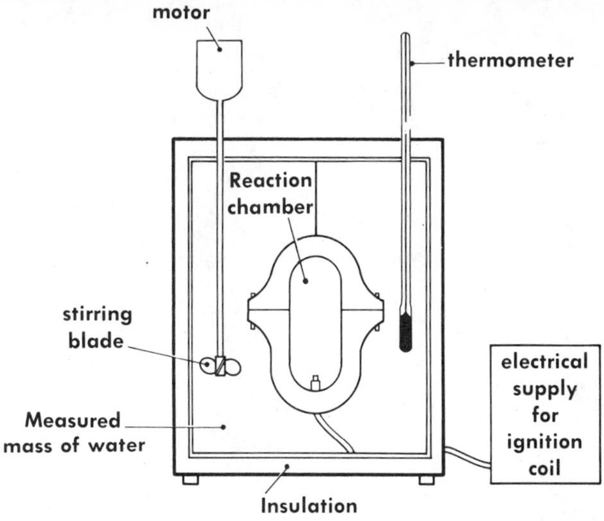

Figure 2–8   Diagrammatic representation of a calorimeter.

For example, if 200 grams of surrounding water in a calorimeter (Fig. 2–8) is raised by 2.5°C, the heat produced by the reaction is:

calories gained by the water = calories lost in the reaction

$$= (m)(\text{sp. ht.})(\Delta t)$$

$$= (200 \cancel{g})\left(\frac{1.0 \text{ cal}}{\cancel{g} \, °\cancel{C}}\right)(2.5°\cancel{C})$$

$$= 500 \text{ cal}$$

For large quantities of heat, the kilocalorie (kcal) is used. One kilocalorie is equal to 1000 calories. The calorimeter is also used to obtain data for the calculation of the specific heat of various substances.

Example 2.3

How many calories are required to raise 5.0 grams of water from 10°C to 25°C?
1. Organize the data:

$$\text{cal} = ?$$

$$m = 5.0 \text{ g}$$

$$\text{sp. ht.} = 1.0 \, \frac{\text{cal}}{\text{g}°\text{C}}$$

$$\Delta t \, (\text{temperature change}) = 25° - 10° = 15°\text{C}$$

2. Write the equation, substitute the data, and perform the calculation, using dimensional analysis.

$$\text{cal} = (m)(\text{sp. ht.})(\Delta t)$$

$$\text{cal} = 5.0 \cancel{g} \times 1.0 \, \frac{\text{cal}}{\cancel{g}°\cancel{C}} \times 15°\cancel{C}$$

$$\boxed{75 \text{ cal}}$$

Example 2.4

What temperature rise can be expected if 15.0 grams of glycerol absorb 210.0 cal of heat? The specific heat of glycerol is $0.314 \dfrac{\text{cal}}{\text{g}^\circ\text{C}}$.

1. Organize the data:

$$\text{energy absorbed} = 210.0 \text{ cal}$$

$$m = 15.0 \text{ g}$$

$$\text{sp. ht.} = 0.314 \dfrac{\text{cal}}{\text{g}^\circ\text{C}}$$

$$\Delta t = ? \text{ degrees}$$

2. Solve:

$$\Delta t = 210.0 \text{ cal} \times \dfrac{1 \text{ g}^\circ\text{C}}{0.314 \text{ cal}} \times \dfrac{1}{15.0 \text{ g}}$$

$$\boxed{\Delta t = 44.6^\circ\text{C rise}}$$

TABLE 2–2 **Conversion Factors**

$$1 \text{ cal} = 4.18 \times 10^7 \text{ ergs}$$
$$1 \text{ cal} = 4.18 \text{ joules}$$
$$1 \text{ joule} = 1 \times 10^7 \text{ ergs}$$

Exercise 2.1

1. What is the kinetic energy in ergs of a 10.0 mg particle moving with a velocity of 50.0 meters per second? (Hint: convert units to grams and centimeters.)
2. Express the answer to the above problem in joules.
3. Calculate the velocity in $\dfrac{\text{cm}}{\text{sec}}$ of a particle having a kinetic energy of 3.0×10^{-12} erg and a mass of 6.0×10^{-26} g.
4. How many calories are required to raise the temperature of 45.0 g of water from 3.2°C to 17.4°C?
5. What is the specific heat of a substance if 2.1 kcal (kilocalories) raises 50.0 g of the substance by 72.0°C?

Exercise 2.2

Perform the following energy conversions using Table 2–2:
1. 2.0×10^4 ergs = _____ cal.
2. 4.5×10^{-5} J = _____ erg.
3. 3.4×10^{-5} cal = _____ erg.
4. 7.2×10^7 ergs = _____ J.
5. 620 cal = _____ J.
6. 8.7×10^6 J = _____ cal.

When a pure solid is heated gently, its temperature rises until liquid appears and melting begins.* During the melting process, the solid and liquid states coexist, and the temperature remains constant. After all solid has melted, continued heating produces a rise in temperature until the boiling point is reached. During the process of boiling the temperature again remains constant until all liquid has been converted to gas.

Pure solids have definite and constant melting points. In fact, since every pure substance melts and freezes at a characteristic temperature, melting point determinations are frequently used in the identification of substances. Similarly, pure liquids have a definite and constant boiling point.

---

*A few solid substances pass directly from the solid to the gaseous state when heated, without forming a liquid. This phenomenon is called sublimation, and such substances are said to sublime. For example, when solid iodine is heated slowly it converts, without melting, to dense violet fumes. More commonly, at $-78°C$ solid carbon dioxide ("dry ice") sublimes to form gaseous carbon dioxide.

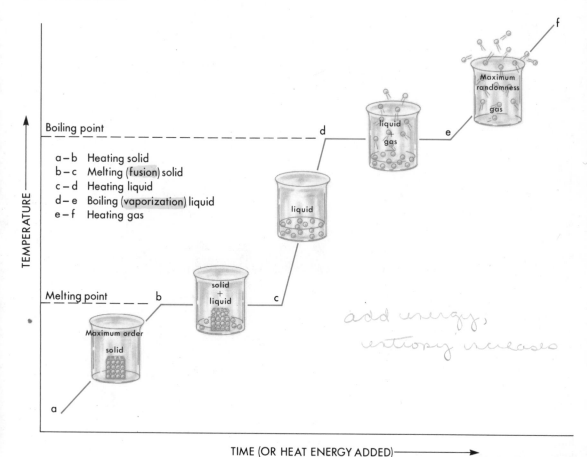

a–b   Heating solid
b–c   Melting (fusion) solid
c–d   Heating liquid
d–e   Boiling (vaporization) liquid
e–f   Heating gas

Figure 2–9   Heating curve showing temperature and changes of state for the heating of a pure substance.

A *heating curve* such as that shown in Figure 2–9 illustrates the relationship of temperatures and changes of state when a pure solid is slowly and uniformly heated. When heat is added at a constant rate, the distance on the time coordinate is also a measure of the quantity of heat energy added. Horizontal portions of a heating curve show periods of melting (b–c) and vaporization (d–e). The sloping lines represent times during which the solid is being heated (a–b), liquid is being heated (c–d), and gas is being heated (e–f).

Figure 2–10 shows the heating curve and energy calculations pertaining to the heating of a 100 gram mass of ice from an initial temperature of $-30°C$ to steam at a final temperature of $150°C$. A 100 gram piece of ice with a thermometer frozen in it at $-30°C$ is placed in a beaker and heat added at a prescribed rate. Temperature readings are taken every few seconds and the data plotted to form the heating

Line Segment	Process	Heat Energy Required	
a to b	Heating solid from $-30°C$ to $0°C$	$(100 \text{ g}) (0.50 \text{ cal/g-°C}) (30°C) =$	1,500 cal
b to c	Melting solid	$(100 \text{ g}) (80 \text{ cal/g})$ $=$	8,000 cal
c to d	Heating liquid from $0°C$ to $100°C$	$(100 \text{ g}) (1.0 \text{ cal/g-°C}) (100°C) =$	10,000 cal
d to e	Vaporizing liquid	$(100 \text{ g}) (540 \text{ cal/g})$ $=$	54,000 cal
e to f	Heating gas from $100°C$ to $150°C$	$(100 \text{ g}) (0.48 \text{ cal/g-°C}) (50°C) =$	2,400 cal

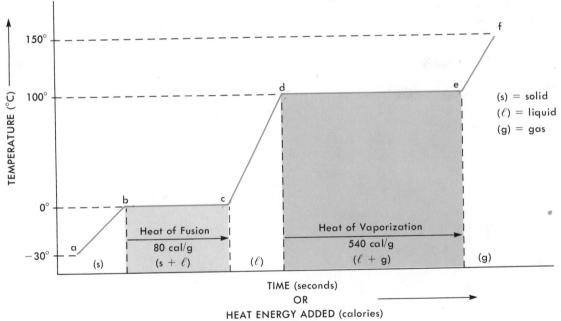

Figure 2–10   Heating curve showing energy requirements for the conversion of 100 grams of ice at an initial temperature of $-30°C$ to steam at $150°C$.

MATTER, ENERGY, AND CHANGE

curve. Specific heats are taken from Table 2–1 for ice (0.50 cal/g°C), liquid water (1.0 cal/g°C), and steam (0.48 cal/g°C).

Along segment (a–b) of Figure 2.10, the temperature of the solid rises with time because of an increase in kinetic energy of the solid ↑KE ↑T particles in ice. At 0°C, some of the solid particles attain enough energy to overcome the attractive forces holding them in a rigid lattice. The crystal begins to melt. During melting (b–c), the temperature remains unchanged because the kinetic energies of the solid and liquid states remain constant. The amount of solid gradually decreases and the amount of liquid increases. When all solid has changed to liquid (c), the average kinetic energy of the liquid increases (c–d), and the temperature rises until the liquid begins to boil (d). At the boiling point, some liquid particles become sufficiently energetic to escape as a gas. Vaporization (d–e) takes place at a constant temperature so long as any liquid remains. After all liquid has been converted to a gas, the temperature of the gaseous state starts going up, and the kinetic energy of the gas particles increases. This is shown by a rising temperature along (e–f).

The quantity of heat absorbed by 1 gram of a solid during melting is called its *heat of fusion.* The heat of fusion of ice is 80 cal/g. This means that it takes 80 calories of heat to melt 1 gram of ice and that 80 calories of heat is liberated when 1 gram of liquid water freezes. The quantity of heat needed to vaporize 1 gram of liquid without changing its temperature is known as the *heat of vaporization.* The heat of vaporization of water is 540 cal/g. This means that 540 calories of heat must be added to water at its boiling point to convert it to steam. Also, the condensation of steam at its boiling point will release 540 calories per gram.

Example 2.5

The specific heat of lead is 0.031 cal/g°C. What mass of steam at 100°C must be condensed to provide enough heat to raise 3.0 g of lead from 80°C to 100°C?

1. Organize the data:

$$\text{sp. ht. of lead} = 0.031 \frac{\text{cal}}{\text{g}^\circ\text{C}}$$

$$\text{Mass of lead} = 3.0 \text{ g}$$

$$\Delta t \text{ of lead} = (100^\circ\text{C} - 80^\circ\text{C}) = 20^\circ\text{C}$$

2. heat of fusion of steam $= 0.48 \frac{\text{cal}}{\text{g}}$ heat of vaporization 540 cal/g

$$\text{Mass of steam} = ?$$

---

2. Solve:

Let m = mass of steam

Heat lost by condensing steam = Heat gained by heating lead

$$\left(540 \frac{cal}{g}\right)m = (3.0 \text{ g})\left(0.031 \frac{cal}{g°C}\right)(20°C)$$

$$m = \frac{3.0 \times 0.031 \times 20}{540} \text{ g}$$

$$\boxed{m = 0.0034 \text{ g}}$$

QUESTIONS & PROBLEMS

2.1 What is matter? What are types, states, and changes of matter?

2.2 What are the two fundamental driving forces behind the changes of matter?

2.3 Name three observations that identify a chemical reaction as being stoichiometric.

2.4 What is a catalyst?

2.5 Label the following changes as physical or chemical: rusting iron, souring milk, dissolving sugar, baking a cake, changing carbon dioxide gas to dry ice.

2.6 Identify the following as pure substances or mixtures: silver, table salt, air, milk, sulfur dioxide, sulfur, tea, brass, nitric acid, ice.

2.7 How fast must a 2.0 g object move in order to have a kinetic energy of 8.0×10^6 ergs?

2.8 If the energy absorbed in a chemical change is 3×10^{-5} J, what is the equivalent energy in ergs?

2.9 How many calories are required to raise the temperature of 140 g of a substance by 22.0° Celsius if the specific heat is 0.232 $\frac{cal}{g°C}$?

2.10 Convert 150 calories to joules and ergs.

2.11 Define:
(a) kinetic energy
(b) potential energy
(c) calorie
(d) specific heat
(e) heat of fusion
(f) heat of vaporization

2.12 How many calories of heat are required to convert 16.0 g of ice at $-18°C$ to water at 90°C?

2.13 Calculate the amount of heat required to change 3.0 g of ice at 0°C to steam at 100°C.

2.14 How many calories are needed to convert 50 g of ice at $-40°C$ to steam at 160°C?

2.15 How many grams of steam at 100°C must be added to 200 g of water at 23°C to raise the water temperature to 75°C?

2.16 How many grams of liquid water can be cooled from 50°C to 0°C by adding 2.0 g of ice?

2.17 How many kilocalories of heat must be removed from 20 kg of water at 30°C to convert it all to ice at 0°C?

2.18   The specific heat of aluminum is 0.215 cal/g°C. How many grams of steam at 100°C must be condensed to release sufficient heat to raise the temperature of 7.18 g of aluminum from 23.0°C to 96.0°C?

2.19   Four grams of steam at 100°C is bubbled through 500 g of water initially at 23.0°C. Calculate the final water temperature.

2.20   What is the specific heat of a metal if 18.9 g of steam at 100°C must be condensed to provide sufficient heat to raise 2.0 kg of the metal from 55°C to 100°C?

CHAPTER THREE

LEARNING OBJECTIVES

At the completion of this chapter, you should be able to:

1. Define elements and compounds, and give examples of each.
2. List several distinguishing characteristics of metallic and nonmetallic elements.
3. Write correct symbols for many elements.
4. Define the law of definite composition.
5. Define the law of multiple proportions.
6. List the three principal subatomic particles by name, symbol, and electrical charge.
7. Define *atomic mass unit.*
8. Calculate atomic numbers, atomic masses, and numbers of each type of subatomic particle from given data.
9. Define the term *isotope.*
10. Calculate average atomic masses from natural isotope distribution.
11. Define the term *mole.*
12. Calculate the molar masses of elements and compounds.
13. Find the number of moles of a substance when the mass is given.
14. Calculate the number of particles in a given number of moles.
15. Express moles of solid and liquid elements and compounds in mass units.
16. Explain the significance of Avogadro's number.

ELEMENTS, COMPOUNDS, AND THE MOLE CONCEPT

ELEMENTS, COMPOUNDS, AND THE MOLE CONCEPT

In Chapter 2 a sample of matter was classified as being either a pure substance or a mixture. Pure substances may be subdivided into *elements* and *compounds*. Because the chemist is concerned primarily with the properties and behavior of pure substances, answers to some fundamental questions must be considered. What are elements and compounds? What are they composed of? How are the descriptions of their properties, composition, and changes quantitatively handled? Is there some central and unifying concept that can provide us with a solid basis as we attempt to interrelate quantitative observations? The elements of matter provide the logical starting point for such inquiries.

3.1 THE ELEMENTS

Elements are pure substances, each of which is composed of one type of atom only. Elements cannot be subdivided into simpler substances. We can think of elements as the "building blocks" of compounds.

Elements are commonly described as **metallic** or **nonmetallic,** although a number of them fall into a gray region that indistinctly separates the two. Elements that exhibit the properties of both metals and nonmetals are called **metalloids.** Further reference to metalloids will be made in Chapter 7, which deals with the periodic chart.

Of the 103* recorded elements, only about 20 per cent are nonmetals. The general properties that differentiate metals from nonmetals are summarized in Table 3–1.

All elements have distinct names and symbols. Many of the symbols come from the ancient Latin or Greek names of elements and thus are not abbreviations of the common English names. This is especially true of metals such as copper, tin, lead, gold, and others that were known to ancient alchemists.

---

*Actually, 105 elements have been reported and named! However, kurchatovium (Ku) or rutherfordium (Rf) 104 and hahnium (Ha) 105, have not yet been officially recognized.

TABLE 3-1 **General Characteristics of Metallic and Nonmetallic Elements**

Metals	Nonmetals
Solid (except mercury)	Gas, liquid, or solid
Silvery color (except copper and gold)	Variety of colors or colorless
Lustrous	Most solid nonmetals are brittle and therefore are not malleable or ductile
Most are ductile	Poor conductors of heat and electricity (except carbon, which is a good conductor of electricity)
Some are malleable	
Fair to good conductors of heat and electricity	
Common Examples	*Common Examples*
gold	oxygen
silver	sulfur
zinc	chlorine
iron	carbon

Symbols for elements are shown by either one or two letters. Some, like fluorine (F), potassium (K), and vanadium (V), are represented by a single capital letter, but most elements have symbols comprised of two letters. In such cases, the first letter is a capital and the second is a lowercase letter. Thus sodium is Na, copper is Cu, and zinc is Zn. It is important that the second letter be lowercase and not a reduced or tiny capital letter. It is incorrect to write Cu for copper or Zn for zinc.

Examples:

uranium	U
sulfur	S
oxygen	O
calcium	Ca (not CA)
iron	Fe (not FE)
zinc	Zn (not ZN)

Table 3-2 lists several elements with their symbols and names.

TABLE 3-2 **Symbols and Derivations of Selected Elements**

Element	Symbol	Derivation
bromine	Br	Greek *bromos* = stench
cobalt	Co	German *Kobald* = goblin
copper	Cu	Latin *cuprum* = the island of Cyprus
gold	Au	Latin *aurum* = shining dawn
iron	Fe	Latin *ferrum* = iron
lead	Pb	Latin *plumbum* = lead
mercury	Hg	Latin *hydrargyrum* = mercury
phosphorus	P	Greek *phosphoros* = light-bearing
potassium	K	Latin *kalium*
sodium	Na	Latin *natrium*
silver	Ag	Latin *argentum*
tin	Sn	Latin *stannum*

ELEMENTS, COMPOUNDS, AND THE MOLE CONCEPT

Most of the elements are solids at room temperature. Only mercury and bromine are liquid. Some common elemental gases such as oxygen, nitrogen, fluorine, chlorine, and hydrogen occur naturally in diatomic form; that is, two atoms combine to form a gaseous molecule. These diatomic elements are written O_2, N_2, F_2, Cl_2, and H_2, respectively. Two other diatomic elements are bromine (Br_2) and iodine (I_2).

Diatomic
O_2 N_2 F_2 Cl_2 H_2
Br_2 I_2

A familiarity with the names and symbols of the elements is very important for the sake of working efficiently in chemistry. Constant reference to a periodic chart or a table of the elements helps to make the recommended memorization of the most commonly used elements a rather painless experience.

3.2 UNIT STRUCTURE OF ELEMENTS

While the concept of the atom as a microscopic unit structure of matter dates from the ancient Greeks, it remained for John Dalton at the beginning of the nineteenth century to propose a remarkable theory of the structure of matter. Despite very limited experimental evidence, Dalton's logic suggested the following:

1. Elements are composed of indivisible solid spheres called atoms (from the Greek *atomos,* meaning indivisible).
2. All atoms of a single element have the same mass and size. *— false*
3. Different elements have atoms that differ in mass and size.
4. Molecules of a compound are formed by the union of two or more atoms of different elements.
5. The atoms that combine to form molecules do so in simple whole-number ratios. For example:

   2 H atoms and 1 O atom form one molecule of water, H_2O
   1 N atom and 3 H atoms form one molecule of ammonia, NH_3
   2 C atoms and 6 H atoms and 1 O atom form one molecule of ethyl alcohol, C_2H_5OH

6. Atoms of some elements may combine in different whole-number ratios to form different molecules. For example:

   1 C atom and 1 O atom form carbon monoxide, CO
   1 C atom and 2 O atoms form carbon dioxide, CO_2
   1 N atom and 1 O atom form nitrogen monoxide, NO
   2 N atoms and 3 O atoms form dinitrogen trioxide, N_2O_3
   1 N atom and 2 O atoms form nitrogen dioxide, NO_2
   2 N atoms and 5 O atoms form dinitrogen pentoxide, N_2O_5

Law of def comp.
Law of mult prop.

Today, of course, our model of the atom is vastly more sophisticated than Dalton's. The atom is more complex than his solid-sphere concept, which he developed without knowledge of the subatomic particles called protons, neutrons, and electrons. Furthermore, atoms can be decomposed by the extraordinary methods used in making atomic energy available, and not all the atoms of a particular element have the same mass. (See Isotopes, 3.6.) Dalton's efforts, however, did give rise to two fundamental laws of chemistry that still hold true. They are the **law of definite composition** and the **law of multiple proportions.**

3.3 THE LAW OF DEFINITE COMPOSITION

This law says, in effect, that the mass proportions of elements in a given compound are always the same. If a sample of pure water is analyzed and found to be 88.8 per cent oxygen by mass and 11.2 per cent hydrogen, this is now and forever the mass ratio of oxygen to hydrogen. Any 100.0 g mass of water, be it from the laboratory, the North Pole, or Mars, must contain the chemical combination 88.8 g of oxygen and 11.2 g of hydrogen. If the mass of what is supposed to be water does not have the specified mass ratio, then it is, in fact, not water.

3.4 THE LAW OF MULTIPLE PROPORTIONS

This law states that when two elements can combine to form more than one compound, the mass ratio of one of the elements will vary by small whole-number multiples when compared to a fixed mass of the other element. For example, nitrogen combines with oxygen to form several compounds. Table 3–3 lists the masses of oxygen in those compounds that combine with 100.0 g of nitrogen.

TABLE 3–3 **Oxygen Mass in Compounds Combining with 100.0 g Nitrogen**

Compound	Nitrogen (g)	Oxygen (g)	Whole-Number Multiples of Oxygen Mass
N_2O	100.0	57.0	$\dfrac{57.0}{57.0} = 1$
NO	100.0	114.0	$\dfrac{114.0}{57.0} = 2$
N_2O_3	100.0	171.0	$\dfrac{171.0}{57.0} = 3$
NO_2	100.0	228.0	$\dfrac{228.0}{57.0} = 4$
N_2O_5	100.0	285.0	$\dfrac{285.0}{57.0} = 5$

The data in Table 3–3 illustrate the law of multiple proportions insofar as the masses of oxygen that combine with 100.0 g of nitrogen are (1×57.0)g, (2×57.0)g, (3×57.0)g, (4×57.0)g, and (5×57.0)g.

ELEMENTS, COMPOUNDS, AND THE MOLE CONCEPT

One of the first significant advances after the Dalton atom was made by J. J. Thomson in 1897, when he discovered the charge to mass ratio and the units of negative electrical charge called **electrons** (Fig. 3–1). In 1907, R. A. Millikan found the charge on the electron to be 1.6×10^{-19} coulomb* (Fig. 3–2). This value, used in connection with Thomson's discovery of the charge-to-mass ratio for the electron (e/m), 1.8×10^8 coul/g (coulombs per gram), indicates that the apparent mass of the electron is 9.1×10^{-28}g. This is about as close to nothing as you can get.

$$charge = 1.6 \times 10^{-19}$$
$$\frac{e}{m} = 1.8 \times 10^8 \frac{coul}{g}$$
$$e = 9.1 \times 10^{-28} g$$

This information led Thomson to postulate that an atom is a sphere of positive charge in which electrons are distributed much like raisins in a pudding. Thomson's "plum pudding" model did not last very long. It is interesting to note that the nature of the electron is less clearly understood today than it was thought to be understood at the turn of the century. Although the electron is commonly pictured as a particle having a specific charge and mass, its mass is not a very meaningful value, and its size (after all, a particle of matter should have physical dimensions) is practically impossible to specify. That is to say, the electron is described as an approximate *point* of charge, and a point has no dimensions! But the electron sometimes behaves as though it had size. Depending on the conditions under which electrons are observed, the size of a single electron may seem to be as large as a whole atom or even larger. This recitation of the mysterious aspects of

---

*A coulomb is a unit of electrical charge. This unit, which has been standardized by international agreement, can be visualized very roughly if you try to picture the total amount of static electricity picked up when a hundred people run plastic combs through their dry hair.

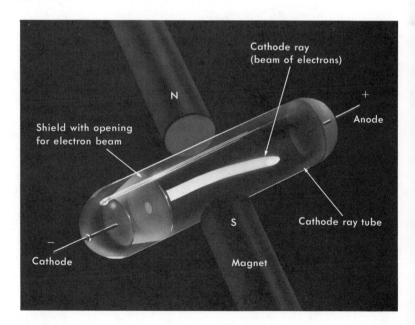

Figure 3–1 Thomson used magnetic and electrical fields in such a way as to determine that the cathode ray was composed of negatively charged particles (electrons), which had specific charge to mass ratio (e/m). The degree to which the cathode ray bends as it moves perpendicular to the magnetic lines of force is dependent on the charge and mass of the electrons.

Cathode ray
(beam of electrons)

N

+

Anode

Shield with opening
for electron beam

S

Cathode ray tube

Cathode

Magnet

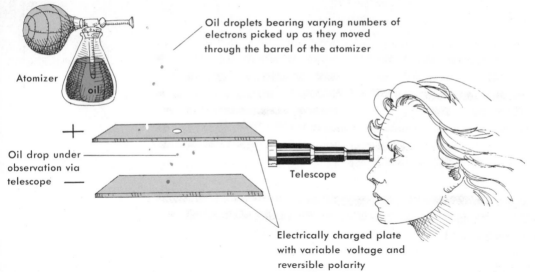

Oil droplets bearing varying numbers of electrons picked up as they moved through the barrel of the atomizer

Atomizer

oil

Oil drop under observation via telescope

Telescope

Electrically charged plate with variable voltage and reversible polarity

Figure 3—2　Simplified diagrammatic representation of Millikan's "oil drop" apparatus. The measurement of voltage applied to the plates as it affected the rate of fall of oil drops enabled Millikan to find the charge on the single electron.

the electron is meant to emphasize the need to distinguish between models and reality. It is very useful to think of the electron as a particle when it *behaves* like a particle. An analogy, however, may be found in the fact that while a man can swim and may sometimes behave like a fish, this does not make him a fish in reality.

In 1911, Ernest Rutherford proved experimentally that the atom has an extremely small and dense core. This discovery led to the acceptance of the nuclear model of the atom, in which a dense nucleus is

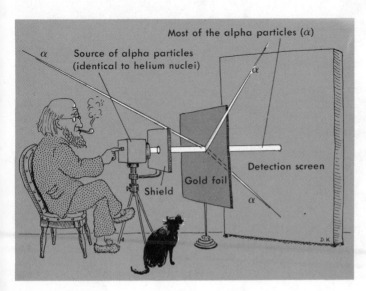

Most of the alpha particles (α)

α　Source of alpha particles (identical to helium nuclei)

α

Detection screen

Gold foil

Shield

α

Figure 3—3　Rutherford's "gold foil" experiment showed that the occasional scattering of alpha particles must be due to an extraordinary concentration of mass (and positive charge) in the core (nucleus) of the gold atoms. Alpha particles, which are identical to helium nuclei (two protons and two neutrons), are produced by the nuclear disintegration of various atoms.

　ELEMENTS, COMPOUNDS, AND THE MOLE CONCEPT

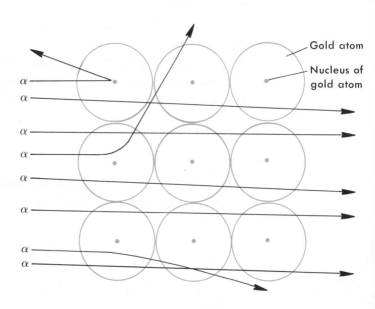

Figure 3–4 Diagrammatic representation of alpha-particle scattering as the gold atom nuclei are approached by alpha particles. Because alpha particles and gold nuclei are both positively charged, the alpha-particle scattering serves as an example of the axiom "like charges repel." The alpha particles are scattered because the much more massive gold nuclei remain fixed, as a large rock would remain fixed when bombarded by Ping-Pong balls.

surrounded by a complement of electrons. Although the electrons constitute the greater volume of the atom (very close to 100 per cent), the remaining subatomic particles compose the nucleus and contain almost all of the mass. See Figures 3–3 and 3–4 for a simplified diagrammatic representation of Rutherford's "gold foil" experiment.

Wilhelm Wein and J. J. Thomson identified the subatomic particle of positive charge, which was given the name **proton.** It remained for Rutherford to establish that the proton is a nuclear particle. The charge on the proton is the same as that of the electron but opposite in sign, and its mass is early 2000 times greater than that of the electron. The proton is assigned the arbitrary mass of 1 **atomic mass unit** (1 amu), because actual masses are difficult to measure. Today, an atomic mass unit is alternately defined as 1/12 the mass of the carbon-12 isotope. The electron, by comparison, is 0.00055 amu.

$1 p = 1 amu$

$1.6 \times 10^{-24} g$

3.6 ISOTOPES

Discussion of the isotope carbon-12 raises questions about the definition and significance of isotopes in general. Isotopes of an element are atoms that have the same number of protons (atomic number) but differing numbers of neutrons (nuclear particles having no electrical charge but with a mass nearly equal to the proton). The variation in number of neutrons does not make different elements. The atoms have different masses and occasionally have different degrees of stability, but they have the same number of electrons and the same chemical behavior. Some isotopes are notoriously unstable. They tend to undergo nuclear decomposition at rates that vary from microseconds to centuries, emitting radioactive particles in the process. Table 3–4

$1 N \cong 1 amu$

lists some isotopes that may be familiar because their uses or dangers have often been discussed in newspapers and magazines.

The reason that most elements have atomic masses that are not perfectly whole numbers is because they have two or more isotopes, of which the naturally occurring element is a mixture. Any natural sample of chlorine gas would be found to be 75.53 per cent Cl-35 and 24.47 per cent Cl-37 by mass, and the resultant atomic mass is 35.45 amu. The average atomic mass has to be weighed toward the Cl-35 because approximately 3/4 of the naturally occurring chlorine has that mass. The weighted average is obtained by multiplying each value by its percent abundance and calculating the sum.

Example 3.1

Find the average atomic mass of chlorine from its natural distribution:

$$\text{chlorine-35} = 75.53\%$$

$$\text{chlorine-37} = 24.47\%$$

1. Convert the percentage to a decimal value and multiply by the more accurate isotopic mass obtained from a reference table:

$$34.97\ (0.7553) = 26.41\ \text{amu}$$

$$36.97\ (0.2447) = 9.046\ \text{amu}$$

2. The sum of the products is the average atomic mass. This is called a *weighted average*.

$$\begin{array}{r} 26.41\ \text{amu} \\ + \ \ 9.046\ \text{amu} \\ \hline 35.46\ \text{amu} \end{array}$$

To account for the difference between the number of protons and the larger value for the mass of the entire atom, Rutherford, in 1920, predicted the existence of an electrically neutral particle in the nucleus that would have about the same mass as the proton. In 1932, James Chadwick discovered the **neutron** and thus validated Rutherford's prediction. Table 3-5 summarizes the characteristics of the principal subatomic particles.

We noted previously that the mass of a proton is about 1.6×10^{-24} g. In other words, if atomic mass units are converted to grams, 1.6×10^{-24} g $= 1$ amu. This system would lead to the expression of a carbon-12 isotope mass as $12 \times 1.6 \times 10^{-24}$ g $= 1.92 \times 10^{-23}$ g. To describe this method of expression atomic masses as cumbersome is something of an understatement. The alternative that chemists have adoped is to establish a system of *relative atomic masses*. Atomic masses are based on a comparison of the mass of a species to that of carbon-12. Magnesium, for example, has an atomic mass of 24 because a magnesium atom is about twice as massive as a carbon-12 atom.

ELEMENTS, COMPOUNDS, AND THE MOLE CONCEPT

TABLE 3–4 Isotopes

Isotope	Reason for Fame or Infamy
Carbon-14	Dating of objects belonging to ancient civilizations
Cobalt-60	Cancer therapy
Strontium-90	Radioactive fallout from atomic bombs
Iodine-131	Thyroid therapy
Uranium-235	Atomic energy
Hydrogen-3 (tritium)	Active substance of hydrogen bombs
Phosphorus-32	Biological tracer

TABLE 3–5 Subatomic Particles

Particle	Symbol	Electrical Charge	Mass (amu)	Approximate Mass (g)
proton	p+	+1	1.0073	1.6×10^{-24}
neutron	n⁰	0	1.0087	1.6×10^{-24}
electron	e⁻	−1	0.0005	9.1×10^{-28}

A convenient method for establishing a comparative system of masses when it is impractical or impossible to compare single units is to compare *equally large numbers* of the units. A comparison of the approximate masses of a penny, a nickel, and a quarter is illustrated in Figure 3–5.

In the same manner that 1000 of each coin produces the same mass ratio as single coins, very large but equal numbers of atoms may be compared. The specific large number used is 6.02×10^{23}, known

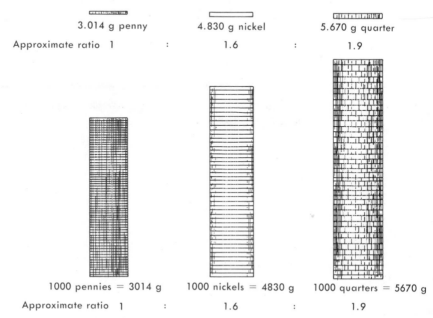

3.014 g penny	4.830 g nickel	5.670 g quarter
Approximate ratio 1 :	1.6 :	1.9

1000 pennies = 3014 g	1000 nickels = 4830 g	1000 quarters = 5670 g
Approximate ratio 1 :	1.6 :	1.9

Figure 3–5 The ratio of the masses of three individual coins is the same as the ratios of a thousand of each coin.

6.02×10²³ (handwritten)

as **Avogadro's number.** Avogadro's number provides the basis for the mole concept, which is one of the most important quantitative tools in chemistry.

3.7 ATOMIC MASS

$p^+ = e^-$ (handwritten)

A look at the periodic chart of the elements might show carbon listed as in Figure 3–6. In the figure the **atomic number, 6,** indicates the number of protons in the nucleus of the atom or the number of electrons in a neutral atom. The atomic mass, 12.011, is the mass of the weighted average of the naturally occurring carbon isotopes C-12 and C-13. The mass number of an isotope minus its atomic number equals the number of neutrons. The conventional way of indicating the atomic mass number and the atomic number of an atom is to write the mass number as a *superscript* to the left of the symbol and the atomic number as a *subscript* to the left.

Atomic # (handwritten)
MASS# (handwritten)

$$^{12}_{6}C$$

6 protons
6 electrons
$12 - 6 = 6$ neutrons

$$^{56}_{26}Fe$$

26 protons
26 electrons
$56 - 26 = 30$ neutrons

$$^{238}_{92}U$$

92 protons
92 electrons
$238 - 92 = 146$ neutrons

The rounded-off atomic mass is the sum of the number of protons and neutrons, because both nuclear particles have the approximate value of 1 amu. This mass of the electron is ignored because it is so small by comparison. Remember that it would require nearly 2000 electrons to have a mass of 1 amu and that the largest atoms have only about 100 electrons.

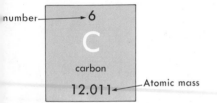

number

6

C

carbon

12.011 — Atomic mass

Figure 3–6 Carbon, as it might be shown on a common form of the periodic table.

ELEMENTS, COMPOUNDS, AND THE MOLE CONCEPT

Exercise 3.1

1. Write the symbol, with atomic numbers as subscripts and mass numbers as superscripts, for the following:
   (a) sodium $^{23}_{11}NA$ $^{96}_{42}Mo$
   (b) phosphorus $^{31}_{15}P$
   (c) nickel $^{207}_{82}Pb$
   (d) molybdenum
   (e) lead $^{59}_{28}Ni$
2. Complete the following chart:

Isotope	Atomic Number	Number of Neutrons	Number of Electrons	Number of Protons	Atomic Mass
$^{27}_{13}Al$	13	14	13	13	27
Chromium	24	28	24	24	52
$_{35}Br$	35	45	35	35	80
Au	79	118	79	79	197
^{32}S	16	16	16	16	32

We mentioned earlier that an organized table of relative atomic masses can be obtained by weighing equally large numbers of atoms of different elements. The validity of larger mass in the same ratio was illustrated in Figure 3–5.

The specific large number used by chemists is 6.02×10^{23}. This number, known as Avogadro's number and symbolized N, can hardly be described as being merely "large." The term "large" does not begin to describe the enormity of 10^{23}. An analogy that might suggest a picture of Avogadro's number would be to try to imagine how much of the earth's surface could be covered by 6.02×10^{23} marbles. The answer is that the entire surface of our planet could be covered by Avogadro's number of marbles to a depth of more than 50 miles!

Avogadro's number of particles is called a **mole,** from the Greek word meaning "pile" or "mound." Although Amadeo Avogadro did not experimentally determine the number that bears his name, he has been so honored because his studies of gas behavior indirectly led to the finding of the number. Several experimental methods can be used to verify that Avogadro's number is indeed 6.02×10^{23}, but discussions of electrodeposition, X-ray diffraction, and monomolecular layers as examples of these experimental techniques are beyond the scope of this text.

Avogadro's number of anything may be specified as one mole; 6.02×10^{23} atoms of lead is a mole of lead *atoms;* 6.02×10^{23} molecules of carbon dioxide is a mole of carbon dioxide *molecules;* and

6.02 × 10²³ sneakers (perish the thought) is a mole of sneakers. How-ever, the *internationally recognized standard for a mole is the number of atoms in 12.000 grams of carbon-12*. This number of atoms is, of course, Avoga-dro's number.

3.9 MOLES OF ELEMENTS

The approximate values for 40 g of calcium, 108 g of silver, and 207 g of lead are each the mass of one mole of the respective element. Why? The answer is discovered by another look at the carbon-12 atom, which is the basis for the atomic mass scale. Observe that the mass number of the carbon-12 atom is the sum of the 6 protons and 6 neu-trons that compose its nucleus. The sum of the nuclear units is equal to 12 atomic mass units. This value cannot easily be measured. But the mass of 6.02 × 10²³ atoms of carbon-12 can be measured with common devices. It is found to be 12 grams. Notice that the sum of protons and neutrons yields a numerical value in arbitrary units that is exactly the same as the numerical mass value of the mole of atoms of C-12 in grams (Fig. 3–7). Comparing other atoms in the same way, it can be demonstrated that the atomic mass in atomic mass units can be multi-plied by Avogadro's number to provide a table of relative molar masses in grams, where each value is the mass of a mole of atoms. Figure 3–8 illustrates this comparison.

It is often convenient to use words that relate a number of things to an equivalent mass. The words "gross" (144), "dozen" (12), and "mole" (6.02 × 10²³) may be used in such a manner. For example, a gross of dimes weighs 349.82 grams, and a gross of quarters weighs 898.20 grams. Thus, we may say that a gross of dimes (144 dimes) is equal to 349.82 grams of dimes; similarly, a gross of quarters is equal to 898.20 grams.

In chemistry the mole is frequently used to relate Avogadro's number to an equivalent mass. When we speak of a mole of chlorine molecules (Cl_2), we mean 6.02 × 10²³ molecules weigh 71.0 grams. A mole of water molecules is 6.02 × 10²³ molecules and weighs 18.0 grams. The mass of any substance that is composed of Avogadro's num-ber of units is called the *molar mass*.

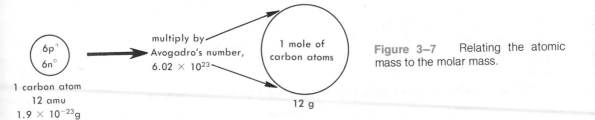

1 carbon atom
12 amu
1.9 × 10⁻²³g

multiply by Avogadro's number, 6.02 × 10²³

1 mole of carbon atoms

12 g

Figure 3–7 Relating the atomic mass to the molar mass.

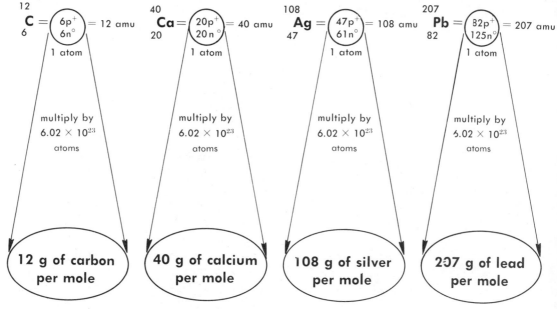

Figure 3–8   Relating the mass numbers of several isotopes to their molar masses.

3.10   MASS, MOLES, AND NUMBERS OF ATOMS

The beauty and utility of the mole concept lies in the fact that once we know how many moles of a substance we have, we can express the amount in units of mass and in numbers of particles. Later we shall see how this can also be done with volumes of gases. We can just as profitably work in reverse as we solve problems involving the mole concept in which we need to take measured masses, volumes, and numbers of particles and express these amounts in moles. We must decide what we are looking for, what data are available, and what kind of an answer makes sense. Dimensional analysis is the surest method to avoid the mistakes that have perennially made the mole concept a horror story for students instead of a simplification of the problem-solving process. Consider the following examples.

Example 3.2

How many moles of calcium atoms have a mass of 200 grams?
1. Find the mass of one mole of calcium from a chart of atomic masses:

$$40 \frac{g}{mole}$$

---

2. Using dimensional analysis:

$$\text{number of moles} = 200 \not{g} \times \frac{1 \text{ mole}}{40 \not{g}}$$

$$\boxed{\text{number of moles} = 5.0}$$

Example 3.3

The value 30.6 mg of sodium is the mass of how many *moles*?
1. Express the mass of sodium in grams so that the unit of mass is compatible with the mass of a mole, which is in grams.*

$$30.6 \not{mg} \times \frac{g}{10^3 \not{mg}} = 3.06 \times 10^{-2} \text{ g}$$

2. Find the mass of one mole of sodium from an atomic mass chart:

$$\text{Na} = 23.0 \frac{g}{\text{mole}}$$

3. Solving:

$$n = 30.6 \times 10^{-3} \not{g} \times \frac{1 \text{ mole}}{23.0 \not{g}}$$

$$\boxed{n = 1.33 \times 10^{-3} \text{ mole}}$$

Example 3.4

What is the mass, in grams, of 0.0024 mole of carbon?
1. Find the mass of one mole of carbon:

$$12 \frac{g}{\text{mole}}$$

2. Use scientific notation to express 0.0024 as 2.4×10^{-3}.
3. Using dimensional analysis:

$$\text{mass} = 2.4 \times 10^{-3} \not{\text{mole}} \times \frac{12 \text{ g}}{1 \not{\text{mole}}}$$

$$\boxed{\text{mass} = 2.9 \times 10^{-2} \text{ g}}$$

Example 3.5

How many atoms are there in a 20.0 g sample of silver?
1. Find the mass of one mole of silver:

$$108 \frac{g}{\text{mole}}$$

*An alternative method is to relate milligram masses to *millimoles*. This method, however, is not recommended until you have gained considerable experience.

ELEMENTS, COMPOUNDS, AND THE MOLE CONCEPT

2. Using dimensional analysis:

$$\text{number of atoms} = 20.0 \text{ g} \times \frac{1 \text{ mole}}{108 \text{ g}} \times \frac{6.02 \times 10^{23} \text{ atoms}}{1 \text{ mole}}$$

$$\boxed{\text{number of atoms} = 1.11 \times 10^{23}}$$

Example 3.6

particle → # moles → g

What is the mass of 1 million atoms of chromium?
1. Find the mass of one mole of chromium:

$$52 \frac{\text{g}}{\text{mole}}$$

2. Using dimensional analysis:

$$\text{mass} = 10^6 \text{ atoms} \times \frac{1 \text{ mole}}{6.02 \times 10^{23} \text{ atoms}} \times \frac{52 \text{ g}}{1 \text{ mole}}$$

$$\boxed{\text{mass} = 8.6 \times 10^{-17} \text{ g}}$$

Note: Be careful with arithmetical detail. An answer of 8.6×10^{17} g looks like a tiny mistake (10^{17} instead of 10^{-17}). However, it is an error of 34 orders of magnitude, and it says in effect that the mass of a million chromium atoms begins to approach the mass of our planet, instead of being microscopic.

Exercise 3.2

(a) Calculate the average atomic mass of silicon from the data below:

Isotope	Isotopic Mass	Per cent of Natural Occurrence
silicon-28	27.98 amu	92.2
silicon-29	28.98 amu	4.7
silicon-30	29.97 amu	3.1

(b) What is the mass of 0.0600 mole of iron in grams?
(c) How many moles of aluminum have a mass of 60.0 mg?
(d) What is the mass of one lead atom in grams?
(e) How many atoms of barium are there in 1.2×10^{-2} mole?

3.11   MOLES OF COMPOUNDS

Because a mole of an element is composed of Avogadro's number of atoms, it follows that a mole of a compound is made of the Avogadro number of molecules, or the smallest number of **ions** in an ionic crystal that expresses the fundamental ratio of the elements in the compound. An ion is an electrically charged particle—usually atoms or related groups of atoms having an unequal total number of electrons and protons. There is a difference in structure between molecular and ionic

---

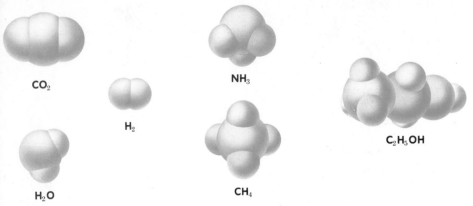

Figure 3–9 Examples of molecules.

substances. Some chemists emphasize this difference by making a distinction between **molecular** masses and **formula** masses. The molecular mass designation is thus reserved for compounds made of molecules (which are defined as discrete, uncharged particles, each composed of a precise number of atoms). A molecule has a describable size, geometry, and mass. We also describe a molecule as the smallest identifiable unit of a substance that can exist as an individual particle under ordinary conditions. A molecule of a compound is composed of more than one kind of atom (CO_2, H_2O, CH_4), whereas molecules of elements consist of one or more atoms of the same kind (H_2, He). Some examples of molecules are illustrated in Figure 3–9.

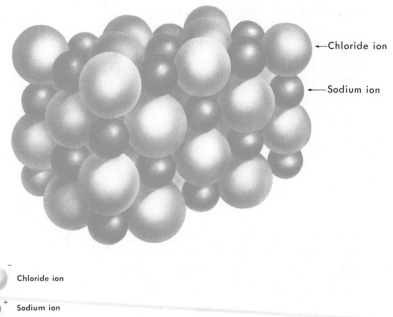

←Chloride ion

←Sodium ion

Chloride ion

Sodium ion

Figure 3–10 The crystal structure of sodium and chloride ions in table salt.

ELEMENTS, COMPOUNDS, AND THE MOLE CONCEPT

The crystal structure formed by the forces of attraction that exist between *ions* is not molecular. For example, table salt (sodium chloride) is a structure composed of sodium ions and chloride ions in a one-to-one ratio, held together by electrostatic forces of attraction. This crystalline compound can form a crystal of almost any reasonable size or mass. Individual sodium chloride units of two ions do not ordinarily exist alone; any visible crystal fragment would contain thousands of ions. The formula, NaCl, represents only the simplest ratio of the ions. It is not meant to suggest a molecule. Figure 3–10 illustrates the arrangement of ions in NaCl.

A more systematic discussion of molecular and ionic compounds will be taken up in the chapter dealing with chemical bonds. However, the application of the mole concept to compounds is uniform regardless of any distinction made between molecular and ionic structures.

Because a molecule of carbon dioxide, for example, is made of one carbon atom and two oxygen atoms (represented by the formula CO_2), it follows that one mole of CO_2 is composed of molecules made from one mole of carbon atoms and two moles of oxygen atoms. If a mole of carbon atoms has a mass of 12 grams, and two moles of oxygen atoms have a mass of 2×16 g, the total mass of a mole of CO_2 is 44 g. We concluded from these calculations that the mass of a mole of a compound is equal to the sum of the atomic masses expressed in grams. This is what we call the **molar mass** of the compound. The term "molar mass" is equally applicable to a nonmolecular compound such as sodium chloride. The one-to-one ratio indicated by the formula NaCl means that the mass of one mole of sodium ions plus the mass of one mole of chlorine ions is equal to the mass of one mole of NaCl units. The molar mass of sodium ions, 23.0 g, added to the molar mass of chlorine ions, 35.5 g, equals the molar mass of NaCl, 58.5 g. In other words, 58.5 g of NaCl is the mass of a mole of table salt. Table 3–6 lists the values for the molar masses of some common compounds.

Some compounds have more complicated formulas than those listed in Table 3–6. Calculations of the mass of a mole of such compounds are illustrated by the following examples.

TABLE 3–6 **Molar Masses**

Compound	Formula	Molar Mass $\left(\dfrac{g}{mole}\right)$
water	H_2O	18
ammonia	NH_3	17
benzene	C_6H_6	78
oxygen	O_2	32
sucrose (table sugar)	$C_{12}H_{22}O_{11}$	342
silver nitrate	$AgNO_3$	170

Example 3.7

What is the mass of a mole of $Ca(NO_3)_2$?
1. The subscript 2 outside the parenthesis means that the number of atoms *within* the parenthesis is doubled. Hence, the formula indicates 2 moles of nitrogen atoms and 6 moles of oxygen atoms.
2. The mass of 1 mole of calcium is added to the total masses of 2 moles of nitrogen and 6 moles of oxygen.

3. The molar mass of $Ca(NO_3)_2$ = $164 \dfrac{g}{mole}$

Example 3.8

Calculate the mass of a mole of $Fe_4[Fe(CN)_6]_3$.
1. Find the total number of each atom per formula. The subscript outside the bracket multiplies the subscript outside the enclosed parenthesis.

$$
\begin{aligned}
7 \text{ moles of Fe} &= 392 \text{ g} \\
18 \text{ moles of C} &= 216 \text{ g} \\
18 \text{ moles of N} &= 252 \text{ g} \\
\hline
\text{Total} &= 860 \text{ g}
\end{aligned}
$$

2. The molar mass of $Fe_4[Fe(CN)_6]_3$ = $860 \dfrac{g}{mole}$

Example 3.9

How many moles of NaOH have a mass of 5.00 g?
1. Find the mass of one mole (i.e., the molar mass) of NaOH from the formula:

$$NaOH = 40.0 \dfrac{g}{mole}$$

2. Using dimensional analysis:

$$n = \dfrac{5.00 \text{ g}}{1} \times \dfrac{1 \text{ mole}}{40.0 \text{ g}}$$

$$n = 0.125 \text{ mole}$$

Example 3.10

What is the mass of 0.042 mole of H_2SO_4?
1. Find the mass of 1 mole of H_2SO_4 from the formula:

$$H_2SO_4 = 98 \dfrac{g}{mole}$$

ELEMENTS, COMPOUNDS, AND THE MOLE CONCEPT

2. Using dimensional analysis:

$$\text{mass} = \frac{0.042 \text{ mole}}{1} \times \frac{98 \text{ g}}{1 \text{ mole}}$$

$$\boxed{\text{mass} = 4.1 \text{ g}}$$

Example 3.11

How many *molecules* are there in 4.0 g of CO_2?
1. Find the mass of a mole of CO_2.

$$CO_2 = 44 \frac{\text{g}}{\text{mole}}$$

2. Using dimensional analysis:

$$\text{number of molecules} = \frac{4.0 \text{ g}}{1} \times \frac{1 \text{ mole}}{44 \text{ g}} \times \frac{6.02 \times 10^{23} \text{ molecules}}{1 \text{ mole}}$$

$$\boxed{\text{number of molecules} = 5.5 \times 10^{22}}$$

Exercise 3.3

1. Calculate the mass of a mole of each of the following:
   (a) $CaBr_2$
   (b) $(NH_4)_2S$
   (c) CH_3COOH
   (d) $Al_2(SO_4)_3$
   (e) $CuSO_4 \cdot 5 \, H_2O$ (Hint: include the mass of 5 moles of water) (Hydroling d)
2. What is the mass of 2.5×10^{-3} mole of H_2O?
3. How many moles are there in 60.0 mg of NH_3?
4. What is the mass of one molecule of CO_2?
5. How many molecules are there in 6.0 mL of mercury if the density of mercury is 13.6 g/mL?

3.12 MOLES IN SOLUTION

Traditionally, the subject of solution "strength" is taken up as part of the general topic of solutions, which appears, more logically, a bit later in a course. Indeed, we shall investigate systematically the nature of the solution process along with related questions in Chapter 9. However, because of the practical advantages gained in laboratory work, we will define solutions operationally and express solution "strengths" in terms of the mole concept so that useful calculations may be performed.

A **solution** is obtained when a substance, called a **solute**, is dispersed evenly throughout the volume of another substance, called a **solvent.** The states of the solutes and solvents may be solid, liquid, or gas; and when both the solute and solvent are in the same state, the

Heterogeneous + mixtures

substance in greater amount is referred to as the solvent. Another requirement of a solution is that the dispersed (or dissolved) particles of solute be sub-microscopic to such an extent that light cannot be reflected. A simple and rather obvious example of a solution is when table sugar is dissolved in water. The sugar is the solute and the water is the solvent. Indeed, water is used so commonly as a solvent that it is often referred to as a "universal" solvent. Of course, there are substances that will not dissolve in water or other solvents, and these substances are described as being **insoluble** in the specified solvent. For example, chalk is insoluble in water. However, we shall restrict our discussion in this chapter to water solutions, also called **aqueous solutions** and abbreviated **(aq)**. Furthermore, although we shall eventually study several methods of describing solution **concentrations** (i.e., "strength" in terms of how much solute is dissolved in a given volume of solvent), our discussion here will deal only with moles of solute per liter of solution, called **molar** concentrations. The **molarity** of a solution is the best means by which we can apply the mole concept to the solving of problems in which some or all of the participating species are in solution. The symbol for the dimension of molarity is **M**.

$$M = \frac{\text{moles of solute}}{\text{liter of solution}}$$

Notice that molarity relates to liters of *solution* rather than liters of solvent. The purpose of this convention is to ensure a particular kind of uniformity: If you have 1 mole of a solute added to *enough* solvent so that a liter of solution is obtained, then you know that 1 mL contains 0.001 mole of solute. Similarly, you know that 100 mL contains 0.1 mole of solute. In order to solve this type of problem, the number of moles must be known, and that information is available only if the volume and molarity of a solution are known.

Example 3.12

How would you prepare 500.0 mL of a 0.200 M aqueous solution of glucose, $C_6H_{12}O_6$?
1. Calculate the molar mass of glucose from the formula, $C_6H_{12}O_6$. Since we are dealing with moles of solute, that is essential information.

$$C_6H_{12}O_6 = 180 \text{ g/mole.}$$

2. The question is: How many grams of solute are needed so that the addition of water to a final solution volume of 500.0 mL will produce the desired concentration of 0.200 mole/liter? Let us use dimensional analysis in order to find the mass of glucose needed.

$$\text{mass of glucose} = \frac{180 \text{ g } C_6H_{12}O_6}{1 \text{ mole } C_6H_{12}O_6} \times \frac{0.200 \text{ mole } C_6H_{12}O_6}{1 \text{ l sol'n}} \times 0.500 \text{ l sol'n}$$

$$\boxed{\text{mass of glucose} = 18.0 \text{ g}}$$

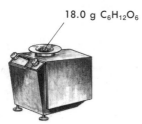

18.0 g $C_6H_{12}O_6$

500.0 mL mark

Figure 3–11 Preparation of 500 mL of 0.200 M solution of glucose. (A) Proper mass of $C_6H_{12}O_6$ weighed. (B) Glucose dissolved in less than 500 mL water. (C) Water added to bring volume of solution up to 500.0 mL.

A   B   C

The actual procedure used to prepare a solution of a specified molarity is to weigh-out the mass of solute we have calculated, transfer the solute to an insufficient volume of solvent, and then use volumetric glassware to bring the volume up to the predetermined level. For example, the 500.0 mL of 0.200 M glucose would be prepared by adding the 18.0 g of solute to about 400 to 450 mL of water, dissolving it, and finally adding water to the 500.0 mL mark on some appropriate vessel such as a volumetric flask (Fig. 3–11).

The next task is to calculate the number of moles of solute from a given molar concentration or from information about the mass of solute and the volume of solution. Examples 3.13 and 3.14 will illustrate this process.

Example 3.13

Find the number of moles of KCN in 400.0 mL of a 1.20 M solution.

For this type of problem we need to determine only the molarity and volume. Since we want moles of KCN, the mass is not needed, and, hence, the molar mass of KCN is irrelevant. Express 400.0 mL in liters and perform the dimensional analysis. Let the symbol, n, stand for moles.

$$n = \frac{1.20 \text{ moles}}{L \text{ soln}} \times 0.4000 \ L \text{ soln} = \boxed{0.480 \text{ mole of KCN}}$$

Example 3.14

If 5.00 g of NaOH (s) is dissolved in water for a total volume of 35.0 mL of solution, what is the molarity of the solution and how many moles of NaOH are there in the 35.0 mL volume?

1. Calculate the molar mass of NaOH from the formula:

molar mass = 40.0 g/mole

2. Express the 35.0 mL volume in liters, since molarity is defined as moles/L:

$$35.0 \text{ mL} \times \frac{L}{10^3 \text{ mL}} = 0.0350 \text{ L}$$

3. Use dimensional analysis to calculate the molarity (moles/L).

$$\text{molarity} = \frac{1 \text{ mole NaOH}}{40.0 \text{ g NaOH}} \times \frac{5.00 \text{ g NaOH}}{0.0350 \text{ L sol'n}} = 3.57 \text{ M}$$

4. The number of moles in 0.0350 L is:

$$n = \frac{3.57 \text{ mole NaOH}}{1 \text{ L sol'n}} \times 0.0350 \text{ L sol'n} = 0.125 \text{ mole}$$

Additional topics dealing with mass relationship in chemical formulas include percentage composition of compounds and empirical formulas. These concepts will be developed in Chapter 4, Chemical Formulas and Nomenclature.

QUESTIONS & PROBLEMS

3.1   List five properties of metals that serve to contrast them with nonmetals.

3.2   Write the symbols for the following elements:
(a) lithium
(b) barium
(c) arsenic
(d) tin
(e) bismuth
(f) fluorine
(g) molybdenum
(h) radium
(i) iodine
(j) krypton

3.3   Name the following elements:
(a) Sc
(b) Hf
(c) Cd
(d) Ar
(e) U
(f) K
(g) Sb
(h) Mn
(i) Sr
(j) V

3.4   Complete the following chart:

Isotope	Atomic Number	Number of Neutrons	Number of Electrons	Number of Protons	Isotopic Mass
$^{91}_{40}Zr$					
$^{66}_{30}Zn$					
$^{195}_{78}Pt$			78		
^{88}Sr	38				
		74			127

ELEMENTS, COMPOUNDS, AND THE MOLE CONCEPT

3.5   Define *mole*.

3.6   What are isotopes of an element?

3.7   How many moles of magnesium are there in 6.0 g?

3.8   How many moles of copper are there in 400.0 mg?

3.9   How many grams are there in 0.0026 mole of silicon?

3.10   What is the mass of an atom of gold?

3:11   What is the total mass of 5.2×10^{-2} mole of iron and 1.2×10^{22} molecules of H_2O?

3.12   How many moles of iron are there in 0.22 g?

3.13   What is the mass of 0.045 mole of NO_2?

3.14   How many molecules of H_2O are needed to produce a mass of 200 g?

3.15   Calculate the molar mass of:
(a) K_2CrO_4
(b) $KMnO_4$
(c) $Na_2S_2O_3$
(d) $Ca_3(PO_4)_2$?, *see* 3.18
(e) H_2SO_3
(f) $Mg(IO_3)_2$

3.16   How many ions are there in 0.20 mole of NaCl?

3.17   What is the mass of 1×10^{18} molecules of CH_4?

3.18   Calculate the molar mass of $Ca_3(PO_4)_2$.

3.19   How many moles of oxygen atoms are there in 600 g of $CaCO_3$?

3.20   How many moles of atoms are there in 49.0 g of H_2SO_4?

3.21   How many oxygen atoms are there in 3.60 moles of CH_3COOH?

3.22   How many nitrogen atoms are there in one kg of NH_3?

3.23   How many grams of silver are there in 0.93 g of $AgNO_3$?

3.24   How many grams of $AgNO_3$ contain 0.060 mole of oxygen atoms?

3.25   Calculate the number of moles of atoms in a mixture of 9.0 g H_2O, 22.0 g CO_2, 68.0 g NH_3 and 8.0 g CH_4.

3.26   Find the moles of Fe atoms in
(a) 1.22 kg of Fe_2O_3.
(b) 454 mg of FeS.

3.27   How many moles of molecules are there in a mixture of 1.0 kg of CO and 220 g of N_2? How many moles of atoms?

3.28   Which contains the greater number of atoms: 10 g of K, 136 g of Br_2, or 5.6 g of Fe?

3.29   What is the molar mass of a compound containing
(a) 0.050 mole in a mass of 3.69 g?
(b) 6.02×10^{22} molecules in a mass of 92.1g?

3.30   Which contains the greater mass?
(a) 3.9 moles of Ba
(b) 3.0 moles of AgCl
(c) 2.4×10^{24} molecules of CH_4

3.31   Calculate the mass of solute needed to prepare each of the following solutions:
(a) 2.00 L of 0.350 M KCN
(b) 50.0 mL of 0.0275 M $CaSO_4$
(c) 450 mL of 2.40×10^{-2} M $H_2C_2O_4$

3.32   How many moles of nitrate ion, NO_3^- (aq), are there in 100.0 mL of 0.500 M solution of $Ca(NO_3)_2$?

3.33   Calculate the molarities resulting from the following preparations:
(a) 20.0 g of NaOH in 250 mL of solution
(b) 5.78 mg of NaBr in 35.0 mL of solution
(c) 72.0 per cent by mass HNO_3 (aq) having a density of 1.42 g/cm^3

CHAPTER FOUR

LEARNING OBJECTIVES

At the completion of this chapter, you should be able to:

1. Name binary and ternary compounds when their formulas are given.
2. Use classical, trivial, and Stock systems of nomenclature.
3. Name hydrated compounds.
4. Write formulas from given names of compounds.
5. Use tables of common oxidation numbers (valences) in writing formulas.
6. Apply the rules for correct formula writing.
7. Use the special nomenclature of acids to write and name their formulas.
8. List the prefixes used in the descriptive nomenclature of nonmetallic compounds.
9. Use the descriptive and Stock systems of nomenclature in naming and writing the formulas of nonmetallic compounds.
10. Calculate the percentage composition of compounds from their chemical formulas.
11. Determine the empirical formulas of compounds from experimental data.
12. Find the true (molecular) formulas of compounds from empirical formulas and molecular masses.

FORMULAS AND NOMENCLATURE

Knowing the names of formulas and being able to write the formulas of compounds correctly are critical skills in chemistry. The efficient use of formulas requires a thorough understanding of the accepted rules and conventions, plus a great deal of practice. A chemical formula is much more than a shorthand method of writing the names of compounds. The formula of a compound specifically indicates the number of each type of atom in a molecule or the ratios of the ions or atoms in those substances that are not molecular. Indirectly, the atomic masses of the constituent atoms in a compound allow the mass ratios to be calculated, as was described in Chapter 3 dealing with the law of definite composition. For example, the formula for water, H_2O, shows that one mole of water molecules is composed of two moles of hydrogen atoms and one mole of oxygen atoms. Furthermore, an 18 gram sample of water is composed of 2 grams of hydrogen and 16 grams of oxygen.

Two related types of calculations in which the mole concept is applied to compounds are **percentage composition** and **empirical formulas.** The percentage composition of a compound expresses the mass ratios of elements or the chemically combined water in hydrated compounds (called water of hydration). Empirical formulas express the simplest ratios of the numbers of each type of atom in a compound. The term *empirical* suggests that these mass data are obtained as a result of direct analysis in the laboratory. The topics of percentage composition and empirical formulas will be discussed in the latter part of this chapter, following a systematic study of chemical nomenclature.

The balancing of chemical equations is vitally dependent upon the accuracy of the formulas of all the reactants and products. There is no such thing as a trivial error in formula writing. For example, $FeCl_2$ is the formula of a compound that is very different from $FeCl_3$, although the error in the subscript appears to be minor. The formula $Ca(OH)_2$ shows that one mole of the compound is composed of one mole of calcium and two moles each of oxygen and hydrogen. The mass of a mole of $Ca(OH)_2$ is quite different from the mass suggested by the incorrect formula $CaOH_2$, where the absence of the parentheses

means that there is only one mole of oxygen in a mole of compound. The preferred rules that govern the writing and naming of formulas were developed by the German inorganic chemist Albert Stock and are known as the **Stock system of nomenclature.** This system is approved by the International Union of Pure and Applied Chemistry (IUPAC). Some chemists and commercial chemical supply companies continue to use the older **classical** and **trivial** (common) names for compounds, but this practice is slowly disappearing. The Stock system is preferred by most chemists today.

A tabulation of some trivial names is presented in Table 4–1. Many trivial names such as water and ammonia are so deeply ingrained by continued usage that they are generally accepted.

4.2 THE BASIS FOR FORMULA WRITING

Success in the skill of formula writing requires a good knowledge of the application of internationally accepted rules and conventions that govern the writing and naming of chemical formulas. These rules and conventions are discussed in the following pages.

Table 4–3 lists essential data for formula writing. The names in parentheses are those using the older classical method of applying the suffix *-ous* or *-ic* to identify ions having multiple oxidation numbers. This older method of nomenclature is not adequate for the unambiguous identification of elements having more than two oxidation states. The "*-ous, -ic*" method is cited in the IUPAC rules as "in use but not recommended." It is strongly urged that the contents of Table 4–2 be memorized much as you would the vocabulary in a language course.

TABLE 4-1 Trivial Names of Common Compounds

Formula	Trivial Name	Chemical Name
$NaHCO_3$	Baking soda	Sodium bicarbonate
$Na_2B_4O_7 \cdot 10\,H_2O$	Borax	Sodium borate 10-water
$MgSO_4 \cdot 7\,H_2O$	Epsom salt	Magnesium sulfate 7-water
CH_3CH_2OH	Grain alcohol	Ethanol
CH_3OH	Wood alcohol	Methanol
N_2O	Nitrous oxide ("laughing gas")	Nitrogen (I) oxide
CaO	Quick lime	Calcium oxide
$Ca(OH)_2$	Slaked lime	Calcium hydroxide
$NaOH$	Lye	Sodium hydroxide
$NaNO_3$	Saltpeter	Sodium nitrate
$C_{12}H_{22}O_{11}$	Table sugar	Sucrose

4.3 THE CONCEPT OF VALENCE

Historical

Writing chemical formulas requires a systematic application of some related and organized table of combining values possessed by atoms and ions. Historically, the combining values, or **valences,** of interacting species (atoms and ions) were determined experimentally. For many years, the "valence" of an element was based on its **combining mass;** also known as the **equivalent mass.** Using oxygen as a standard for comparison, the definition of the equivalent mass of a substance was *that mass which could combine with or replace 8.00 g of oxygen in a chemical reaction.* Although we have little use for the concept of equivalent masses in modern chemistry, it does demonstrate the relationship between the atomic mass, the combining mass, and the small whole number multiple, which we call the "valence," that relates them. For example, relating the atomic mass of oxygen (16.00 g/mole) to its combining mass (8.00 g/equivalent mass) we see the small whole number, 2, which relates them when the combining mass is divided into the atomic mass.

$$\text{Valence} = \frac{16.00\ \dfrac{g}{\text{mole}}}{8.00\ \dfrac{g}{\text{equiv. mass}}} = 2\ \frac{\text{equiv. masses}}{\text{mole}}$$

TABLE 4–2 Valences for selected elements as they relate molar mass to combining weights

Compound	Element	M molar mass (g/mole)	C combining mass from compound analysis (g/equiv. mass)	ratio $\dfrac{M}{C}$ = Valence
Al_2O_3	Al	26.98	8.99	$\dfrac{26.98}{8.99} = 3$
$MgCl_2$	Mg	24.31	12.16	$\dfrac{24.31}{12.16} = 2$
$FeCl_2$	Fe	55.85	27.93	$\dfrac{55.85}{27.93} = 2$
$FeCl_3$	Fe	55.85	18.62	$\dfrac{55.85}{18.62} = 3$
PbO	Pb	207.19	103.60	$\dfrac{207.19}{103.60} = 2$
PbO_2	Pb	207.19	51.80	$\dfrac{207.19}{51.80} = 4$
Cu_2S	Cu	63.54	63.54	$\dfrac{63.54}{63.54} = 1$
CuS	Cu	63.54	31.77	$\dfrac{63.54}{31.77} = 2$

This whole number, 2, in the case of oxygen was known as its "valence," or combining value. Look at Table 4–2 for other examples of empirically derived valences for some selected elements in various compounds.

From Table 4–2, we can see that Fe, Pb, and Cu (three examples among many possibilities) exhibit more than one common valence.

Why any element should have more than one combining value was not understood less than a century ago. Earlier, chemists recognized empirically that iron, for example, had common valences of 2 and 3, which were associated with different chemical properties. Hence, when iron combined with chlorine to form $FeCl_2$ (called ferrous chloride since the suffix -ous is related to the lower valence of 2), the properties of the compound were found to be different as compared to $FeCl_3$ (called ferric chloride since the suffix -ic is related to the higher valence of 3).

But our chemistry forefathers could be just as practical as any of us today. In effect, their attitude was that since the concept of valence worked, it must be true. Therefore, chemists did indeed use valences to write formulas and name compounds. Progress was made with the hope and expectation that deeper understanding would follow. And so it was that with our clearer insights into the structure of the atom, and particularly with our developing ability to describe the distribution of electrons in atoms, a modern concept of valence evolved.

FORMULAS AND NOMENCLATURE

The Modern Concept of Valence

One of the delightful aspects of almost any scientific inquiry is the likelihood that more new questions will be raised than answers found. This is true now because a truly meaningful rationale for our modern system of valence, which is based on **oxidation numbers** and **ionic charges,** is based ultimately on understandings that we shall not come to until the chapters on atomic structure and chemical bonding have been studied. But once again our motive is to speak the language of chemistry now with the understanding that we shall learn the grammar later. Let us start by distinguishing between ionic charges and oxidation numbers (since they are not synonymous) and then relating the two.

Ionic Charges.   Ionic charges, which can be verified experimentally, constitute the most useful pool of factual data from which a systematic table of valences can be made. At this point, we need not be concerned with how many electrons are lost or gained by atoms and groups of atoms. All we need to know is the net charge on the ion. And *for all practical purposes the net charge on the ion constitutes its valence.* Hence, where the classical valences of iron were 2 and 3, we now say in terms of ionic charge that the positively charged ions (cations) Fe^{2+} and Fe^{3+} have valences of $2+$ and $3+$ (say, *plus* two and *plus* three). This assignation of ionic charge to the statement of valence marks a departure from the historical system that includes no recognition of positive or negative signs in association with the combining values of elements. Table 4–3 will present a fairly extensive list of commonly used ionic species to be used as our practical table of valences. While Table 4–3 is meant to serve as a reference table, we suggest that greater speed in formula writing will occur as students set out to memorize the contents as aggressively as possible.

In terms of modern valence, the simple imperative of utmost importance (i.e., the cardinal rule of writing formulas of compounds) is that *the sum of the valences of all the species composing a compound must equal zero.* Thus far we have considered valence in terms of ionic charge, but what about compounds or parts of compounds where the constituents are not in the ionic "state"?

Oxidation Numbers.   The concept of oxidation numbers is an artificial device chemists have constructed in order to describe the "apparent" valence of atoms in both free and combined conditions. Basically, oxidation number relates to the number of electrons *lost* (in reality or hypothetically) by a neutral atom. For example, when an iron atom loses two electrons $Fe - 2e^- \rightarrow Fe^{2+}$, we say that the process of losing electrons constitutes **oxidation** and that the **oxidation number** of the iron has gone from 0 to $+2$. Any free element is said to be in the zero **oxidation state,** and the oxidation number on each atom of that element is zero. Hence, we can see that the ionic charge on the resulting iron ion is the same as the oxidation number. But it is not always true that the oxidation number reflects an ionic charge and for that reason a distinction has to be made. For example:

a — ion has always same charge (can have diff oxidation #'s)

elements are species

→ # of e⁻ lost (can be diff for same element when combined w/ diff. subs.

Fe^{2+} symbolizes an iron ion having a *charge* of "plus" two. We call this an iron (II) ion. Note: the symbol is written with the charge (number written before sign) at the upper right.

$\overset{+2}{Fe}$ symbolizes an iron atom having an *oxidation number* of "plus" two. We say this characterizes iron in the $+2$ oxidation state. Note: this symbol is written with the oxidation number (sign written before number) above.

In the case of iron, the examples are rather straightforward and nonambiguous. However, let us examine chlorine (with which iron may combine) in order to see how the concept of oxidation numbers becomes an invaluable method for "bookkeeping" as the "apparent" valence of an atom demonstrates variability.

TABLE 4–3 A table of modern valences based on commonly used ionic species, alphabetically arranged

Cations		Anions	
Aluminum	Al^{3+}	Acetate	$(CH_3COO^-$ or, $OAc^-)$
Ammonium	NH_4^+	Arsenate	AsO_4^{3-}
Barium	Ba^{2+}	Arsenite	AsO_3^{3-}
Cadmium	Cd^{2+}	Bromide	Br^-
Calcium	Ca^{2+}	Carbonate	CO_3^{2-}
Chromium (II)	Cr^{2+}	Chlorate	ClO_3^-
Chromium (III)	Cr^{3+}	Chloride	Cl^-
Cobalt (II)	Co^{2+}	Chlorite	ClO_2^-
Cobalt (III)	Co^{3+}	Chromate	CrO_4^{2-}
Copper (I)	Cu^+ (cuprous)	Cyanate	OCN^-
Copper (II)	Cu^{2+} (cupric)	Cyanide	CN^-
Hydrogen	H^+	Dichromate	$Cr_2O_7^{2-}$
Iron (II)	Fe^{2+} (ferrous)	Dihydrogen phosphate	$H_2PO_4^-$
Iron (III)	Fe^{3+} (ferric)	Fluoride	F^-
Lead (II)	Pb^{2+} (plumbous)	Hydride	H^-
Lithium	Li^+	Hydrogen carbonate	HCO_3^- (bicarbonate)
Magnesium	Mg^{2+}	Hydrogen sulfate	HSO_4^- (bisulfate)
Manganese (II)	Mn^{2+}	Hydrogen sulfite	HSO_3^- (bisulfite)
*Mercury (I)	Hg_2^{2+} (mercurous)	Hypochlorite	ClO^-
Mercury (II)	Hg^{2+} (mercuric)	Hydroxide	OH^-
Nickel (II)	Ni^{2+}	Iodide	I^-
Nickel (III)	Ni^{3+}	Monohydrogen phosphate	HPO_4^{2-}
Potassium	K^+	Nitrate	NO_3^-
Silver	Ag^+	Nitride	N^{3-}
Sodium	Na^+	Nitrite	NO_2^-
Strontium	Sr^{2+}	Oxalate	$C_2O_4^{2-}$
Tin (II)	Sn^{2+} (stannous)	Oxide	O^{2-}
Zinc	Zn^{2+}	Perchlorate	ClO_4^-
		Permanganate	MnO_4^-
		†Peroxide	O_2^{2-}
		Phosphate	PO_4^{3-}
		Phosphide	P^{3-}
		Sulfate	SO_4^{2-}
		Sulfide	S^{2-}
		Sulfite	SO_3^{2-}
		Thiosulfate	$S_2O_3^{2-}$

*The mercury (I) ion normally occurs as a diatomic ion. Each ion of the diatomic unit has the valence, $1+$. Imagine the structural model ($Hg^+ \cdot Hg^+$).

†The peroxide ion is analogous to the mercury (I) diatomic configuration. Imagine the structural model, ($O^- \cdot O^-$).

First the oxidation number changes as chlorine forms chloride ions:

(diatomic)

$$Cl_2 + 2e^- \rightarrow 2\,Cl^-$$

Here we see a molecule of chlorine *gaining* electrons (one electron for each atom) in the formation of chloride ions. The oxidation number changes from zero to -1 ("minus" one). The minus sign indicates both the ionic charge and the fact that chlorine has undergone a process *opposite* of oxidation and, hence, its oxidation number is negative. Now, let us compare chlorine as we did in the case of iron:

Cl^- symbolizes a chloride ion having a charge of "minus" one. (Conventionally, the numeral 1 is not included with ionic charges.)

$\overset{-1}{Cl}$ symbolizes a chlorine atom having an oxidation number of "minus" one. This characterizes chlorine in the "minus" one oxidation state.

Let us see why the difference between ionic charges and oxidation numbers is more serious than a mere difference in definitions. Although a hard-line distinction between the two expressions of valence is not especially relevant to our current problem (i.e., writing and naming formulas) when we come to study the phenomenon of oxidation and reduction later, we shall be able to appreciate the distinctions. Look at Tables 4–4 and 4–5 for an intriguing insight into some differences between ionic charge and oxidation numbers.

TABLE 4–4 Ionic charges of selected ions containing chlorine or sulfur

Ions	Symbol and Ionic Charge
Chloride	Cl^-
Chlorate	ClO_3^-
Perchlorate	ClO_4^-
Sulfide	S^{2-}
Sulfate	SO_4^{2-}
Hydrogen Sulfate	HSO_4^-
Sulfite	SO_3^{2-}
Thiosulfate	$S_2O_3^{2-}$

polyatomic ions

TABLE 4–5 **Various oxidation numbers of chlorine and sulfur**

Substances	Name	Oxidation numbers of atoms (or ions)
Cl_2	Chlorine	$\overset{0}{Cl_2}$
$NaCl$	Sodium Chloride	$\overset{+1}{Na}\overset{-1}{Cl}$
$NaClO$	Sodium Hypochlorite	$\overset{+1}{Na}\overset{+1}{Cl}\overset{-2}{O}$
$NaClO_3$	Sodium Chlorate	$\overset{+1}{Na}\overset{+5}{Cl}\overset{-2}{O_3}$
$NaClO_4$	Sodium Perchlorate	$\overset{+1}{Na}\overset{+7}{Cl}\overset{-2}{O_4}$
Cl_2O_7	Dichlorine Heptoxide	$\overset{+7}{Cl_2}\overset{-2}{O_7}$
S	Sulfur	$\overset{0}{S}$
Na_2S	Sodium Sulfide	$\overset{+1}{Na_2}\overset{-2}{S}$
SO_2	Sulfur Dioxide	$\overset{+4}{S}\overset{-2}{O_2}$
SO_3	Sulfur Trioxide	$\overset{+6}{S}\overset{-2}{O_3}$
H_2SO_3	Sulfurous Acid	$\overset{+1}{H_2}\overset{+4}{S}\overset{-2}{O_3}$
$Na_2S_2O_3$	Sodium Thiosulfate	$\overset{+1}{Na_2}\overset{+2}{S_2}\overset{-2}{O_3}$

Summarizing Table 4–5 briefly (aware of many unanswered questions —as promised), we see that while the chloride ion and the sulfide ion each have ionic charges and oxidation numbers that are identical (1− and −1 for chloride; 2− and −2 for sulfide), the sampling of a few substances illustrates that a variety of oxidation numbers are possible.

4.4 FORMULA-WRITING RULES

The rules for the writing of formulas are summarized as follows:
1. Careful attention must be paid to capital and lowercase letters in the symbols used.
   Co is the *element* cobalt
   CO is the *compound* carbon monoxide
2. Roman numerals are not used with a formula.
   $FeCl_2$ is iron(II) chloride
   $Fe(II)Cl_2$ is redundant and unconventional and therefore is not to be used
3. Parentheses are used when *more* than one polyatomic ion is indicated

FORMULAS AND NOMENCLATURE

in the formula. Parentheses around a *single* polyatomic ion usually
are not necessary.

Ca(OH)$_2$ is calcium hydroxide; representing

    1 calcium atom

    ② oxygen atoms

    2 hydrogen atoms

CaOH$_2$ is incorrectly written; representing

    1 calcium atom

    ① oxygen atom

    2 hydrogen atoms

KOH is potassium hydroxide

K(OH) contains unnecessary parentheses and is therefore considered wrong by convention

4. In the writing of **binary** (two-element) compounds, the symbol for
the metal is usually written first, followed by the symbol for the
nonmetal.

    Na Cl   is correct

    ↑ ↖

  metal nonmetal

    Cl Na   is incorrect

The same rule for the ordering of symbols applies to **ternary** (three-
element) compounds.

5. When a compound is composed of nonmetals exclusively, the symbol of the more metallic element is written first. This is the element
to the left in a period, or, as a secondary choice, lower in a group
when both elements belong to the same group. Consider examples
taken from a small section of the periodic table (Fig. 4–1). The
following examples will serve to illustrate the rule:

SiC   below carbon in group IVA

NO$_2$ to the left of oxygen in the second period

SO$_3$ below oxygen

P$_4$O$_{10}$ to the left of and below oxygen

SiO$_2$ far to the left of and below oxygen

CS$_2$ far to the left of sulfur

GROUP		
IV A	V A	VI A
second period C	N	O
third period Si	P	S

Figure 4–1   A small section of the periodic table.

6. As with many rules, there are exceptions. In the case of binary
compounds composed of nonmetals, the element that appears first
in this sequence is placed first: B, Si, C, Sb, As, P, N, H, Te, Se, S,
At, I, Br, Cl, O, F.

    Examples: NH$_3$, H$_2$S, S$_2$Cl$_2$, ClO$_2$, OF$_2$.

7. An important rule in formula writing is that *the sum of the positive and negative valences (ionic charges or oxidation numbers) in a compound must be equal to zero.* The rules for formula writing will now be applied and emphasized in a series of examples.

Example 4.1

Write the formula for potassium hydroxide.

1. Set down the components with their respective ionic charges.

   $K+$ $OH-$

2. See whether the sum of the electrical charges equals zero.

   | $1+$ | and | $1-$ | $= 0$ |
   | potassium | | hydroxide | |

3. The original number of potassium and hydroxide ions is accepted as correct. The formula is KOH.

Example 4.2

Write the formula for copper (I) chloride.

1. Set down the components with their respective ionic charges.

   $Cu+$ $Cl-$

2. See whether the sum of the electrical charges equals zero.

   | $1+$ | and | $1-$ | $= 0$ |
   | copper (I) | | chloride | |

3. The original number of ions is accepted as correct. The formula is CuCl.

Example 4.3

Write the formula for copper(II) chloride.

1. Set down the appropriate symbols for the ions involved.

   Cu^{2+} $Cl-$

2. In order to balance the electrical charges, two chloride ions are required.

   $Cu^{2+}, 2Cl-$

3. The formula is $CuCl_2$.

Example 4.4

Write the formula for potassium sulfate.

1. $K+$ SO_4^{2-}
2. Two potassium ions are needed for balancing.

   $2K+, SO_4^{2-}$
3. The formula is as follows:

   K_2SO_4

Note: There are no parentheses used in the case of a single polyatomic group in the formula.

FORMULAS AND NOMENCLATURE

Example 4.5

Write the formula for iron(III) oxalate.
1. Fe^{3+} $C_2O_4^{2-}$
2. The balancing of electrical charges requires a total of six positive and negative charges (the lowest common denominator).
   $2Fe^{3+}, 3(C_2O_4^{2-})$
3. Parentheses *must* be placed around the oxalate ion. The subscript 3 multiplies the number of both the carbon atoms and oxygen atoms composing the polyatomic ion.
4. The completed formula is as follows: $Fe_2(C_2O_4)_3$.

Example 4.6

Write the formula for mercury(I) chloride.
1. Set down the appropriate symbols, remembering that the mercury(I) ion exists as a diatomic unit.
   Hg_2^{2+} Cl^-
2. Balancing of the charge requires two chloride ions.
   $Hg_2^{2+}, 2Cl^-$
3. The charges *appear* to be unbalanced. However, the Hg_2^{2+} notation means that there is in reality only one unit of $(Hg-Hg)^{2+}$. If it were customary to write the formula,
   $(Hg-Hg)Cl_2$
   the appearance of the formula might be less confusing.
4. The completed formula is
   Hg_2Cl_2
   and the subscripts are *not* reduced to make the formula HgCl.

Example 4.7

Write the formula for sodium peroxide.
1. Set down the components, bearing in mind that the peroxide ion is a diatomic species.
   Na^+ O_2^{2-}
2. The peroxide ion may be somewhat misleading as in the case of the Hg_2^{2+} ion. The formula for sodium peroxide follows the same pattern as was illustrated in Example 4.6
3. The balanced formula clearly indicates the need for two sodium ions $2Na^+$, O_2^{2-}.
4. While the formula might be interpreted as $Na_2(OO)$, the conventional form is Na_2O_2
5. Notice once again that the formula is *not* reduced to the simplest ratio, NaO.

Example 4.8

Write the formula for lead(IV) sulfide.
1. Set down the symbols and oxidation numbers for the lead(IV) and sulfur atoms:
   $\overset{+4}{Pb}\ \overset{-2}{S}$
2. Notice that two sulfur atoms are sufficient to make the sum of the oxidation numbers equal to zero.
   $\overset{+4}{Pb}, \overset{-2}{2S}$
3. The correct formula, then, is PbS_2 and *not* Pb_2S_4.

The possibility of writing Pb_2S_4 for lead(IV) sulfide as indicated in Example 4.8 is a more likely mistake when a shortcut method is used for the writing of formulas. Use of the quicker method (see below) assumes that a fundamental understanding of formula writing has been developed and applied in the manner of the previous examples.

The Shortcut Method of Formula Writing

The rapid technique for writing formulas consists of "switching" the common oxidation numbers of the atoms in a compound. When the numerical values *without* the positive and negative signs are exchanged between the atoms and then written as *subscripts*, the formula is usually completed. But you must be careful to check the sum of the product of the oxidation numbers if there is any uncertainty. There is nothing scientific about the "switching" device—it is merely a fast-action "gimmick" that gets the job done most of the time.

Example 4.9

Use the shortcut method to write the formula for sodium chromate.
1. Set down the appropriate symbols.

$$Na^{\oplus} \quad CrO_4{}^{2\ominus}$$

2:4
ratio
usually d
reduced

2. Switch the valences, and the final form of the formula is

$$Na_2CrO_4$$

(cannot reduce
number of
atoms needed
for compound
- diatomic)

Example 4.10

Use the shortcut method to write the following formulas:
1. cobalt(III) sulfide
2. tin(IV) oxide
3. ammonium monohydrogen phosphate
4. nickel(III) hydroxide

1. $Co^{3+} \quad S^{2\ominus}$ = Co_2S_3
? 2. $Sn^{4+} \quad O^{2\ominus}$ = Sn_2O_4

The ratio 2:4 is usually reduced,

$$Sn_2O_4 \quad = \quad SnO_2$$

?, 3. $NH_4{}^{\oplus} \quad HPO_4{}^{2\ominus}$ = $(NH_4)_2HPO_4$
4. $Ni^{3+} \quad OH^{\ominus}$ = $Ni(OH)_3$

Note: In samples 3 and 4, parentheses are needed to indicate the presence of more than one polyatomic ion in the formula.

FORMULAS AND NOMENCLATURE

Exercise 4.1

Write the formulas for the following compounds.
(a) copper(II) sulfate
(b) magnesium hydroxide
(c) aluminum arsenite
(d) cobalt(III) bromide
(e) barium chlorate
(f) sodium peroxide
(g) mercury(I) nitrate
(h) potassium permanganate
(i) calcium oxalate
(j) nickel(II) sulfate
(k) tin(IV) nitrite
(l) cadmium nitride
(m) chromium(II) hydroxide
(n) silver acetate
(o) iron(III) monohydrogen phosphate
(p) ammonium sulfide
(q) potassium thiocyanate
(r) strontium hydrogen carbonate
(s) lead(IV) chromate
(t) mercury(II) hydroxide

4.5 NOMENCLATURE OF BINARY AND TERNARY COMPOUNDS

Naming formulas correctly is most easily accomplished by starting with the subscripts and switching them back to their original total oxidation number (or ionic charge) values. This process is exactly the reverse of the shortcut method for writing formulas, previously described. Remember, too, we mentioned earlier that -*ide* is the binary-compound suffix, except for acids.

- ide binary suffix (except for acids)

Example 4.11

What is the name of the compound $Mg(CN)_2$?
1. Because magnesium has only one common valence, there is no need to make any special distinction such as calling it magnesium(I) or magnesium(II). The inclusion of a Roman numeral when it is unnecessary is wrong, by convention.
2. The name of the formula is simply the combination of the names of the ionic constituents, magnesium cyan*ide*.

Example 4.12

Write the name of the compound $PbCl_2$.
1. Lead has more than one possible common oxidation number. The actual oxidation number of the lead in the $PbCl_2$ compound *must* be included in the name of the compound.
   Stock: lead(II) or lead(IV)
   Classical: plumb*ous* or plumb*ic*

   $PbCl_2$ (Cl^{-1}), Pb^{+2}
2. By switching the subscripts to their oxidation number positions, the lead is observed to be in the $+2$ oxidation state.

$$PbCl_2$$

3. The name of the compound is lead (II) chlor*ide* (plumbous chloride).

Example 4.13

What is the name of the compound having the formula $Fe_3(AsO_4)_2$?
1. Separating the formula into two identifiable components, the name will be as follows:
   Stock: iron(II) or iron(III)
   Classical: ferr*ous* or ferr*ic*
2. AsO_4^{3-} is the arsen*ate* ion.
3. Switching the subscripts to their total oxidation number positions will indicate the valence of the metal atom and hence the correct name of the compound.

$$Fe_3(AsO_4)_2$$

4. The name of the compound is iron(II) arsenate (ferrous arsenate).

Example 4.14

Write the name of the compound $Sn(SO_3)_2$.
1. The metallic part of this ternary compound may be:
   Stock: tin(II) or tin(IV)
   Classical: stannous or stannic
2. The SO_3^{2-} is the sulf*ite* ion.
3. If the sulfite is correctly expressed as SO_3^{2-}, the charge on the tin ion must also be doubled:

$$Sn(SO_3)_2$$

4. The name of the compound is tin(IV) sulfite (stannic sulfite).

Summary of Binary and Ternary Compound Nomenclature

1. Lowercase letters are used throughout. Names of compounds should not be capitalized.
2. When the metallic component of a formula has more than one common valence, it is usually indicated by a Roman numeral in parentheses or by the appropriate *-ous* or *-ic* suffix if the classical system is used. However, it is emphasized that the Stock system is preferred.
3. The suffix on the nonmetallic member of the formula depends on the formal name of the ion or **oxyanion** (complex negative ions containing oxygen). *radicals*
4. It is difficult to generalize oxyanion nomenclature except to say that the ones with the *lower* number of oxygen atoms per ion are likely to be linked to the *-ite* suffix, whereas the ones with more oxygen atoms have the *-ate* suffix.

FORMULAS AND NOMENCLATURE

anion (- ions)
cations (+ ions)

oxyanions
(- ion w/o)

TABLE 4–6 Anion and Oxyanion Nomenclature

	Cl^-	chloride
Lower Number of Oxygens	ClO^-	hypochlorite*
	ClO_2^-	chlorite
Higher Number of Oxygens	ClO_3^-	chlorate
	ClO_4^-	perchlorate**

⎱ oxyanions

Hypo- means less or lower than usual.
**Per-* (from *hyper*) means more or higher than usual.

5. Simple nonoxyanions usually have the suffix *-ide*. Table 4–6 lists some samples of anion and oxyanion nomenclature.

Exercise 4.2

Write the names of the following compounds according to the Stock system.
(a) KF
(b) $NaHCO_3$
(c) $FeCl_2$
(d) $MgSO_3$
(e) $Ca(OH)_2$
(f) Li_3AsO_4
(g) NH_4Cl
(h) HgO
(i) K_2HPO_4
(j) PbS_2
(k) $K_2Cr_2O_7$
(l) AlN
(m) CsCN
(n) $KMnO_4$
(o) CdI_2
(p) Hg_2I_2
(q) $Ni(OCN)_2$
(r) $Sn(CO_3)_2$
(s) $Co_2(C_2O_4)_3$
(t) $Mn(NO_2)_2$

4.6 NOMENCLATURE OF ACIDS

The special nomenclature of acids is derived from the standard nomenclature of binary and ternary compounds and is expressed by a firm set of rules.

Although the topic of acids will be more systematically investigated in a later chapter, they can usually be recognized by their chemical formulas. Acids generally have a formula beginning with *hydrogen*, as in HCl, HNO_3, and H_2SO_4 (hydrochloric acid, nitric acid, and sulfuric acid, respectively), and they are usually dissolved in water.

First Rule.   When the compound has the suffix *-ide* in the standard nomenclature, the corresponding acid name begins with the prefix *hydro-*, followed by the *root* word, and finally by the suffix *-ic* and the terminal word *acid*.

Example 4.15

Name the compound HBr as an acid.
1. The formula begins with hydrogen, indicating an acid.
2. The standard (i.e., nonacid compound) nomenclature would be hydrogen brom*ide*
3. Because the suffix is -*ide,* the root *brom* is preceded by *hydro-* and followed by -*ic.* The name is:

hydro bromic acid
prefix   root   suffix

Table 4–7 lists common roots of some of the nonmetals most frequently included in acids.

TABLE 4–7 **Root Names of Some Nonmetals**

Nonmetal	*Root*
Fluorine	*fluor*
Chlorine	*chlor*
Bromine	*brom*
Iodine	*iod*
Sulfur	*sulfur* or *sulf*
Phosphorus	*phosph* or *phosphor*
Nitrogen	*nitr*

Examples of the application of the first rule are listed in Table 4–8.

TABLE 4–8 **Application of the First Rule in Acid Nomenclature**

Formula	*Standard Nomenclature*	*Acid Nomenclature*
HCl	hydrogen chlor*ide*	*hydrochloric* acid
HCN	hydrogen cyan*ide*	*hydrocyanic* acid
HI	hydrogen iod*ide*	*hydriodic* acid
HF	hydrogen fluor*ide*	*hydrofluoric* acid

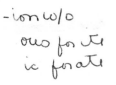

Second Rule.   When the suffix of the compound is -*ite* or -*ate,* the corresponding name of the **oxyacid** begins with the root and is followed by the suffix -*ous* in place of -*ite* and -*ic* in place of -*ate.* Examples of the application of the second rule are listed in Table 4–9.

Example 4.16

Name the compound HNO_2 as an acid.
1. The standard nomenclature is hydrogen nitr*ite.*
2. The root is *nitr-* and the corresponding acid suffix is -*ous.*
3. The name of the compound is:

nitr ous acid
root   suffix related to -*ite*

FORMULAS AND NOMENCLATURE

Example 4.17

Name the compound H_2SO_4 as an acid.
1. The standard name is hydrogen sulf*ate*.
2. The root is *sulfur-* and the corresponding acid suffix is *-ic*.
3. The name of the compound is:

sulfur ic acid

root   suffix related to *-ate*

TABLE 4–9 **Application of the Second Rule in Acid Nomenclature**

Formula	Standard Nomenclature	Acid Nomenclature
HNO_2	hydrogen nit*rite*	nit*rous* acid
H_2SO_3	hydrogen sulf*ite*	sulfur*ous* acid
H_3AsO_3	hydrogen arsen*ite*	arsen*ous* acid
HNO_3	hydrogen nit*rate*	nit*ric* acid
H_2CrO_4	hydrogen chrom*ate*	chrom*ic* acid
H_2CO_3	hydrogen carbon*ate*	carbon*ic* acid
H_2SO_4	hydrogen sulf*ate*	sulfur*ic* acid

Third Rule. When there are more than two oxyacids in a series, as in ClO^- (hypochlorite) or ClO_4^- (perchlorate), the prefixes *hypo-* and *per-* are maintained in the acid nomenclature. Table 4–10 lists some selected examples of *hypo-* and *per-* acids. (w/and rule)

TABLE 4–10 **Application of the Third Rule in Acid Nomenclature**

Formula	Standard Nomenclature	Acid Nomenclature
HClO	hydrogen *hypo*chlor*ite*	*hypo*chlor*ous* acid
HBrO	hydrogen *hypo*brom*ite*	*hypo*brom*ous* acid
$HClO_4$	hydrogen *per*chlor*ate*	*per*chlor*ic* acid
$HMnO_4$	hydrogen *per*mangan*ate*	*per*mangan*ic* acid

Example 4.18

Write the formula for oxalic acid.
1. Note the suffix, *-ic*. This corresponds to the standard name, hydrogen oxalate.
2. Write the symbols and switch the appropriate valences to subscripts.

H⊕ $C_2O_4{}^{2-}$

3. The formula is: $H_2C_2O_4$.

Example 4.19

Write the formula for sulfurous acid.
1. Note the suffix *-ous*. This corresponds to the standard name, hydrogen sulfite.
2. Write the symbols and switch the valences to the subscript position.

3. The formula is H_2SO_3.

Exercise 4.3

Write the formulas for the following acids:
(a) cyanic acid   (f) nitrous acid
(b) acetic acid   (g) arsenic acid
(c) hypoiodous acid   (h) hydrocyanic acid
(d) permanganic acid   (i) chromous acid
(e) hydrobromic acid   (j) hydrofluoric acid

Example 4.20

What is the name of the acid $HClO_2$?
1. Identify the compound as a possible classical acid since the formula begins with hydrogen.
2. Note the standard name of the compound as hydrogen chlor*ite*.
3. The suffix *-ite* corresponds to the acid name chlor*ous* acid.

hydrochlorous acid

Example 4.21

What is the name that corresponds to the formula $HPO_4{}^{2-}$?
1. *Caution:* Observe the notation of electrical charge on the formula. This is *not* a neutral compound, and therefore it should not be named according to the rules for acid nomenclature.
2. The formula should be identified as the monohydrogen phosphate ion.
3. This example serves to illustrate the importance of being concerned with detail in the writing and naming of formulas. Although a hydrogen sulfate ion, $HSO_4{}^-$, may behave as an acid (as we will learn in the chapter on acids and bases), it is not subject to the classical acid nomenclature designed for neutral compounds.

Exercise 4.4

Name the following according to the nomenclature for acids and ions:
(a) $HBrO$ (f) $HOAc$
(b) H_2CrO_4 (g) $H_2PO_4{}^-$
(c) HCN (h) $HClO_4$
(d) $HCO_3{}^-$ (i) HNO_2
(e) H_3PO_4 (j) $HSO_3{}^-$

4.7 NOMENCLATURE OF NONMETALLIC COMPOUNDS

In naming of compounds in which no metals are present, or in cases where there is a metal (or metalloid)* that has an oxidation number of $+4$ or higher, an especially descriptive system of nomenclature is often used. The method involves the use of Greek or Latin prefixes that describe the number of atoms present in the molecule. The prefixes are illustrated by the examples in Table 4–11. Also listed are the alternative names according to the Stock system. The oxidation number of the more metallic element (the first symbol in the formula) is calculated on the basis of the common oxidation number of the more nonmetallic element (the second symbol in the formula).

Example 4.22

Name the compound SO_2 according to the Stock system.

1. Because each oxygen atom has an oxidation number of -2, the total oxidation number for 2 oxygen atoms is -4.
2. Axiomatically, the sum of the oxidation numbers in a formula of a compound must equal zero. Therefore, the single sulfur atom must have an oxidation number of $+4$.

$$\overset{+4}{S}\ \overset{-2}{O_2}$$

3. The Stock name of the compound is sulfur(IV) oxide.

---

*For the time being, we can describe metalloids as those elements falling into a kind of gray area; that is, they are not distinctly metals or nonmetals but have properties, varying in degree and depending on external conditions, of both.

TABLE 4–11 **Descriptive and Stock Nomenclature**
for Selected Compounds

Prefix	Example	Descriptive Name	Stock Name
mono = 1	CO	carbon **mon**oxide	carbon(II) oxide
di = 2	SiO_2	silicon **di**oxide	silicon(IV) oxide
tri = 3	SO_3	sulfur **tri**oxide	sulfur(VI) oxide
tetra = 4	CCl_4	carbon **tetra**chloride	carbon(IV) chloride
penta = 5	N_2O_5	**di**nitrogen **pent**oxide	nitrogen(V) oxide
hexa = 6	XeF_6	xenon **hexa**fluoride	xenon(VI) fluoride
hepta = 7	Cl_2O_7	**di**chlorine **hept**oxide	chlorine(VII) oxide
octa = 8	Cl_2O_8	**di**chlorine **oct**oxide	chlorine(IV) oxide
nona = 9	I_4O_9	**tetra**iodine **non**oxide	iodine(V) oxide
deca = 10	P_4O_{10}	**tetra**phosphorus **dec**oxide	phosphorus(V) oxide (dimer)*

Example 4.23

Name the following compounds according to the descriptive and Stock systems:

1. N_2O $\overset{+1}{N_2}\overset{-2}{O}$
   descriptive: dinitrogen monoxide
   Stock: nitrogen(I) oxide

2. MnO_2 $\overset{+4}{Mn}\overset{-2}{O_2}$
   descriptive: manganese dioxide
   Stock: manganese(IV) oxide

3. P_4O_6 $\overset{+3}{P_4}\overset{-2}{O_6}$
   descriptive: tetraphosphorous hexoxide
   Stock: phosophorus(III) oxide dimer

4. Cl_2O $\overset{+1}{Cl_2}\overset{-2}{O}$
   descriptive: dichlorine monoxide
   Stock: chlorine(I) oxide

5. V_2O_5 $\overset{+5}{V_2}\overset{-2}{O_5}$
   descriptive: divanadium pentoxide
   Stock: vanadium(V) oxide

Example 4.24

Write the formulas for the following compounds:

1. phosphorus pentachloride
   PCl_5
2. disulfur dichloride
   S_2Cl_2
3. nitrogen(III) oxide
   N_2O_3
4. silicon(IV) bromide
   $SiBr_4$
5. diarsenic pentasulfide
   As_2S_5

---

*A **dimer** is a molecule composed of two identical simpler molecules. Thus, P_4O_{10} is a dimer of P_2O_5 (diphosphorus pentoxide).

1. Write the Stock or descriptive name of the following compounds:
(a) NO_2 (d) N_2O_5 (g) I_2Cl_7
(b) CF_4 (e) VCl_4 (h) PbO_2
(c) I_4O_9 (f) BF_3 (i) CrO_3
                           (j) Sb_2O_5

2. Write the formulas for the following compounds:
(a) diarsenic trisulfide   (f) sulfur (VI) fluoride
(b) uranium hexachloride   (g) xenon hexaiodide
(c) selenium dioxide   (h) carbon(IV) fluoride
(d) dinitrogen trioxide   (i) diantimony trioxide
(e) carbon(II) oxide   (j) bismuth(V) oxide

4.8 NOMENCLATURE OF HYDRATES

A hydrated compound is one in which a specific number of moles of water, called water of hydration, are chemically bound into the crystal structure of a compound in a definite molar ratio of water to compound. When the water of hydration is removed, the compound is said to be **anhydrous.** The nomenclature is quite descriptive. For example, $CuSO_4 \cdot 5\,H_2O$ means that there are 5 moles of water of hydration per mole of salt in the crystal.

Example 4.25

Write the alternate names of $CuSO_4 \cdot 5\,H_2O$.
1. Copper(II) sulfate 5-water
2. Copper(II) sulfate 5-hydrate
3. Copper(II) sulfate pentahydrate
4. Cupric sulfate pentahydrate

Example 4.26

Write the alternate names of $MgSO_4 \cdot 7\,H_2O$
1. Magnesium sulfate 7-water
2. Magnesium sulfate 7-hydrate
3. Magnesium sulfate heptahydrate

Example 4.27

Write the formulas for the following hydrates.
1. Calcium chloride 6-water
   $CaCl_2 \cdot 6\,H_2O$
2. Ferrous sulfate heptahydrate
   $FeSO_4 \cdot 7\,H_2O$
3. Calcium sulfate 1-hydrate
   $CaSO_4 \cdot H_2O$
4. Iron(II) phosphate 8-water
   $Fe_3(PO_4)_2 \cdot 8\,H_2O$

Exercise 4.6

1. Write two alternate names for each of the following hydrates:

(a) $Ca(NO_3)_2 \cdot 3\ H_2O$
(b) $Sn(NO_3)_2 \cdot 20\ H_2O$
(c) $Na_2SO_3 \cdot 7\ H_2O$
(d) $U(SO_4)_2 \cdot 4\ H_2O$
(e) $ZnSO_4 \cdot 7\ H_2O$
(f) $NaI \cdot 2\ H_2O$

2. Write formulas for the following hydrates:

(a) Aluminum oxide 3-hydrate
(b) Cobalt(II) perchlorate 5-water
(c) Thallium(III) sulfate heptahydrate
(d) Sodium sulfate decahydrate
(e) Sodium acetate 3-water

4.9 PERCENTAGE COMPOSITION OF COMPOUNDS

The calculation of the mass composition of a compound by percentage is a useful skill in analytical chemistry. Because the formula mass of a compound in grams equals the mass of a mole of the compound, the individual molar masses of the elements in the compound can be compared to the total mass and expressed as a per cent.

Example 4.28

Calculate the percentage composition of ammonium sulfide.

1. Write the formula:
$(NH_4)_2S$
2. Calculate the mass of a mole of $(NH_4)_2S$, which is composed of the following:

(a) 2 moles of nitrogen atoms

$(NH_4)_2S$

$$2\ \text{moles N} \times \frac{14.00\ \text{g}}{1\ \text{mole N}} = 28.00\ \text{g N}$$

(b) 8 moles of hydrogen atoms

$(NH_4)_2S$

$$8\ \text{moles H} \times \frac{1.01\ \text{g}}{1\ \text{mole H}} = 8.08\ \text{g H}$$

(c) 1 mole of sulfur atoms = 32.06 g S

Add a, b, and c

(d) 1 mole of $(NH_4)_2S$ = 68.14 g ← total

3. The percentage of each element is determined by calculating its fraction of the whole.

$$\text{per cent N} = \frac{28.00 \text{ g N}}{68.14 \text{ g (NH}_4)_2\text{S}} \times 100 = 41.09\%$$

$$\text{per cent H} = \frac{8.08 \text{ g H}}{68.14 \text{ g (NH}_4)_2\text{S}} \times 100 = 11.9\%$$

$$\text{per cent S} = \frac{32.06 \text{ g S}}{68.14 \text{ g (NH}_4)_2\text{S}} \times 100 = \underline{47.05\%}$$

$$100.0\% \text{ total}$$

4. The sum of the percentages should equal 100%. (Rounding errors may make the sum slightly lower or higher than 100%.)

Exercise 4.7

1. Calculate the percentage composition of the following compounds:
   (a) KBr   (d) $Na_2S_2O_3$
   (b) $Ca(OH)_2$ (e) $Fe_2(HPO_4)_3$
   (c) H_2SO_4

2. Find the percentage of water in the compound $MgSO_4 \cdot 7 H_2O$.
3. What is the percentage composition of indium chloride if 0.50 g in solution is added to enough silver nitrate solution so that 0.99 g of silver chloride is produced?

Example 4.29

Formula by % comp. (analytical)

Find the empirical formula of a compound that was analyzed to contain 32.4% sodium, 22.5% sulfur, and 45.1% oxygen.

1. The first step is to express the percentages of the constituent atoms. This is easily accomplished if exactly 100 g of a compound is assumed. The numerical values are unchanged.

$$32.4\% \text{ of } 100 \text{ g} = 32.4 \text{ g of Na}$$

$$22.5\% \text{ of } 100 \text{ g} = 22.5 \text{ g of S}$$

$$45.1\% \text{ of } 100 \text{ g} = 45.1 \text{ g of O}$$

2. The second step is to calculate the comparative number of moles of Na, S, and O atoms:

$$\text{moles of Na} = 32.4 \text{ g} \times \frac{1 \text{ mole}}{23.0 \text{ g}} = 1.41 \text{ moles}$$

$$\text{moles of S} = 22.5 \text{ g} \times \frac{1 \text{ mole}}{32.1 \text{ g}} = 0.70 \text{ mole}$$

$$\text{moles of O} = 45.1 \text{ g} \times \frac{1 \text{ mole}}{16.0 \text{ g}} = 2.82 \text{ moles}$$

3. The formula $Na_{1.41}S_{0.70}O_{2.82}$ is not acceptable. The ratios of the atoms in a formula must be in terms of small *whole* numbers. This is done by dividing all the mole values by the *smallest* number present (in this case, the sulfur). The results will then contain no number less than 1.

$$\text{Na} = \frac{1.41}{0.70} = 2.01 \approx 2$$

$$S = \frac{0.70}{0.70} = 1.00$$

$$O = \frac{2.82}{0.70} = 4.03 \approx 4$$

4. The simplest whole number ratio of the atoms in the compound is observed to be,
$Na_2S_1O_4$
which is finally written as:
Na_2SO_4
(sodium sulfate)

4.10 EMPIRICAL AND TRUE FORMULAS OF COMPOUNDS

The *empirical* formula shows the simplest relationship of atoms in a compound as obtained from analytic data. The *true* formula shows not only the ratio between atoms but also the total number of each atom in a molecule (or each ion in a formula unit).

The true formula for mercury (I) chloride is Hg_2Cl_2, although its empirical formula is $HgCl$. Similarly, CH_2 is the simplest or empirical formula for a compound having a 1:2 ratio of carbon to hydrogen atoms. The true formula may indeed be CH_2, but it may also be other compounds having a 1:2 ratio of C to H.

Two different compounds may have the same empirical formula but different true formulas. Benzene, for example, has an empirical formula of CH, but its true formula is C_6H_6. Acetylene has the same empirical formula as benzene, but its true formula is C_2H_2. The true formula can be determined from an empirical formula if we know the molar mass. For example, the empirical formula for hydrogen peroxide is HO, but, since its molar mass is 34 g/mole, the true formula is H_2O_2. Conversely, water has an empirical formula of H_2O and a molar mass of 18 g/mole, so the true formula is also H_2O.

Some examples showing calculations of empirical and true formulas follow. Bear in mind that the true formula will be either the same as the empirical formula or some multiple of the empirical formula.

Example 4.30

A 1.39 g sample of *hydrated* copper(II) sulfate is heated until all the water of hydration is driven off. The anhydrous salt has a mass of 0.89 g. Determine the formula of the hydrate.
1. Determine the mass of the water by calculating the difference between the mass of the hydrate and the anhydrous salt (residue).

$$\begin{array}{r} 1.39 \text{ g hydrate} \\ -0.89 \text{ g residue} \\ \hline 0.50 \text{ g water} \end{array}$$

2. Calculate the comparative number of moles of $CuSO_4$ and H_2O:

$$CuSO_4 = 63.5\ g + 32.0\ g + (4 \times 16.0\ g) = \frac{159.5\ g}{1\ mole}$$

$$H_2O = (2 \times 1.0\ g) + 16.0\ g = \frac{18.0\ g}{1\ mole}$$

$$CuSO_4 = 0.89\ g \times \frac{1\ mole}{159.5\ g} = 0.0056\ mole$$

$$H_2O = 0.50\ g \times \frac{1\ mole}{18.0\ g} = 0.028\ mole$$

Hint: The arithmetic can be made a little simpler and less liable to computational error if the mole numbers are made larger without changing their ratios. We can do this by multiplying both mole numbers by 100 or 1000 or whatever is most convenient. For example, the calculated ratio $\frac{0.0056}{0.028}$ is exactly the same as $\frac{56}{280}$, where both numbers were multiplied by 10^4.

3. Use the simplified molar ratios to find the whole-number values. Divide both numbers by the smaller value:

$$CuSO_4 = \frac{56}{56} = 1$$

$$H_2O = \frac{280}{56} = 5$$

4. The formula of the hydrate is:
$$CuSO_4 \cdot 5\ H_2O$$
$$[copper(II)\ sulfate\ 5\text{-}water]$$

Occasionally, calculation of the empirical formula needs a little extra refinement before it works out to a small whole-number answer, as illustrated in Example 4.31.

Example 4.31

A 200.0 mg sample of a K, Cr, and O compound was analyzed and found to contain 70.8 mg of chromium and 76.0 mg of oxygen. Find the empirical formula.
1. Find the mass of potassium in the sample by subtracting the Cr and O sum from the total:

$$\begin{aligned}
total\ mass\ of\ sample &= 200.0\ mg \\
mass\ of\ Cr\ and\ O &= \underline{146.8\ mg} \\
K &= 53.2\ mg
\end{aligned}$$

2. Calculate the comparative number of moles of each element, after converting the milligram ratio to an equivalent gram ratio for convenience. Then divide all the mole numbers by the smallest value:

$$K = 53.2\ g \times \frac{1\ mole}{39.1\ g} = 1.36\ moles,\ \frac{1.36}{1.36} = 1$$

$$Cr = 70.8 \not{g} \times \frac{1 \text{ mole}}{52.0 \not{g}} = 1.36 \text{ moles}, \quad \frac{1.36}{1.36} = 1$$

$$O = 76.0 \not{g} \times \frac{1 \text{ mole}}{16.0 \not{g}} = 4.75 \text{ moles}, \quad \frac{4.75}{1.36} = 3.49 \approx 3.5$$

3. The formula $K_1Cr_1O_{3.5}$ is not acceptable. The entire formula must be multiplied by the smallest number possible to produce a whole-number ratio. Obviously, 3.5 moles of oxygen must be *doubled*. The empirical formula becomes:

$$K_2Cr_2O_7$$
(potassium dichromate)

Example 4.32

A compound containing 92.3% carbon and 7.7% hydrogen has a molar mass of 26 g/mole. Calculate its true formula.
1. Find the empirical formula (assume 100 g of compound).

$$C = 92.3 \text{ g} \times \frac{1 \text{ mole}}{12.0 \text{ g}} = 7.69 \text{ moles} \approx 7.7 \text{ moles}$$

$$H = 7.7 \text{ g} \times \frac{1 \text{ mole}}{1.0 \text{ g}} = 7.7 \text{ moles}$$

Dividing by the smaller number gives a 1:1 ratio of C atoms to H atoms. The empirical formula is C_1H_1, or simply CH.
2. Calculate the true formula:
The true formula will be CH or some multiple thereof. The molar mass of CH is 13 g/mole. Since the given molar mass of the compound is 26 g/mole, the true formula is $(CH)_2$, written properly as C_2H_2.

Exercise 4.8

1. Find the empirical formulas of compounds having the following percentage composition:
(a) 75.0% C, 25.0% H
(b) 72.4% Fe, 27.6% O
(c) 21.8% Mg, 27.9% P, 50.3% O
(d) 70.0% Fe, 30.0% O
(e) 59.30% C, 4.55% H, 23.00% N, 13.15% O
2. A 10.0 g sample of a compound is analyzed and found to be composed of 4.21 g sodium, 1.89 g phosphorus, and 3.90 g oxygen. Find the empirical formula.

QUESTIONS & PROBLEMS

4.1   Write the names of the following ions:
(a) S^{2-} (f) HPO_4^{2-}
(b) SO_3^{2-} (g) HSO_3^-
(c) $S_2O_3^{2-}$ (h) SCN^-
(d) ClO^- (i) $H_2PO_4^-$
(e) ClO_2^- (j) HCO_3^-

4.2   Write the formulas for the following ions:
(a) phosphide   (f) hydrogen sulfate
(b) iodide   (g) bicarbonate
(c) cyanide   (h) ferrous
(d) dichromate   (i) ammonium
(e) arsenate   (j) cyanate

4.3 Explain what is *wrong* with each of the following formulas:
(a) Cr(III)Cl$_3$ (d) KCn
(b) FE(NO$_3$)$_2$ (e) H$_3$PO$_4^-$
(c) MgOH$_2$

4.4 Name the following compounds according to the Stock system:
(a) Na$_2$SO$_3$ (k) CuSO$_4$ · 3 H$_2$O
(b) ZnHPO$_4$ (l) Rb$_3$AsO$_3$
(c) Fe(OAc)$_3$ (m) Bi(OCN)$_3$
(d) NaNO$_2$ (n) BaC$_2$O$_4$
(e) Hg$_2$(NO$_3$)$_2$ (o) KMnO$_4$
(f) Ba(OH)$_2$ (p) Be$_3$N$_2$
(g) AgHSO$_4$ (q) ZnS
(h) FeBr$_2$ (r) Co$_2$(Cr$_2$O$_7$)$_3$
(i) Na$_3$N   (s) Pb(HCO$_3$)$_2$
(j) K$_2$O$_2$ (t) Sn(SO$_4$)$_2$

4.5 Name the following compounds according to the classical *(ous, ic)* system:
(a) CuCl$_2$ (f) Fe(ClO$_4$)$_3$
(b) HgO   (g) Ni(OH)$_2$
(c) Fe$_2$(C$_2$O$_4$)$_3$ (h) Cu$_2$S
(d) PbS$_2$ (i) Sn(CO$_3$)$_2$
(e) CoHPO$_4$ (j) Sb(ClO)$_2$

4.6 Name the following nonmetallic compounds by the Stock or the descriptive *(mono-, di-, tri-,* etc.) method:
(a) PCl$_3$ (f) S$_2$Cl$_2$
(b) SiF$_4$ (g) P$_4$O$_{10}$
(c) Bi$_2$O$_5$ (h) Br$_3$O$_8$
(d) NO$_2$ (i) VF$_6$
(e) P$_4$S$_7$ (j) As$_4$O$_6$

4.7 Name the following as acids:
(a) HBr   (f) HCN
(b) HNO$_2$ (g) HOCN
(c) H$_3$PO$_4$ (h) H$_2$S
(d) HIO$_3$ (i) HOAc
(e) HClO$_4$ (j) H$_2$C$_2$O$_4$

4.8 Write the formulas for the following compounds:
(a) copper(II) sulfate
(b) carbon disulfide
(c) magnesium hydroxide
(d) aluminum arsenate
(e) phosphoric acid
(f) cobalt(II) bromide
(g) barium chlorate
(h) zinc iodide
(i) acetic acid
(j) mercurous iodide
(k) diarsenic pentoxide
(l) potassium permanganate
(m) sodium peroxide
(n) nickelous sulfate
(o) stannic nitrate
(p) cadmium hydrogen carbonate
(q) carbon(II) oxide
(r) mercuric cyanide
(s) chromium(II) hydroxide
(t) ammonium carbonate

4.9 Name the following compounds:
(a) NH$_4$OH
(b) HCN (as an acid)
(c) As$_2$S$_3$
(d) SiC
(e) NaMnO$_4$
(f) CdI$_2$
(g) Ag$_2$Cr$_2$O$_7$
(h) Ba(ClO$_3$)$_2$
(i) SiF$_4$
(j) Na$_2$SO$_4$ · 10 H$_2$O
(k) Ni(OCN)$_2$
(l) XeCl$_6$
(m) CoCl$_3$ · 3 H$_2$O
(n) HI (as an acid)
(o) BiCl$_3$
(p) AlN
(q) Li$_3$AsO$_4$
(r) H$_2$C$_2$O$_4$ (as an acid)
(s) Fe(SCN)$_3$
(t) CuHPO$_4$

4.10   Write the formulas for the following compounds and ions:
(a) sulfur trioxide
(b) diiodine pentoxide
(c) iron(II) hydrogen carbonate
(d) boron trifluoride
(e) potassium sulfide
(f) stannic oxide
(g) hydrocyanic acid
(h) sulfite ion
(i) cuprous thiosulfate
(j) ammonia
(k) cobalt(II) ion
(l) hydrosulfuric acid
(m) antimony(V) perchlorate
(n) carbon monoxide
(o) zinc sulfate 20-water
(p) dichlorine heptoxide
(q) strontium hydroxide
(r) barium hypochlorite 2-hydrate
(s) ammonium ion
(t) cupric phosphate

4.11   Complete the following table by writing the acceptable names and formulas:

	Br^-	CO_3^{2-}	PO_4^{3-}
K^+	KBr potassium bromide		
Fe^{2+}			
Ni^{3+}			
Pb^{2+}			
NH_4^+			

4.12   Calculate the percentage composition of the following:
(a) KOCN   (c) Fe_3O_4
(b) $Fe(NO_3)_2$ (d) NH_4HCO_3

4.13   If 3.20 g of $Na_2SO_4 \cdot 10 H_2O$ is heated to complete dehydration, what mass of residue remains?

4.14   What is the empirical formula of a compound that is analyzed to contain 75.8% As and 24.2% oxygen?

4.15   A compound has the following composition by mass: carbon 52.1%; hydrogen 13.1%; oxygen 34.8%. What is the simplest (empirical) formula of the compound?

4.16   Complete the following table by writing acceptable names and formulas.

	OH^-	Cl^-	NO_3^-	SO_4^{2-}	ClO_3^-	$C_2O_4^{2-}$	S^{2-}
H^+							
Zn^{2+}							
Fe^{3+}		$FeCl_3$ iron(III) chloride					
Al^{3+}							
Ag^+							
Ba^{2+}							
Cu^{2+}							
Ca^{2+}							
Pb^{2+}							

4.17   Find the true formulas of the compounds having the following percentage compositions and molar masses:
(a) 40.00% C, 6.71% H, 53.29% O; 175 g/mole
(b) 43.6% P, 56.4% O; 284 g/mole
(c) 39.65% C, 1.65% H, 58.70% Cl; 182 g/mole

4.18   A compound has an empirical formula C_2H_6O. Its molar mass is 138 g/mole. What is its true formula?

4.19   A binary (only 2 elements) compound contains 38.9% chlorine with the remainder being oxygen. Its molar mass is 182.9 g/mole. What is the true formula of the compound?

4.20   An acidic compound was analyzed and found to be composed of 54.3% O, 40.7% C, and 5.1% H.
(a) Calculate its empirical formula.
(b) Calculate its true formula if its molar mass is 59.0 g/mole.

CHAPTER FIVE

LEARNING OBJECTIVES

At the completion of this chapter, you should be able to:

1. Define the term *stoichiometry* and describe a stoichiometric reaction.

2. List four experimental observations that tentatively identify a reaction as being stoichiometric.

3. Balance equations by inspection.

4. Interpret the information that a balanced equation conveys.

5. Classify chemical equations as representative of the following types of reactions: decomposition, combination, single replacement, double replacement.

6. Use the activity series to predict the products of single replacement reactions.

7. Perform stoichiometric calculations from balanced equations when given data in the form of moles or masses.

8. Use the factor-label method to perform calculations.

9. Distinguish between theoretical and actual product yields in reactions.

10. Identify action-limiting species in stoichiometric calculations.

CHEMICAL EQUATIONS AND STOICHIOMETRY

The chemical equation is a very special and valuable shorthand that we use to describe the quantitative relationships among the **reactants** and **products** of a chemical change. The reactants, conventionally written to the left of a **yield** sign (an arrow, $\rightarrow$), show the substances that interact to produce new substances called products. A chemical reaction is often evidenced by the production of a gas, the formation of a precipitate, an emission of heat and/or light, or possibly a color change. A reaction is said to be complete when any one of the reactants has been completely converted to a product. Most chemical reactions do not go to completion, that is, they reach a point at which the reactants and products coexist in a dynamic balance. These are called *equilibrium reactions*. An equilibrium reaction is commonly distinguished from a stoichiometric reaction by use of a double-arrow yield sign ($\rightleftharpoons$), in which the arrow pointing to the right ($\rightarrow$) indicates the *forward reaction*, and the arrow pointing to the left ($\leftarrow$) indicates the *reverse reaction.*

The factor that makes a chemical equation a special kind of shorthand is its *quantitative* nature. An equation describes not only what reacts and what is produced but also *how much* of each participating species is involved. The law of conservation of matter is paramount in this regard. When a chemist begins with a given number of grams of reactant, he expects to end up with the same number of grams of product, regardless of how dramatic the chemical change may be. In a chemical change, the amount of matter that may be converted to radiant energy is not detectable by ordinary laboratory balances. Chemical reactions are most conveniently described in terms of the mole concept.

Before analyzing chemical equations as mole-to-mole ratios of reactants and products, some useful notations should be pointed out. Although an equation does not necessarily have to indicate more than the molar ratios of original reactants and final products, special symbols may be used to give a fuller picture of the conditions under which reactants combine and products emerge. For example, substances may be annotated to indicate whether they are solids, liquids, gases, or in solution. The overall reaction may be described quantitatively as **exothermic** (heat-producing) or **endothermic** (heat-absorbing) by using a **ΔH** convention, called **enthalpy** change. The ΔH represents a heat loss or gain at a constant pressure and temperature. An exothermic reaction is represented by a negative ΔH value and, conversely, a positive ΔH value indicates an energy-absorbing reaction. For example, ΔH = −12.5 kcal refers to a change in which 12.5 kcal of heat energy is liberated in an exothermic reaction. The presence of a catalyst can be indicated also by noting it over the "yield" sign. Heating of reactants is noted by a capital Greek letter delta (Δ). A summary of these notations is provided in Table 5–1.

5.2 BALANCING EQUATIONS

In the process of balancing chemical equations, a careful distinction must be made between the subscripts and the **coefficients.** The coefficients in chemical equations are the large numbers written *in front* of the formulas (never in the middle of a formula), which tell us how many moles of each substance are used or formed. The subscripts illustrate the *law of definite composition;* the coefficients are governed by the *law of conservation of matter.* For example, the chemical combination of aluminum and oxygen produces aluminum oxide. If conservation were the only factor to deal with, an incorrect equation could be written:

$$Al(s) + O_2(g) \rightarrow AlO_2(s)$$

wrong

TABLE 5–1 **Notations in Chemical Equations**

Symbol	Meaning	Examples
(s)	Solid reactant or insoluble solid product	$Zn(s)$ $PbCrO_4(s)$ $H_2O(s)$
(*l*)	Liquid reactant or liquid product	$H_2O(l)$ $Hg(l)$ $Br_2(l)$
(g)	Gaseous reactant or product	$CO_2(g)$ $H_2(g)$ $H_2O(g)$
(aq) aqueous	Reactants or products in water solutions	$NaCl(aq)$ $KBr(aq)$ $Fe(NO_3)_2(aq)$
$\xrightarrow{\Delta}$	Reactants heated	$A + B \xrightarrow{\Delta} C + D$
$\xrightarrow{Pt}$	Platinum used as a catalyst	$A + B \xrightarrow{Pt} C + D$
ΔH = −	Exothermic reaction	$CH_4(g) + 2 O_2(g) \rightarrow$ $2 H_2O(l) + CO_2(g)$ ΔH = − 212 kcal
ΔH = +	Endothermic reaction	$2 HCl(g) \rightarrow Cl_2(g) + H_2(g)$ ΔH = + 44.1 kcal

The formula AlO_2 does balance the equation insofar as it results in the same number of Al and O units on both sides of the equation. However, the equation is wrong! The law of definite composition clearly states that there is only one correct formula for aluminum oxide, Al_2O_3, based on the common oxidation numbers of aluminum and oxygen:

$$\overset{+3}{Al}, \overset{-2}{O} = Al_2O_3$$

The problem can be dealt with most efficiently by establishing a few rules for balancing equations by **inspection.**

Rules

1. *Write the correct formula for every reactant and product.* This formula is based on the oxidation numbers of the species involved. The result is a **skeleton equation,** or unbalanced equation. Example:

$$Al(s) + O_2(g) \rightarrow Al_2O_3(s)$$

2. *The subscripts must not be changed* (law of definite composition). Example:

$$Al_2(s) + O_3(g) \rightarrow Al_2O_3(s)$$

cannot be used. There is no evidence to support the existence of aluminum as diatomic molecules, and O_3 is the formula for ozone, not oxygen.

3. The balancing coefficients used should be the smallest numbers possible. The use of a fractional coefficient is perfectly satisfactory if you find it helpful in the balancing procedure. For example, in the equation

$$Al(s) + O_2(g) \rightarrow Al_2O_3(s)$$

the oxygen atoms can be directly balanced by multiplying the O_2 formula by the fraction $\frac{3}{2}$:

$$2\, Al(s) + \tfrac{3}{2}O_2(g) \rightarrow Al_2O_3(s)$$
or, if you prefer, whole number coefficients:
$$4\, Al(s) + 3\, O_2(g) \rightarrow 2\, Al_2O_3(s)$$

4. It must be remembered that the coefficient multiplies *every* atom in the formula.

$$2\, Al_2O_3 \text{ means}$$
4 aluminum atoms and
6 oxygen atoms

The coefficient means that in 2 moles of Al_2O_3, there are 4 moles of aluminum atoms and 6 moles of oxygen atoms.

Example 5.1

In the formula 2 $Fe_2(SO_4)_3$, how many iron, sulfur, oxygen, and sulfate units are there?

(a) 2 Fe_2 $(SO_4)_3$

→ 4 iron atoms

(b) 2 Fe_2 $(S_1O_4)_3$

→ 2 × 3 × 1 sulfur atoms = 6 sulfur atoms

(c) 2 Fe_2 $(S_1O_4)_3$

→ 2 × 3 × 4 oxygen atoms = 24 oxygen atoms

(d) 2 Fe_2 $(SO_4)_3$

→ 2 × 3 sulfate ions = 6 sulfate ions

5.3 THE MEANING OF THE BALANCED EQUATION

The balanced equation indicates the proportionate number of moles of each reactant and product involved. Because a number of moles can be expressed as grams, volume of gas (which we will cover in the next chapter), and number of particles, a great deal of information is available. For example, the equation 2 Al + $\frac{3}{2}O_2$ → Al_2O_3 means the following:

1. 2 moles of aluminum atoms + $1\frac{1}{2}$ moles of oxygen molecules produces 1 mole of aluminum oxide.
2. Since 1 mole of aluminum atoms = 27.0 g and 1 mole of O_2 molecules = 32.0 g, then:

	2 Al(s)	+	$\frac{3}{2}O_2$(g)	→	Al_2O_3(s)
moles:	2		$\frac{3}{2}$		1
mass:	54.0 g		48.0 g		102.0 g
molecules:	(2 × 6.02 × 10^{23})		($\frac{3}{2}$ × 6.02 × 10^{23})		(6.02 × 10^{23})

Example 5.2

Balance the following equation, and discuss the information that it conveys:

$$C_3H_8(g) + O_2(g) \rightarrow CO_2(g) + H_2O(l)$$

1. Check the moles of carbon atoms on each side of the equation:

$$C_3H_8(g) + O_2(g) \rightarrow CO_2(g) + H_2O(l)$$

3 moles of carbon atoms   1 mole of carbon atoms

CHEMICAL EQUATIONS AND STOICHIOMETRY

(a) Balance the moles of carbon atoms:

$$C_3H_8(g) + O_2(g) \rightarrow 3\ CO_2(g) + H_2O(l)$$

2. Check the hydrogen atoms:

$$C_3H_8(g) + O_2(g) \rightarrow 3\ CO_2(g) + H_2O(l)$$

8 moles of hydrogen atoms

2 moles of hydrogen atoms

(a) Balance the hydrogen atoms:

$$C_3H_8(g) + O_2(g) \rightarrow 3\ CO_2(g) + 4\ H_2O(l)$$

3. Check the oxygen atoms:

$$C_3H_8(g) + O_2(g) \rightarrow 3\ CO_2(g) + 4\ H_2O(l)$$

2 moles of oxygen atoms

6 moles of oxygen atoms + 4 moles of oxygen atoms = 10 moles of oxygen atoms total

(a) Balance the oxygen atoms:

$$C_3H_8(g) + 5\ O_2(g) \rightarrow 3\ CO_2(g) + 4\ H_2O(l)$$

4. The equation is balanced.

	reactants	products
	$C_3H_8(g) + 5\ O_2(g)$	$\rightarrow \quad 3\ CO_2(g) + 4\ H_2O(l)$
moles:	1 mole + 5 moles	3 moles + 4 moles
mass:	44.0 g + 160.0 g	132.0 g + 72.0 g
molecules:	6.02×10^{23}	$(3 \times 6.02 \times 10^{23})$
	$+ (5 \times 6.02 \times 10^{23})$	$+ (4 \times 6.02 \times 10^{23})$

Example 5.3

Write a balanced equation that describes the reaction between solid aluminum and oxygen gas as it produces solid aluminum oxide in an exothermic reaction (380.0 kcal).

$$2\ Al(s) + \tfrac{3}{2}\ O_2(g) \rightarrow Al_2O_3(s)\ (\Delta H = -380.0\ kcal)$$

Example 5.4

Write and balance the equation for the heating of solid mercury(II) oxide to produce oxygen gas and liquid mercury.

1. The skeleton equation:

$$HgO(s) \xrightarrow{\Delta} Hg(l) + O_2(g)$$

2. The oxygen atoms are balanced by placing a coefficient of $\tfrac{1}{2}$ before the O_2.

$$HgO(s) \xrightarrow{\Delta} Hg(l) + \tfrac{1}{2}\ O_2(g)\ \text{or}$$

$$2\ HgO(s) \xrightarrow{\Delta} 2\ Hg(l) + O_2(g)$$

Example 5.5

Write a balanced equation for the reaction between aqueous solutions of iron(III) nitrate and sodium sulfide, assuming that iron(III) sulfide is precipitated and that the sodium nitrate produced remains in water solution.
1. Write the correct formulas for all the reactants and products, and use clarifying symbols.

$$Fe(NO_3)_3(aq) + Na_2S(aq) \rightarrow Fe_2S_3(s) + NaNO_3(aq)$$

2. Check the iron ions:

(a) Balance the iron ions:

$$2\ Fe(NO_3)_3(aq) + Na_2S(aq) \rightarrow Fe_2S_3(s) + NaNO_3(aq)$$

3. 2 $Fe(NO_3)_3$ means that there is a total of 6 moles of nitrate ions. Balance the nitrate ions:

$$2\ Fe(NO_3)_3(aq) + Na_2S(aq) \rightarrow Fe_2S_3(s) + 6\ NaNO_3(aq)$$

4. Balance the sulfide ions and sodium ions simultaneously by inserting a coefficient of 3 before Na_2S. The equation is checked and found to be balanced.

$$2\ Fe(NO_3)_3(aq) + 3\ Na_2S(aq) \rightarrow Fe_2S_3(s) + 6\ NaNO_3(aq)$$

Example 5.6

Balance the following equation: zinc and hydrochloric acid $\rightarrow$ zinc chloride and hydrogen gas.
1. Write the correct formula for each reactant and product:

$$Zn(s) + HCl(aq) \rightarrow ZnCl_2(aq) + H_2(g)$$

2. Notice that a coefficient of 2 before the HCl balances the equation:

$$Zn(s) + 2\ HCl(aq) \rightarrow ZnCl_2(aq) + H_2(g)$$

Exercise 5.1

1. Balance the following skeleton equations, assuming that the reactions do occur.

(a) $KClO_3(s) \xrightarrow{\Delta} KCl(s) + O_2(g)$
(b) $Na(s) + H_2O(l) \rightarrow NaOH(aq) + H_2(g)$
(c) $CuO(s) + H_2(g) \rightarrow Cu(s) + H_2O(l)$
(d) $SO_3(g) + H_2O(l) \rightarrow H_2SO_4(aq)$
(e) $H_2SO_4(aq) + Al(OH)_3(s) \rightarrow Al_2(SO_4)_3(aq) + H_2O(l)$
(f) $Pb(OAc)_2(aq) + K_2Cr_2O_7(aq) \rightarrow PbCr_2O_7(s) + KOAc(aq)$
(g) $(NH_4)_2S(aq) + Fe(NO_3)_3(aq) \rightarrow NH_4NO_3(aq) + Fe_2S_3(s)$

CHEMICAL EQUATIONS AND STOICHIOMETRY

(h) $Ca(s) + H_2O(l) \rightarrow Ca(OH)_2(s) + H_2(g)$

(i) $(NH_4)_2CO_3(s) \xrightarrow{\Delta} NH_3(g) + CO_2(g) + H_2O(l)$

(j) $Al(s) + H_2C_2O_4(aq) \rightarrow Al_2(C_2O_4)_3(s) + H_2(g)$

2. Write balanced equations for the following reactions:
   (a) solid sodium + water → sodium hydroxide solution + hydrogen gas
   (b) solid calcium oxide + water → calcium hydroxide solid
   (c) solid potassium cyanide + nitric acid solution → solutions of potassium nitrate + hydrocyanic acid

   (d) solid silver oxide $\xrightarrow{\Delta}$ solid silver + oxygen gas
   (e) hydrogen gas + oxygen gas → liquid water
   (f) hydrogen peroxide solution → oxygen gas + liquid water
   (g) solid iron + solid sulfur → iron(II) sulfide solid
   (h) gold(III) chloride solution + solid zinc → zinc chloride solution + solid gold
   (i) hydrogen gas + nitrogen gas → ammonia gas
   (j) solutions of phosphoric acid + strontium hydroxide → solid strontium phosphate + liquid water

3. Write descriptive equations for the following reactions:
   (a) Xenon gas reacts with fluorine gas (F_2) to produce crystals of xenon tetrafluoride while giving off 62.5 kcal of heat in the exothermic change.
   (b) Zinc metal reacts with a water solution of copper(II) sulfate to produce a solution of zinc sulfate and solid copper.
   (c) When chlorine gas (Cl_2) is bubbled into a water solution of potassium bromide, a potassium chloride solution and liquid bromine are produced.

4. Given the skeleton equation $C_2H_4(g) + O_2(g) \rightarrow CO_2(g) + H_2O(l)$, answer the following questions after balancing.
   (a) How many moles of oxygen are needed to produce 2 moles of CO_2?
   (b) How many grams of C_2H_4 are required to produce 2 moles of water?

5.4 TYPES OF CHEMICAL REACTIONS

Although it has been mentioned previously that the writing of chemical equations is not a matter of magically manipulating symbols and numbers until a balance is achieved, it is possible to predict the products of certain reactions. The ultimate test of the reality of a chemical reaction, however, must be done by experimentation.

Reactions are conveniently classified as **decomposition** (analysis), **combination** (synthesis), **single replacement** (substitution), or **double replacement** (metathesis). More specialized titles include **neutralization** and **hydrolysis** reactions, which are types of double replacement in one sense. The special cases of **oxidation-reduction** (redox) and **ionic** equations will be taken up in separate chapters.

Decomposition Reactions

Decomposition reactions are ones in which a compound is treated in such a way that it decomposes. The decomposition products of binary compounds are obviously the two elements. The products of ternary compounds are not predictable without additional information. Examples of decomposition reactions:

1. $HgO(s) \overset{\Delta}{\rightarrow} Hg(l) + \frac{1}{2}O_2(g)$
2. $KClO_3(s) \overset{\Delta}{\rightarrow} KCl(s) + \frac{3}{2}O_2(g)$
3. $PbO_2(s) \overset{\Delta}{\rightarrow} Pb(s) + O_2(g)$
4. $NaNO_3(s) \overset{\Delta}{\rightarrow} NaNO_2(s) + \frac{1}{2}O_2(g)$
5. $(NH_4)_2CO_3(s) \overset{\Delta}{\rightarrow} 2\,NH_3(g) + CO_2(g) + H_2O(l)$

2 elements
not ions

Combination Reactions

Combination or synthesis reactions may involve the simple chemical union of two elements to form a compound. Sometimes compounds may be reactants in a combination reaction that produces a more complex compound. Examples of combination reactions:

$$C(s) + O_2(g) \rightarrow CO_2(g)$$
$$P_4(s) + 5\,O_2(g) \rightarrow P_4O_{10}(s)$$
$$Mg(s) + \frac{1}{2}O_2(g) \rightarrow MgO(s)$$
$$K_2O(s) + H_2O(l) \rightarrow 2\,KOH(aq)$$
$$SO_3(g) + H_2O(l) \rightarrow H_2SO_4(aq)$$

+ H₂O

Single Replacement Reactions

Single replacement reactions or substitution reactions occur when a more chemically active free element displaces a less chemically active element from a compound. In reality, this is an oxidation-reduction reaction because there must be a loss and gain of electrons in such a process. A free element is in the "zero" oxidation state. When it becomes chemically combined as a result of the replacement reaction, it necessarily exhibits some "valence" or oxidation number that is above or below zero. A gain (more positive) in oxidation number is by definition *oxidation;* the converse is *reduction.*

However, for the sake of simplicity, replacement reactions can be predicted with a fair degree of reliability by reference to an experimentally obtained **activity series.** The activity series, shown in Table 5–2, indicates that a free metallic element on the list can replace any metallic ion *below* it from a compound in aqueous solution; but there is no reaction when it is added to a water solution of a compound containing an ion *above* it in the activity series.

CHEMICAL EQUATIONS AND STOICHIOMETRY

TABLE 5–2 The Activity Series

Decreasing Metal Activity	Element	Activity Description
	Li $\rightarrow$ Li$^+$ + e$^-$	
	K $\rightarrow$ K$^+$ + e$^-$	
	Rb $\rightarrow$ Rb$^+$ + e$^-$	
	Cs $\rightarrow$ Cs$^+$ + e$^-$	Will replace hydrogen from liquid water
	Ba $\rightarrow$ Ba^{2+} + 2e$^-$	
	Sr $\rightarrow$ Sr^{2+} + 2e$^-$	
	Ca $\rightarrow$ Ca^{2+} + 2e$^-$	
	Na $\rightarrow$ Na$^+$ + e$^-$	
	Mg $\rightarrow$ Mg^{2+} + 2e$^-$	Will replace hydrogen from steam
	Al $\rightarrow$ Al^{3+} + 3e$^-$	
	Mn $\rightarrow$ Mn^{2+} + 2e$^-$	
	Zn $\rightarrow$ Zn^{2+} + 2e$^-$	
	Cr $\rightarrow$ Cr^{3+} + 3e$^-$	Will replace hydrogen from acids
	Fe $\rightarrow$ Fe^{2+} + 2e$^-$	
	Ni $\rightarrow$ Ni^{2+} + 2e$^-$	
	Sn $\rightarrow$ Sn^{2+} + 2e$^-$	
	Pb $\rightarrow$ Pb^{2+} + 2e$^-$	
	H$_2$ $\rightarrow$ 2 H$^+$ + 2e$^-$	
	Cu $\rightarrow$ Cu^{2+} + 2e$^-$	
	2 Hg $\rightarrow$ Hg$_2^{2+}$ + 2e$^-$	Will not replace hydrogen from acids
	Ag $\rightarrow$ Ag$^+$ + e$^-$	
	Au $\rightarrow$ Au^{3+} + 3e$^-$	

Example 5.7

Write the equation for the reaction between Zn(s) and CuSO$_4$(aq).
1. Write the reactants in the equation format:

$$Zn(s) + CuSO_4(aq) \rightarrow \text{?}$$

Can zinc replace the copper in copper(II) sulfate?
2. Observe on Table 5–2 that zinc is *above* copper. Therefore a safe prediction would be that zinc can replace the copper(II) ion.

$$Zn(s) + CuSO_4(aq) \rightarrow ZnSO_4(aq) + Cu(s)$$
replaces

Example 5.8

Write the equation for the reaction between copper and hydrochloric acid.
1. Set up the equation:

$$Cu(s) + HCl(aq) \rightarrow \text{?}$$

Can copper replace the hydrogen in hydrochloric acid?
2. From Table 5–2 it is seen that copper is *below* hydrogen and therefore will not react. A commonly accepted symbol that indicates no reaction is NR.

$$Cu(s) + HCl(aq) \rightarrow NR$$

---

1. An active metal replacing hydrogen from liquid water produces a hydroxide and hydrogen gas.

$$Li(s) + H_2O(\ell) \rightarrow LiOH(aq) + \tfrac{1}{2}H_2(g)$$

Remember that water can be conveniently written as HOH (hydrogen hydroxide). A more pictorial representation is as follows:

$$Li(s) + HOH(\ell) \rightarrow LiOH(aq) + \tfrac{1}{2}H_2(g)$$
replaces

Other examples:

$$Ca(s) + 2\,H_2O(\ell) \rightarrow Ca(OH)_2(s) + H_2(g)$$
$$Cs(s) + H_2O(\ell) \rightarrow CsOH(aq) + \tfrac{1}{2}H_2(g)$$
$$Ba(s) + 2\,H_2O(\ell) \rightarrow Ba(OH)_2(s) + H_2(g)$$

2. Moderately active metals replacing hydrogen from steam produce metal oxides and hydrogen gas.

$$Mg(s) + H_2O(g) \rightarrow MgO(s) + H_2(g)$$
$$2\,Fe(s) + 3\,H_2O(g) \rightarrow Fe_2O_3(s) + 3\,H_2(g)$$

3. A metal above hydrogen in the activity series will replace hydrogen ion from acids:

$$Zn(s) + 2\,HCl(aq) \rightarrow ZnCl_2(aq) + H_2(g)$$
$$Ba(s) + 2\,HCl(aq) \rightarrow BaCl_2(aq) + H_2(g)$$
$$2\,Al(s) + 3\,H_2SO_4(aq) \rightarrow Al_2(SO_4)_3(aq) + 3\,H_2(g)$$

but

$$Ag(s) + HCl(aq) \rightarrow NR$$

4. A metal will replace any other metal ion *below* itself in the activity series.

$$Zn(s) + CuSO_4(aq) \rightarrow ZnSO_4(aq) + Cu(s)$$
$$Fe(s) + Pb(NO_3)_2(aq) \rightarrow Fe(NO_3)_2(aq) + Pb(s)$$
$$Cu(s) + 2\,AgNO_3(aq) \rightarrow Cu(NO_3)_2(aq) + 2\,Ag(s)$$

Double Replacement Reactions

A double replacement, or metathesis, reaction occurs between two compounds in which there is a "switch" of positive and negative ions. We might think of it as a kind of an exchange of partners in which the metal ion (cation) of compound A becomes the positive ion (cation) member of compound B as the original metal ion of compound B switches over and becomes the positive ion member of compound A (Fig. 5–1).

The reason that a double replacement reaction occurs will become more apparent as the topics of electrolyte strength and solubility rules are taken up in Chapter 9. However, it is useful in the meantime to categorize double replacement reactions as those in which more stable products emerge. The more stable compounds are most notably *precipitates* and *water.*

Examples of Double Replacement Reactions

1. Two soluble compounds in aqueous solution interact to form an insoluble precipitate.

$$AgNO_3(aq) + KCl(aq) \rightarrow AgCl(s) + KNO_3(aq)$$

$$3\,NaOH(aq) + FeCl_3(aq) \rightarrow 3\,NaCl(aq) + Fe(OH)_3(s)$$

$$Pb(NO_3)_2(aq) + K_2CrO_4(aq) \rightarrow PbCrO_4(s) + 2\,KNO_3(aq)$$

$$BaCl_2(aq) + ZnSO_4(aq) \rightarrow BaSO_4(s) + ZnCl_2(aq)$$

$$2\,Na_3PO_4(aq) + 3\,CaBr_2(aq) \rightarrow Ca_3(PO_4)_2(s) + 6\,NaBr(aq)$$

2. When an acid and base react to form a salt and water, we have described the reaction as an acid-base neutralization. For the moment, we will characterize a base as an ionic *hydroxide* compound. Just as acids can be recognized by formulas beginning with H, bases have formulas that end with OH.

Neutralization is another example of a double replacement reaction. Sometimes the base is not in the obvious form of a hydroxide. Metallic oxides that react with water to form hydroxides are known as **basic anhydrides.** Thus $MgO(s)$, $BaO(s)$, and $Na_2O(s)$ are basic anhydrides that form hydroxides when they react with water.

For example, $Na_2O(s) + H_2O(\ell) \rightarrow 2\,NaOH(aq)$ illustrates the reaction between the basic anhydride Na_2O and water, which forms

Figure 5–1 Principle of the double replacement reaction.

compound A   compound B

the typically alkaline basic solution, NaOH. The effect of this observation is to provide the basis for understanding how a metallic oxide behaves similarly to metal hydroxides in reacting with acids in acid-base neutralization.

$$NaOH(aq) + HCl(aq) \rightarrow NaCl(aq) + H_2O(l)$$

$$2\,KOH(aq) + H_2SO_4(aq) \rightarrow K_2SO_4(aq) + 2\,H_2O(l)$$

$$2\,HNO_3(aq) + CaO(s) \rightarrow Ca(NO_3)_2(aq) + H_2O(l)$$

$$2\,Al(OH)_3(s) + 3\,H_2SO_4(aq) \rightarrow Al_2(SO_4)_3(aq) + 6\,H_2O(l)$$

$$SrO(s) + 2\,HBr(aq) \rightarrow SrBr_2(aq) + H_2O(l)$$

Acid anhydrides are nonmetallic oxides that react with water to form acids. For example, $CO_2(g)$, $SO_2(g)$, and $P_4O_{10}(s)$ react with water to produce $H_2CO_3(aq)$, $H_2SO_4(aq)$, and $H_3PO_4(aq)$, respectively. Acid anhydrides may react with bases (or basic anhydrides) in acid-base neutralization:

$$2\,NaOH(aq) + CO_2(g) \rightarrow Na_2CO_3(aq) + 2\,H_2O(l)$$

$$2\,KOH(aq) + SO_2(g) \rightarrow K_2SO_3(aq) + 2\,H_2O(l)$$

Exercise 5.2

1. Complete and balance the following reactions, and label each one as *decomposition, combination, single replacement, double replacement*:
   (a) $Ca(s) + O_2(g) \rightarrow$
   (b) $Al(s) + S(s) \rightarrow$
   (c) $Au_2O_3(s) \xrightarrow{\Delta}$
   (d) $P_4(s) + Cl_2(g) \rightarrow$ phosphorus trichloride solid
   (e) $Cs(s) + H_2O(l) \rightarrow$
   (f) $Al(s) + H_2C_2O_4(aq) \rightarrow$
   (g) $N_2O_5(g) + H_2O(l) \rightarrow$
   (h) $NiCl_2(aq) + KOH(aq) \rightarrow$
   (i) $Fe_2O_3(s) + C(s) \rightarrow$
   (j) $CuO(s) + HNO_3(aq) \rightarrow$
2. Complete and balance the following reactions:
   (a) solid strontium + water $\rightarrow$
   (b) solid cadmium + copper(II) sulfate solution $\rightarrow$
   (c) solid copper(II) oxide + hydrogen gas $\rightarrow$
   (d) sulfur trioxide gas + water $\rightarrow$
   (e) solid antimony(III) sulfide + hydrochloric acid solution $\rightarrow$
   (f) solid platinum(IV) oxide $\xrightarrow{\Delta}$
   (g) sulfuric acid solution + solid chrominum(III) oxide $\rightarrow$
   (h) liquid benzene (C_6H_6) + oxygen gas $\rightarrow$
   (i) solid aluminum + liquid bromine $\rightarrow$
   (j) solutions of lead(II) nitrate + potassium chromate $\rightarrow$

The study of the quantitative relationships between substances as represented by formulas and equations is called stoichiometry. We have emphasized that a correctly balanced equation is an absolute necessity if you are to know the mole-to-mole ratios of reactants and products. The importance of calculating the masses and/or volumes of the constituents of a reaction is both an economic and a safety consideration. For example, the chemically pure materials used in the laboratory are expensive. It would be irresponsible to use a large excess of reagents if the precise amount needed were known. If everyone in a freshman chemistry laboratory uses more materials than is necessary, it would result in a waste of dollars and materials.

Suppose, as another example, that a poisonous gas is to be produced and collected in bottles. Unless the expected volume of the gas is calculated, there is no sure way of knowing how many gas-collecting bottles are needed. If several people were to produce more of the gas than could be collected in the available bottles, many students in the laboratory would be subjected to the irritating and toxic overflow. While the empirical (trial and error) approach has its place, the availability of reference texts for data about physical and chemical properties and the performance of stoichiometric calculations describe the practice of chemistry. The try-and-see approach is an unwanted last resort for the *experienced* chemist and then is used only with appropriate safety equipment and fume-exhaust hoods.

In effect, all stoichiometric problems are **mole-to-mole** calculations. However, because a mole can be expressed as a mass, a gas volume, or a number of particles, stoichiometric problems may be classified as moles-to-mass, mass-to-moles, mass-to-mass, mass-to-volume, volume-to-mass, and volume-to-volume types. Problems involving gas volumes will be recovered in the next chapter.

Review for a moment the following equation: $KClO_3(s) \xrightarrow{\Delta} KCl(s) + \frac{3}{2}O_2(g)$. This equation clearly states that the decomposition of 1 mole of $KClO_3$ yields 1 mole of KCl and $1\frac{1}{2}$ moles of O_2. One mole of $KClO_3$ weighs 122.5 g. It is not reasonable, on the basis of economy and safety, to use more than a fraction of a mole of $KClO_3$ to produce oxygen and KCl. The actual amount of $KClO_3$ required should be based on the amount of oxygen that one is prepared to collect. The point is, however, that once the stoichiometry of a reaction is established via the balanced equation, calculation of various amounts of reactants and products can be determined by dimensional analysis, using the *factor-label* method. In order to keep the following examples in proper perspective, it should always be remembered that the balanced equation contains all the stoichiometry and that everything else is arithmetic.

The factor-label method for solving stoichiometric problems is really a useful application of the mole concept to dimensional analysis. The following examples of the types of stoichiometric problems will be generalized in terms of substance A and substance B. In other words, given a specific mass or a number of moles of substance A, what mass or number of moles of substance B will be produced or be needed to react completely? The applications of the generalized procedures to specific problems will be illustrated by examples.

Moles-to-Moles Problems

If a balanced equation indicates that 2 moles of A *produce* 3 moles of B in a particular reaction,

$$2 \text{ moles A} \xrightarrow{\text{product}} 3 \text{ moles B}$$

or,

$$2A \rightarrow 3 \text{ B}$$

the question might be to find the number of moles of B that would be produced by 0.24 mole of A.

The same mathematics (that is, the same sequence for the dimensional analysis of the labeled factors) applies when the equation indicates that 2 moles of A *reacts with* 3 moles of B to produce some other product:

$$2A + 3B \rightarrow \text{a product}$$

The molar ratio $\dfrac{2 \text{ moles A}}{3 \text{ moles B}}$ is true in each case.

To find moles of B, write the number of moles of A and multiply by the molar ratio of B to A:

1. $\boxed{\text{moles A} \xrightarrow{\text{to get}} \text{moles B}}$

2. $\left(\begin{array}{c}\text{actual number}\\ \text{of moles of A}\end{array}\right) \times (\text{molar ratio}) \rightarrow \left(\begin{array}{c}\text{actual number}\\ \text{of moles of B}\end{array}\right)$

3. $\dfrac{0.24 \text{ mole A}}{1} \times \dfrac{3 \text{ moles B}}{2 \text{ moles A}} = \boxed{0.36 \text{ mole B}}$

Example 5.9

How many moles of $KClO_3$ are needed to produce 0.50 mole of oxygen?
1. Write the balanced equation:

$$2\ KClO_3(s) \rightarrow 2\ KCl(s) + 3\ O_2(g) \quad \textit{Decomposition}$$

2. The essential data are as follows:

$$2\ KClO_3 \rightarrow 3\ O_2$$

$$2\ \text{moles} \rightarrow 3\ \text{moles}$$

3. Identify the problem as a *moles-to-moles* type.
4. The sequence is as follows:

$$\text{moles } O_2 \rightarrow \text{moles } KClO_3$$

5. The steps in this sequence are as follows:

$$\text{moles } O_2 \times \text{molar ratio} \rightarrow \text{moles } KClO_3$$

6. Substitute the labeled factors and perform the dimensional analysis:

$$n = \frac{0.50\ \text{mole } O_2}{1} \times \frac{2\ \text{moles } KClO_3}{3\ \text{moles } O_2}$$

$$\boxed{n = 0.33\ \text{mole of } KClO_3}$$

Moles-to-Mass Problems

The sequence of operations in this type of problem is to change moles of A to moles of B and then to grams of B.

$$\boxed{\text{moles A} \xrightarrow{\text{to get}} \text{moles B} \xrightarrow{\text{to get}} \text{mass B}}$$

1. Use the molar ratio to find the actual number of moles of B produced by the given number of moles of A:

$$\left(\begin{array}{c}\text{given number}\\ \text{of moles A}\end{array}\right) \times \left(\frac{3\ \text{moles B}}{2\ \text{moles A}}\right) \rightarrow \left(\begin{array}{c}\text{actual number}\\ \text{of moles of B}\end{array}\right)$$

2. Multiply the actual number of moles of B by the molar mass of B to obtain grams.

$$(\text{moles B}) \times \left(\frac{g}{\text{mole}}\right) = \text{grams of B}$$

3. In sequence:

$$\begin{pmatrix} \text{actual number} \\ \text{of moles of A} \end{pmatrix} \times (\text{molar ratio}) \times \begin{pmatrix} \text{molar mass} \\ \text{of B} \end{pmatrix} = \begin{array}{c} \text{grams} \\ \text{of B} \end{array}$$

Example 5.10

What mass of $KClO_3$ is needed to produce 0.0200 mole of KCl?
1. Write the balanced equation; this time using a fractional coefficient in balancing.

$$KClO_3(s) \rightarrow KCl(s) + \tfrac{3}{2} O_2(g)$$

2. The balanced equation indicates that 1 mole of $KClO_3$ yields 1 mole of KCl.
3. Identify the problem as a *moles-to-mass* type.
4. The sequence is as follows:

$$\text{moles KCl} \rightarrow \text{moles } KClO_3 \rightarrow \text{mass } KClO_3$$

5. The steps are:

(moles KCl) × (molar ratio) × (molar mass of $KClO_3$) = grams of $KClO_3$

6. Perform the calculation after determining the molar mass of $KClO_3$ to be 122.5 g/mole.

$$\text{mass} = \frac{0.0200 \text{ mole } KCl}{1} \times \frac{1 \text{ mole } KClO_3}{1 \text{ mole } KCl} \times \frac{122.5 \text{ g } KClO_3}{1 \text{ mole } KClO_3}$$

mass = 2.45 g $KClO_3$

Mass-to-Moles Problems

Using the same general equation, $2A \rightarrow 3B$, the problem could be to find the number of moles of B produced from a certain mass of A. The approach is to change grams of A to moles of A, and then find moles of B.

$$\text{mass A} \rightarrow \text{moles A} \rightarrow \text{moles B}$$

Therefore, the sequential scheme is:

$$\underbrace{\text{mass of A} \times \frac{1 \text{ mole A}}{\text{molar mass A}}}_{\text{moles of A}} \times \underbrace{\frac{3 \text{ moles B}}{2 \text{ moles A}}}_{\text{molar ratio}} = \text{moles of B}$$

CHEMICAL EQUATIONS AND STOICHIOMETRY

Example 5.11

   How many moles of C_2H_2 are needed to produce 4.0 g of CO_2 when C_2H_2 is burned?
1. Write the balanced equation:

$$C_2H_2(g) + \tfrac{5}{2} O_2(g) \rightarrow 2 \; CO_2(g) + H_2O(l)$$

2. The stoichiometry of the reaction is as follows: 1 mole of C_2H_2 produces 2 moles of CO_2.
3. Calculate the molar mass of CO_2 (44.0 g/mole) and follow the sequence:

$$\text{mass } CO_2 \rightarrow \text{moles } CO_2 \rightarrow \text{moles } C_2H_2$$

4. $\text{moles} = 4.0 \; g \; CO_2 \times \dfrac{1 \text{ mole } CO_2}{44.0 \; g \; CO_2} \times \dfrac{1 \text{ mole } C_2H_2}{2 \text{ moles } CO_2}$

$$\boxed{n = 0.045 \text{ mole } C_2H_2}$$

Example 5.12

   Given the following balanced equation:

$$P_4(s) + 6 \; I_2(s) \rightarrow 4 \; PI_3(s)$$

calculate the number of moles of I_2 used in the formation of 6.0 g of PI_3.
1. The essential stoichiometry of the reaction is that 6 moles of I_2 will produce 4 moles of PI_3. Simplifying, we will see that the 6:4 ratio is the same as 3:2.
2. Write the solution sequence for this mass-to-moles problem:

$$\text{mass } PI_3 \rightarrow \text{moles } PI_3 \rightarrow \text{moles } I_2$$

3. Find the molar mass of PI_3 and follow the sequence:

$$PI_3 = 412 \text{ g/mole}$$

$$n = 6.0 \; g \; PI_3 \times \dfrac{1 \text{ mole } PI_3}{412 \; g \; PI_3} \times \dfrac{3 \text{ moles } I_2}{2 \text{ moles } PI_3}$$

$$\boxed{n = 0.022 \text{ mole } I_2}$$

Mass-to-Mass Problems

   This is a very common type of problem, often referred to as a mass-mass (or "weight-weight") problem. In effect, it asks what mass of B will be produced from a given mass of A? Or, conversely, if a certain mass of B is obtained, what mass of A must have been used? The sequence of steps is:

$$\text{mass A} \rightarrow \text{moles A} \rightarrow \text{moles B} \rightarrow \text{mass B}$$

1. Change mass of A to moles of A.
2. Change moles of A to moles of B by multiplying by the molar ratio.

3. Convert moles of B to mass of B by multiplying by the molar mass of B.

4. In sequence:

$$\text{mass of A} \times \frac{1 \text{ mole A}}{\text{molar mass A}} \times \frac{3 \text{ moles B}}{2 \text{ moles A}} \times \text{molar mass of B}$$

$$= \text{grams of B}$$

$$\left(\begin{array}{c}\text{actual number}\\ \text{of moles of A}\end{array}\right) \times (\text{molar ratio}) \times (\text{molar mass}) = \text{mass of B}$$

Example 5.13

What mass of aluminum would be used in the production of 10.2 g of aluminum oxide if aluminum reacted directly with oxygen?

1. Write the balanced equation to obtain the stoichiometry:

$$2\ Al(s) + \tfrac{3}{2}\ O_2(g) \rightarrow Al_2O_3(s)$$

2. The molar ratio data show us that 2 moles of Al yield 1 mole of Al_2O_3.
3. Write the sequence of steps for this mass-to-mass problem:

$$\text{mass } Al_2O_3 \rightarrow \text{moles } Al_2O_3 \rightarrow \text{moles } Al \rightarrow \text{mass } Al$$

4. Find the molar masses of Al and Al_2O_3 and follow the sequence:

$$Al = 27.0 \text{ g/mole}$$

$$Al_2O_3 = 102.0 \text{ g/mole}$$

$$\text{mass Al} = 10.2 \text{ g } Al_2O_3 \times \frac{1 \text{ mole } Al_2O_3}{102.0 \text{ g } Al_2O_3} \times \frac{2 \text{ moles Al}}{1 \text{ mole } Al_2O_3} \times \frac{27.0 \text{ g Al}}{1 \text{ mole Al}}$$

$$\text{mass} = \text{moles} \times \text{molar ratio} \times \text{molar mass}$$

$$\boxed{\text{mass} = 5.40 \text{ g Al}}$$

Exercise 5.3

Solve the following problems:
1. Given the skeleton equation, $C_3H_8(g) + O_2(g) \rightarrow CO_2(g) + H_2O(l)$
(a) How many moles of C_3H_8 are needed to produce 0.25 mole of water?
(b) What mass of water is produced from 0.050 mole of C_3H_8?
(c) What mass of oxygen is needed to produce 1.4 g of CO_2 in this reaction?
2. Based on the observation that aluminum reacts with hydrochloric acid in a simple replacement reaction, answer the following questions:
(a) What is the stoichiometry of the reaction?
(b) How many moles of aluminum are needed to produce 0.30 mole of gas?
(c) What mass of aluminum is needed to produce 2.0×10^{-3} of hydrogen?
(d) How many moles of hydrochloric acid are needed to react completely with 60.0 mg of aluminum?

CHEMICAL EQUATIONS AND STOICHIOMETRY

It is an experimentally proven fact that many reactions produce less than the calculated stoichiometric amounts of products. This may be due to varying degrees of purity among the reactants, incompleteness of chemical reactions, or other factors. Furthermore, there may be more than one reaction occurring during the operation, so that various by-products are obtained at the expense of additional reactants. All of these effects combine to produce yields that usually are not as high as predicted from the ideal equation.

When the **actual yield** is compared to the **theoretical yield** or calculated yield, the resulting fraction is expressed as a percentage value and is called the **percentage yield.**

Example 5.14

If 6.00 g of carbon were burned to form CO_2, what mass of CO_2 would be expected? If the actual yield of CO_2 was measured to be 21.0 g, what is the percentage yield?

1. Write the balanced equation to determine the stoichiometry.

$$C(s) + O_2(g) \rightarrow CO_2(g)$$

$$1 \text{ mole} \qquad\qquad 1 \text{ mole}$$

2. Find the theoretical yield of CO_2:

$$\text{mass } CO_2 = 6.00 \text{ g } C \times \frac{1 \text{ mole } C}{12.0 \text{ g } C} \times \frac{1 \text{ mole } CO_2}{1 \text{ mole } C} \times \frac{44.0 \text{ g } CO_2}{1 \text{ mole } CO_2}$$

moles of carbon $\times$ molar ratio $\times$ molar mass

mass = 22.0 g CO_2 (theoretical)

3. % yield = $\dfrac{\text{actual}}{\text{theoretical}} \times 100$

% yield = $\dfrac{21.0 \text{ g}}{22.0 \text{ g}} \times 100 =$ 95.4%

5.8   LIMITING FACTOR OF A REACTANT IN A REACTION

When the masses of reactants are known, a calculation must be made to determine whether or not the masses are stoichiometric. For example, if one of two reactants is present in excess, some of the excess reactant will have no part in the reaction. Any calculation based on the excess mass will be, of course, incorrect. Before a percentage yield can be determined, it is necessary to find out which reactant is completely consumed.

Example 5.15

A 6.54 g sample of zinc is added to 17.0 g of $CuSO_4$ in solution. If 5.42 g of copper is produced, what is the percentage yield?
1. Determine the stoichiometry:

$$Zn(s) + CuSO_4(aq) \rightarrow ZnSO_4(aq) + Cu(s)$$

2. Observe that 1 mole of Zn reacts with 1 mole of $CuSO_4$ to form 1 mole of Cu. Are there equal moles of Zn and $CuSO_4$?

$$\text{moles of Zn} = 6.54 \text{ g} \times \frac{1 \text{ mole}}{65.4 \text{ g}} = 0.100 \text{ mole}$$

$$\text{moles of } CuSO_4 = 17.0 \text{ g} \times \frac{1 \text{ mole}}{159.6 \text{ g}} = 0.106 \text{ mole}$$

3. Note the excess of $CuSO_4$. Therefore, the calculation of the mass of copper must be based on the amount of zinc used, because zinc is the limiting reactant.

$$\text{mass of Cu} = \frac{0.100 \text{ mole Zn}}{1} \times \frac{1 \text{ mole Cu}}{1 \text{ mole Zn}} \times \frac{63.5 \text{ g Cu}}{1 \text{ mole Cu}}$$

$$\text{mass} = \text{moles} \times \text{molar ratio} \times \text{molar mass of copper}$$

$$\boxed{\text{mass} = 6.35 \text{ g Cu (theoretical)}}$$

4. The percentage yield of copper can be calculated.

$$\% \text{ yield} = \frac{\text{actual}}{\text{theoretical}}$$

$$\% \text{ yield} = \frac{5.42 \text{ g Cu}}{6.35 \text{ g sample}} \times 100 = \boxed{85.4\%}$$

QUESTIONS & PROBLEMS

5.1 What is the meaning of each of the following symbols in equations: (aq), (s), (l), (g), ($\xrightarrow{\Delta}$)?

5.2 Interpret the following equations:
(a) $K(s) + H_2O(l) \rightarrow KOH(aq) + \frac{1}{2} H_2(g)$
(b) $CH_3OH(l) + \frac{3}{2} O_2(g) \rightarrow$
$\quad CO_2(g) + 2 H_2O(l) (\Delta H = -173.4 \text{ kcal})$

5.3 Balance the following equations:
(a) $Ni(s) + HCl(aq) \rightarrow NiCl_2(aq) + H_2(g)$
(b) $Al(s) + C(s) \rightarrow Al_4C_3(s)$
(c) $O_2(g) \rightarrow O_3(g)$
(d) $MnO_2(s) + HCl(aq) \rightarrow$
$\quad MnCl_2(aq) + Cl_2(g) + H_2O(l)$
(e) $I_2(s) + Cl_2(g) \rightarrow ICl_3(s)$
(f) $As_4(s) + O_2(g) \rightarrow As_4O_6(s)$
(g) $MnO_2(s) \rightarrow Mn_3O_4(s) + O_2(g)$
(h) $NO(g) + Cl_2(g) \rightarrow NOCl(g)$
(i) $SiO_2(s) + H_2F_2(g) \rightarrow SiF_4(g) + H_2O(l)$
(j) $H_2O(l) + NOCl(g) \rightarrow HCl(aq) + HNO_2(aq)$

CHEMICAL EQUATIONS AND STOICHIOMETRY

5.4   Write skeleton equations and then balance the following:
(a) solid cesium + water → cesium hydroxide solution + hydrogen gas
(b) solid gold(III) oxide → solid gold + oxygen gas
(c) solid iron(II) sulfide + oxygen gas → solid iron(III) oxide + sulfur dioxide gas
(d) bismuth(III)chloride solution + sodium sulfide solution → solid bismuth(III) sulfide + sodium chloride solution
(e) solid silicon dioxide + hydrofluoric acid → solid silicon tetrafluoride + water
(f) solid carbon + solid aluminum oxide → solid aluminum carbide + carbon monoxide gas
(g) solid calcium carbonate → solid calcium oxide + carbon dioxide gas
(h) carbon dioxide gas + calcium hydroxide solution → calcium hydrogen carbonate solution
(i) copper(II)phosphate solution + tin(IV) chloride solution → copper(II) chloride solution + solid tin(IV) phosphate
(j) solid cadmium + mercury(I) nitrate solution → cadmium nitrate solution + liquid mercury

5.5   Complete and balance the following equations:
(a) $Rb_2O(s) + H_2O(l) \rightarrow$
(b) $Mg(s) + HClO_4(aq) \rightarrow$
(c) $HOAc(aq) + NaOH(aq) \rightarrow$
(d) $KCN(aq) + H_2SO_4(aq) \rightarrow$
(e) $Na_2S(aq) + Fe_2(Cr_2O_7)_3 \rightarrow$
(f) $P_4O_{10}(g) + H_2O(l) \rightarrow$
(g) $Co(OAc)_3(aq) + H_2S(l) \rightarrow$
(h) $H_2CrO_4(aq) + NiC_2O_4(aq) \rightarrow$
(i) $Al(s) + H_3PO_4(aq) \rightarrow$
(j) $C_5H_{12}(g) + O_2(g) \rightarrow$

5.6   How many moles of zinc are required to react completely with 1.6 moles of $CuSO_4$ in solution?

5.7   Calculate the mass of HgO needed to produce 0.12 mole of mercury upon complete decomposition.

5.8   What mass of hydrogen gas can be collected when 2.2×10^{-2} mole of magnesium is completely reacted with hydrochloric acid?

5.9   Iron(III) oxide reacts with aluminum to produce iron and aluminum oxide. What mass of aluminum is needed to reduce 4.3 g of iron(III) oxide?

5.10   What mass of 80.0 percent pure gold(III) oxide is needed to yield 1.4 g of pure gold upon decomposition?

5.11   Find the percentage yield in a reaction where the actual yield of product is 1.55 g and the theoretical yield is 4.02 g.

5.12   A one gram sample of impure sodium chloride is dissolved in water and allowed to react with silver nitrate solution. The dried precipitate of AgCl formed weighs 1.475 g. Assuming that all of the NaCl reacts completely, calculate the percentage NaCl in the original sample.

5.13   A sample of impure $CaCO_3$ weighing 5.565 grams is thermally decomposed completely to produce 1.500 grams of CO_2. What is the percentage of $CaCO_3$ in the original sample?

5.14   Calculate the grams of:
(a) $BaSO_4$ produced as a result of the reaction of 10.0 grams of sodium sulfate with excess barium chloride.
(b) Silver chloride produced as a result of the reaction of 25.0 grams of aluminum chloride with excess silver nitrate.

5.15   (a) How many grams of $KClO_3$ must be thermally decomposed to produce 12.0 grams of oxygen?
(b) How many grams of zinc metal are required to liberate 10.0 grams of hydrogen by reaction with excess H_2SO_4?

5.16   To produce ferrous sulfide, 140 g of sulfur is heated with 182 g of iron. How many grams of sulfur are left uncombined?

5.17   How many grams of $Ca(NO_3)_3$ could be made from the reaction of 62.6 g of 80% pure $CaCl_2$ with an excess amount of $AgNO_3$?

5.18   Epsom salt ($MgSO_4 \cdot 7H_2O$) may be made by the action of H_2SO_4 on the mineral magnesite. What mass of Epsom salt can be made from 216 g of impure magnesite containing 75 per cent $MgCO_3$? Equations are as follows:

$$MgCO_3(s) + H_2SO_4(aq) \rightarrow$$
$$MgSO_4(s) + H_2O(l) + CO_2(g)$$
$$MgSO_4(s) + 7H_2O(l) \rightarrow MgSO_4 \cdot 7H_2O(s)$$

5.19   A sample of calcium bicarbonate weighing 16.60 g is heated until the compound is completely decomposed into CaO, water, and CO_2. How many grams of CaO are obtained?

5.20   A solution containing 22.2 g of acetic acid is mixed with a solution containing 8.91 g of aluminum nitrate. How many moles of aluminum acetate will result from the complete reaction?

5.21   How many grams of aluminum acetate could be obtained in Problem 5.20 if the reaction were only 60.0% complete?

5.22   The equation for the production of $Cl_2(g)$ by electrolysis of aqueous sodium chloride is:

$$2 NaCl(s) + 2 H_2O(l) \rightarrow$$
$$2 NaOH(aq) + Cl_2(g) + H_2(g)$$

Calculate the mass of $Cl_2(g)$ obtained from the electrolysis of 200 g of 84% pure NaCl(s).

5.23   Calculate the mass of $Mg(OH)_2(s)$ made when 27.1 g of basic anhydride MgO(s) reacts with excess water.

5.24   Calculate the mass of $CaCO_3(s)$ produced by mixing 112.6 g of 85.0% pure BaO(s) with 81.1 g of 65.0% pure $Na_2CO_3(aq)$.

5.25   Silver reacts with dilute $HNO_3(aq)$ as follows:

$$3 Ag(s) + 4 HNO_3(aq) \rightarrow$$
$$3 AgNO_3(aq) + NO(g) + 2 H_2O(l)$$

(a) What mass of Ag is required to prepare 91.6 g of $AgNO_3$?
(b) How many moles of $AgNO_3$ may be produced from the reaction of 176 g of 71.2% pure Ag with 58.8 g of HNO_3?

5.26   One mole of H_2O and 16.9 g of Na are allowed to react.
(a) How many moles of NaOH(aq) are produced?
(b) How many grams of $H_2(g)$ are produced?

5.27   At low pressures the reaction of $NH_3(g)$ with $O_2(g)$ is only 37% complete. How many moles of $NO_2(g)$ will result if 45.8 g of $NH_3(g)$ are mixed with 3.7 moles of $O_2(g)$? The equation for the reaction is:

$$4 NH_3(g) + 7 O_2(g) \rightarrow 6 H_2O(g) + 4 NO_2(g)$$

5.28   Which will produce a larger mass of oxygen per gram of reactant in a complete reaction:
(a) $2 KClO_3(s) \xrightarrow{\Delta} 2 KCl(s) + 3 O_2(g)$
(b) $2 NaNO_3(s) \xrightarrow{\Delta} 2 NaNO_2(s) + O_2(g)$

5.29   EQUATIONS
Complete and balance the following equations:
A. Combination
1. zinc + sulfur
2. iron + sulfur [solid iron(II) sulfide is formed]
3. aluminum + oxygen gas
4. sodium + chlorine gas
B. Decomposition
1. mercury(II) oxide + heat
2. copper(II) carbonate + heat
[solid copper(II) oxide and CO_2 are formed]
3. potassium chlorate + heat
[solid potassium chloride and O_2 are formed]
4. calcium carbonate + heat
[solid calcium oxide and CO_2 are formed]
C. Single Replacement
1. zinc + sulfuric acid
2. aluminum + hydrochloric acid
3. magnesium + silver nitrate solution
4. sodium + water
D. Double Replacement
1. solutions of ammonium carbonate + barium chloride

2. solutions of aluminum sulfate + potassium hydroxide

3. H₂S gas bubbled through a copper(II) chloride solution

4. sulfuric acid + calcium hydroxide solution

5.30 Complete and balance, indicating solids (s), gases (g), and aqueous solutions (aq).

(a) Zinc metal is added to hydrochloric acid

(b) Solutions of aluminum chloride and lead(II) nitrate are mixed to form solid lead(II) chloride and a solution of aluminum nitrate

(c) Solid iron(III) oxide + $H_2(g)$ react to form solid iron and water.

(d) Solid silver oxide is decomposed by heat to form solid silver and oxygen

(e) Solid copper(I) oxide + oxygen forms solid copper(II) oxide

(f) Solutions of barium chloride and sodium sulfate are mixed to produce sodium chloride and a precipitate of barium sulfate

(g) Hydrobromic acid is added to a solution of silver nitrate to produce a precipitate of silver bromide and . . . ? (You complete it.)

(h) Aluminum metal is added to acetic acid

(i) The hydrocarbon $C_6H_{12}(l)$ is burned completely

(j) Solid carbon reacts with steam to form two gases, carbon dioxide and hydrogen

CHAPTER SIX

LEARNING OBJECTIVES

At the completion of this chapter, you should be able to:

1. Describe in general terms the effect of changes of temperature and pressure on gases.

2. Define an ideal or perfect gas.

3. List the principal postulates of the kinetic molecular theory.

4. Find the molar mass of a gas from its density.

5. List the four parameters that affect the behavior of gases.

6. State Boyle's law and Charles's law.

7. Define the terms *torr* and *atmosphere*.

8. Calculate the effect of changes in temperature and pressure on a volume of gas by use of the combined gas laws.

9. State Dalton's law of partial pressure.

10. Find the partial pressures exerted by particular gases in a mixture.

11. Calculate gas volumes collected by water displacement.

12. Use the table of vapor pressure of water in solving gas law problems.

13. Derive the value and proper units of R, the gas constant.

14. Write the equation of state and explain its terms.

15. Solve problems involving the mass, molar mass, and density of gases using the equation of state.

16. Calculate the volume occupied by a given number of moles of gas at various temperature and pressure conditions.

17. Apply the equation of state to the solution of stoichiometric problems involving gases at nonstandard conditions.

THE BEHAVIOR OF GASES

It is fascinating to observe some of the properties of the gaseous state. Let us consider air as a familiar example. The behavior of this invisible mixture (approximately 80 per cent nitrogen and 20 per cent oxygen by volume) is typical of gases. It offers negligible resistance to our daily human activities, yet it is capable of lifting airplanes weighing many tons, it transmits the destructive forces of hurricanes and tornados, and it incinerates objects from outer space as they pass through the earth's atmosphere.

The force of the atmosphere is sufficient to crush an evacuated tin can, to support one end of a slat of wood while the other end is broken off by a "karate chop," and to support water in an inverted tumbler (Fig. 6–1). The mercury barometer is based on the ability of the atmosphere to sustain a column of mercury at a height of 760 mm.

Gases increase in volume when heated and decrease in volume when cooled. If an inflated balloon is removed from a refrigerator and placed in a tub of hot water, the balloon visibly expands to a larger volume (Fig. 6–2). The process is easily reversed by returning the expanded balloon to a refrigerator. While solids and liquids are similarly affected by temperature change, their changes in volume are not nearly so dramatic. Gases exert a markedly increased force (pressure) when heated in a container of fixed volume. We are cautioned not to incinerate aerosol cans because of the danger of explosion. Figure 6–3 illustrates the effect of temperature on the pressure exerted by a gas in a container of fixed volume.

Compressibility is another physical property of gases that has numerous practical applications. For example, cylinders of compressed gases are widely used by laboratories and industrial concerns. Scuba diving offers a more dramatic application of compressed air. Refrigeration and air-conditioning systems take advantage of the temperature-volume-compressibility relationships of gases.

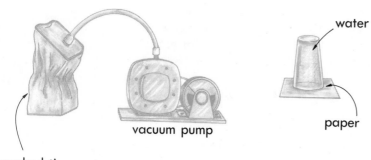

crushed tin can

vacuum pump

water

paper

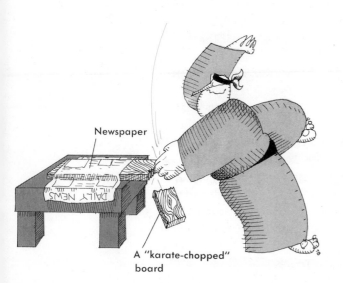

Newspaper

A "karate-chopped" board

Figure 6–1 Demonstrations of atmospheric pressure.

THE BEHAVIOR OF GASES

Figure 6–2   The effect of temperature on a gas volume.

cool balloon

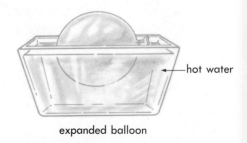

—hot water

expanded balloon

gas pressure gauges

Figure 6–3   The effect of
temperature on the pressure
exerted by a gas in a fixed
volume.

ice water

room temperature

hot water

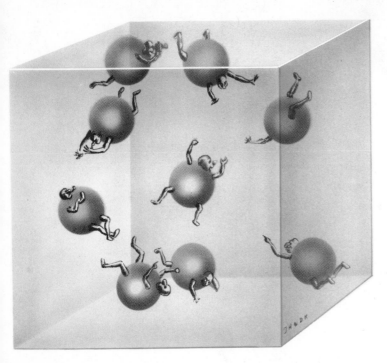

Figure 6—4   Negligible van der Waals forces between tiny particles that are far apart.

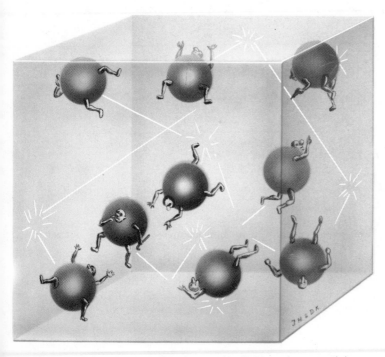

Figure 6—5   The motion, direction, and collisions of gas particles.

THE BEHAVIOR OF GASES

About 100 years ago, James Clark Maxwell and Ludwig Boltzmann suggested a model to explain their observations of the behavior of gases. This explanation is called the kinetic molecular theory. The basic statements of the theory are most accurately applied to **ideal, or perfect, gases.** An ideal gas is one in which no attractive forces exist in the molecules that might prevent them from having perfectly **elastic collisions.** An elastic collision occurs when two objects collide and rebound without causing any net change in their physical structures or total kinetic energies. There is no such thing as an ideal gas in reality, but most gases (especially when they are at low pressures and at high temperatures) behave nearly enough like ideal gases to permit the application of the following basic postulates:

1. Gases consist of atoms or molecules so small that their actual volumes are negligible compared to the spaces between them (Fig. 6–4).
2. Gas particles are constantly moving at high speed at ordinary temperatures. Their motion is straight-line, and the direction is random. The collisions between the particles are perfectly elastic (Fig. 6–5).
3. The collisions between the gas particles result in no net loss of their average kinetic energies. The average kinetic energy of the gas particles is a function of the temperature of the gas (Fig. 6–6).

The model of rapidly moving and colliding molecules presents an excellent explanation of gas behavior. Some illustrative data will be used to emphasize the statistical approach to the description of gas behavior. Words such as "model" and "average" are pointedly used to

ideal law Gas Law

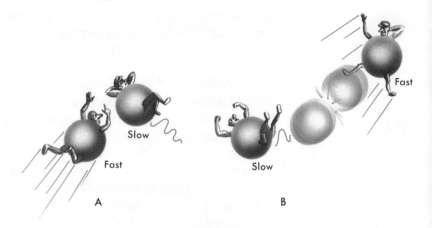

Figure 6–6   Average KE of *A* = average KE of *B* resulting from the transfer of kinetic energy upon collision.

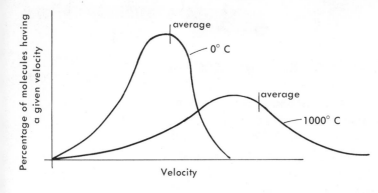

Figure 6–7 Maxwell-Boltzmann molecular velocity distribution curves.

indicate that gas measurements should be made for *moles* of gas rather than for individual molecules.

Maxwell and Boltzmann theorized that a volume of a particular gas placed in a container that does not permit any energy exchange with the outside would lead to some characteristic *average* velocity among the contained molecules. Some molecules in this isolated system will have velocities and related kinetic energies above or below that average. A graphic representation of this theory is presented in Figure 6–7.

It is interesting to compare two dissimilar gases in order to see how their behavior relates to the kinetic molecular theory. The comparison will be between the low density gas hydrogen (2.0 amu) and the much heavier gas carbon dioxide (44.0 amu). The comparisons are at ordinary room temperature and normal barometric pressure.

6.3 AVOGADRO'S HYPOTHESIS AND THE MOLE CONCEPT APPLIED TO GASES

When Avogadro concluded empirically that equal volumes of gases under the same conditions of temperature and pressure would contain equal numbers of molecules, he would have welcomed the knowledge that his observations were consistent with kinetic molecular theory.

Gases that occupy equal volumes at the same temperature and that exert the same force on the walls of the container must contain equal numbers of molecules. If not, these gases would not have the same average kinetic energies. Avogadro's principle states also that equal numbers of moles of gases at the same temperature and pressure will occupy the same volume. This observation is one of the cornerstones of the mole concept, which states that a mole of any ideal gas at *standard conditions* occupies 22.4 liters. The term "standard conditions," or standard temperature and pressure (STP), refers to 0°C (273°K) and one atmosphere pressure (760 torr).

The experimentally supported observation that one mole of any

$KE \propto T$

$1 \, atm = 760 \, torrs$

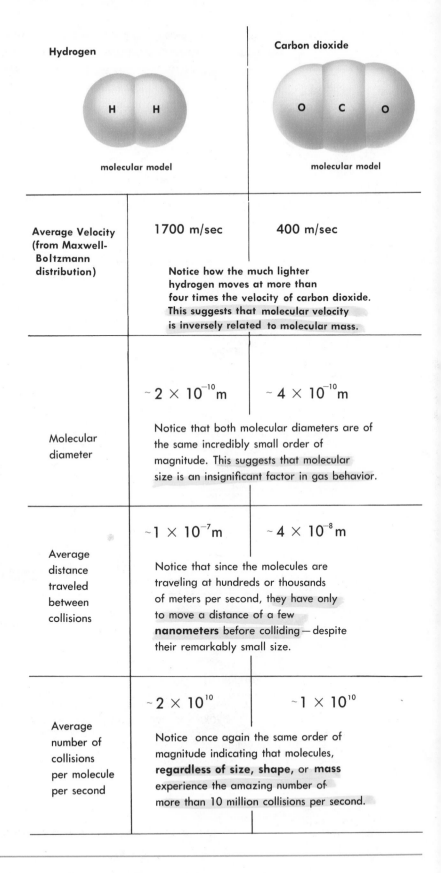

	Hydrogen	Carbon dioxide
	molecular model	molecular model
Average Velocity (from Maxwell-Boltzmann distribution)	1700 m/sec	400 m/sec
	Notice how the much lighter hydrogen moves at more than four times the velocity of carbon dioxide. This suggests that molecular velocity is inversely related to molecular mass.	
Molecular diameter	$\sim 2 \times 10^{-10}$ m	$\sim 4 \times 10^{-10}$ m
	Notice that both molecular diameters are of the same incredibly small order of magnitude. This suggests that molecular size is an insignificant factor in gas behavior.	
Average distance traveled between collisions	$\sim 1 \times 10^{-7}$ m	$\sim 4 \times 10^{-8}$ m
	Notice that since the molecules are traveling at hundreds or thousands of meters per second, they have only to move a distance of a few **nanometers** before colliding — despite their remarkably small size.	
Average number of collisions per molecule per second	$\sim 2 \times 10^{10}$	$\sim 1 \times 10^{10}$
	Notice once again the same order of magnitude indicating that molecules, **regardless of size, shape,** or **mass** experience the amazing number of more than 10 million collisions per second.	

gas at 0°C and standard atmospheric pressure contains 6.02×10^{23} molecules is another aspect of the mole concept. By virtue of this observation, 22.4 liters of gas at STP is called the **molar volume.**

The concept of molar volume allows us to solve a variety of problems in which gas volumes are described in terms of the mole concept.

Example 6.1

How many moles are there in 5.0 liters of oxygen gas at STP?
1. Start with the fact that one mole of any ideal gas occupies 22.4 liters at STP.
2. Using dimensional analysis:

$$n = 5.0 \, \cancel{L} \times \frac{1 \text{ mole}}{22.4 \, \cancel{L}}$$

$$\boxed{n = 0.22 \text{ mole}}$$

Example 6.2

What volume is occupied by 0.17 mole of a gas at STP?

$$V = 0.17 \, \cancel{\text{mole}} \times \frac{22.4 \text{ L}}{1 \, \cancel{\text{mole}}}$$

$$\boxed{V = 3.8 \text{ L}}$$

Example 6.3

Approximately how many *molecules* are there in 4000 ml of a gas at STP?

$$\text{number of molecules} = 4000 \, \cancel{\text{ml}} \times \frac{1 \, \cancel{L}}{10^3 \, \cancel{\text{ml}}} \times \frac{1 \, \cancel{\text{mole}}}{22.4 \, \cancel{L}} \times \frac{6.02 \times 10^{23} \text{ molecules}}{1 \, \cancel{\text{mole}}}$$

$$\boxed{\text{number of molecules} = 1.075 \times 10^{23} \text{ (to 4 sig. figs.)}} \quad ?$$

conversion factors don't count?

Exercise 6.1

1. How many moles of NH_3 gas occupy 300 mL at STP?
2. What volume is occupied by 3.2×10^{-3} mole of N_2 gas at STP?
3. Approximately how many molecules of CO_2 gas are there in a 2.0 liter volume at STP?
4. What is the density of CO_2 gas at STP? (Hint: remember that gas densities are expressed as grams per liter).
5. If 3.0 liters of an unknown gas at STP weighs 5.85 g, what is its molar mass?

$$\frac{44 g}{22.4 \, \ell}$$

$$\frac{3 \, \ell}{22.4 \, \ell} = .134 \text{ moles} \qquad \frac{5.85 g}{.134 \text{ moles}} = 44 \, g/m$$

THE BEHAVIOR OF GASES

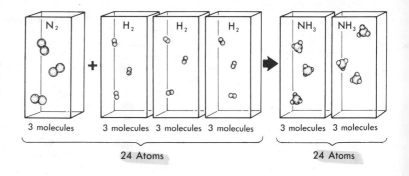

Figure 6–8 Equal numbers of molecules occupy equal volumes at the same temperature and pressure.

An important contribution that immediately preceded Avogadro's hypothesis was the **law of combining volumes,** presented by J. L. Gay-Lussac in 1808. Gay-Lussac found that volumes of gases in chemical reactions were related by small whole-number ratios when measured at fixed temperature and pressure. In terms of his own principle, Avogadro explained that 1 volume of nitrogen could react with 3 volumes of hydrogen to produce 2 volumes of ammonia and still be consistent with the law of conservation of matter. If the equation

$$N_2(g) + 3H_2(g) \rightarrow 2NH_3(g)$$

is visualized by a simple model (shown in Fig. 6–8), the truth of the strange arithmetic 1 + 3 = 2 can be understood.

Notice in Figure 6–8 that conservation of matter is demonstrated in terms of the number of nitrogen and hydrogen atoms at the beginning at the end of the reaction. The only difference is in the arrangement of the atoms in molecules. However, the kinetic molecular theory clearly states that the volume occupied by a gas depends on numbers of molecules and *not* on their size, shape, or mass. The heavier ammonia molecules simply move more slowly than the lighter nitrogen and hydrogen, but the average kinetic energies are all the same.

We discussed the kinetic molecular theory earlier in this chapter with the understanding that temperature and pressure were not variables. The specific value for standard temperature and pressure (STP) was added as another aspect of the mole concept. It is now appropriate to consider the interrelationship of the **gas laws,** where temperature, pressure, and number of molecules are realistically treated as variable factors.

6.4 THE GAS LAWS

The kinetic molecular theory plainly suggests that gases must be measured and described in terms of four **parameters** (variables), organized in Table 6–1.

TABLE 6–1 Parameters

Parameter	Symbol	Brief Description	Unit of Measure
Pressure	P	Force exerted on container walls by molecular collisions	torr atm (atmosphere) mm Hg
Volume	V	Volume of container	L (liters)
Number of moles	n	Number of moles of gas in container	mole
Temperature	T	Absolute temperature of gas	°K

6.5 PRESSURE, VOLUME, AND BOYLE'S LAW

Pressure is defined as the force exerted on a unit area. The pressure exerted by gases is the result of collisions between the rapidly moving molecules and the walls of the container. Gas pressure could be accurately described in dynes per cm², where a dyne is a unit of force.* However, the pressure exerted by gases is more conveniently related to the atmosphere that is pressing in on objects at sea level with a force of 1 million dynes per square centimeter. It so happens that a column of mercury 760 mm high also exerts the same force. If a completely full mercury column (a **barometer**) is inverted into a container of mercury, the relationship between its downward force (mass) and atmospheric pressure can be observed (Fig. 6–9). Naturally, the column of mercury must be at least 760 mm long.

---

*Force is equal to mass times acceleration, F = ma. A dyne is the force needed to change the speed of (accelerate) a mass of 1 gram by 1 cm per second, each second. More concisely expressed,

$$1 \text{ dyne} = \underbrace{g}_{\text{mass}} \times \underbrace{cm/sec^2}_{\substack{\text{cm per sec per sec} \\ \text{(acceleration)}}}$$

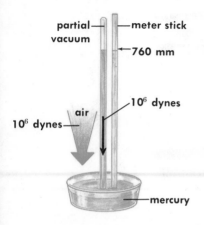

Figure 6–9   Atmospheric pressure equals the force per unit area exerted by a column of mercury 760 mm high.

THE BEHAVIOR OF GASES

Of course, atmospheric pressure changes, but **standard atmospheric pressure** may be defined as that which can support a mercury column 760 mm high. In honor of the inventor of the barometer, Evangelista Torricelli, one mm of mercury in such a column is called a **torr.** The alternative unit of pressure is the **atm** (atmosphere), which is equal to 760 torr.

Robert Boyle in 1662 described the relationship between volume and pressure as being *inverse* (at constant temperature).

$$V \propto \frac{1}{P}$$

Mathematically expressed:

$$V = \frac{k}{P} \quad \text{or,}$$

$$k = PV \quad (T = \text{constant})$$

$$P \times V = \text{constant} \ (T)$$

Looking at Figure 6–10, we can observe the following relationships:

Pressure (atm)	Volume (L)	P × V (L-atm)
1	2	2
2	1	2
4	$\frac{1}{2}$	2
8	$\frac{1}{4}$	2

Although the proportionality constant in the previous example is 2 L-atm, the constant in each individual system depends on the initial volume and the temperature.

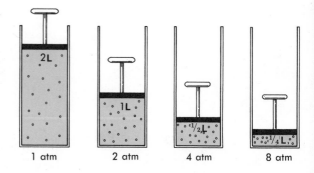

Figure 6–10   A cylinder and piston system showing the relationship between P and V at constant temperature. As the pressure increases, the volume decreases.

Example 6.4

If a gas occupies 250 mL at 600 torr, what volume will it occupy at standard pressure if the temperature remains constant?

1. Organize the data:

initial volume	$V_1 = 250$ mL
final volume	$V_2 = ?$
initial pressure	$P_1 = 600$ torr
final pressure	$P_2 = 760$ torr

2. *Reason* that the increase in pressure will cause a corresponding decrease in volume. Simply multiply the original volume by the pressure change in the form of a *common fraction* (that is, a number less than 1). Any number multiplied by a common fraction will provide an answer that is less than the original number.

Boyles

$$\frac{V_1 P_1}{T_1} = \frac{V_2 P_2}{T_2}$$

$$V_2 = 250 \text{ mL} \times \frac{600 \text{ torr}}{760 \text{ torr}}$$

final volume = 197 mL

There is one other problem that must be considered before leaving the discussion of pressure. It is the question of the pressure exerted by a mixture of gases.

6.6 DALTON'S LAW OF PARTIAL PRESSURES

In 1801, John Dalton reported that the total pressure exerted by a nonreacting mixture of gases is equal to the sum of the pressures that each gas in the mixture exerts individually. Figure 6–11 illustrates this simple cause-and-effect relationship.

pressure (atm)

pressure (atm)

pressure (atm)

Figure 6–11   The total pressure exerted by a mixture of gases is equal to the sum of the individual pressures.

1 mole of gas A

1 mole of gas A and
1 mole of gas B

1 mole of gas A and
1 mole of gas B and
1 mole of gas C

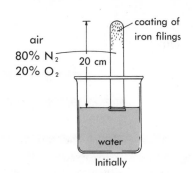
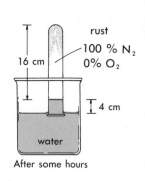

Figure 6–12 Dalton's law of partial pressures. The removal of oxygen from air reduces the pressure by about 20 per cent.

air
80% N₂
20% O₂

coating of
iron filings

20 cm

Initially

16 cm

rust
100 % N₂
0% O₂

4 cm

After some hours

water

water

A fairly simple demonstration of Dalton's law can be effected by wetting the inside of a 20 cm test tube with water so that a thin layer of iron filings can adhere to the inner wall. Invert the tube in a beaker of water so that the pressure in the tube nearly equals the room pressure. After the iron filings combine chemically with the oxygen in the air (the rust is obvious), the water level will be observed to have risen about 4 cm. This 20 per cent reduction of the gas volume in the tube corresponds nicely to the fact that air contains about 20 per cent oxygen by volume. Since the iron filings remove 20 per cent of the original gas volume, the total pressure is reduced by the same fraction. Figure 6–12 illustrates the 20 per cent reduction in pressure.

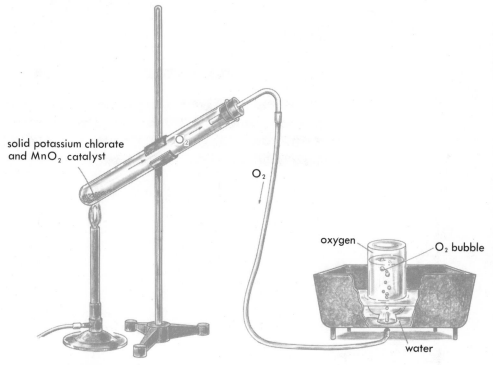

solid potassium chlorate
and MnO₂ catalyst

O₂

O₂

oxygen

O₂ bubble

water

Figure 6–13 Collection of oxygen gas by water displacement.

In summary, Dalton's law of partial pressures may be expressed as follows:

$$\overbrace{P_{total} = P_a + P_b + P_c + \ldots}^{\text{sum of the partial pressures of each gas}}$$

One of the most useful applications of Dalton's law is made when a gas is collected over water. Collection of gas over water is a common laboratory method of collecting a volume of a water-insoluble gas by the displacement of water from a bottle. Figure 6–13 illustrates how oxygen may be collected by water displacement.

The bottle of oxygen pictured in Figure 6–14 may be the subject of some useful calculations. One essential measurement is the pressure of the contained gas or gases. We can determine this quickly by equalizing the level of the water in the bottle with the water level in the trough (Fig. 6–14).

If the barometer reading is 760 torr and the levels are equal, the gas pressure in the bottle must also be 760 torr. But the gas in bottle contains water vapor in addition to oxygen, because water has evaporated into the gas space. The oxygen is exerting a pressure of something less than 760 torr, because the water vapor is also exerting pressure. A correction must be made if the true pressure of dry oxygen gas is to be obtained. The equation applicable in this case is:

$$P_{total} = P_{oxygen} + P_{water\ vapor}$$

To obtain P_{oxygen}, the equation is rearranged:

$$P_{oxygen} = P_{total} - P_{water\ vapor}$$

The pressure exerted by water vapor depends only on the temperature; the hotter the water, the more molecules there are that are able to exist in the vapor state in a particular container and, consequently, the higher is its vapor pressure.

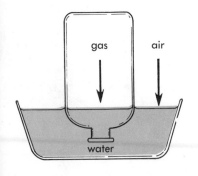

Figure 6–14   Method of equalizing internal and external gas pressure.

TABLE 6–2 **Vapor Pressure of Water**

Temp. (°C)	Torr	Temp. (°C)	Torr
0	4.6	28	28.3
5	6.5	29	30.0
10	9.2	30	31.8
15	12.8	31	33.7
16	13.6	32	35.7
17	14.5	33	37.7
18	15.5	34	39.9
19	16.5	35	42.2
20	17.5	40	55.3
21	18.6	50	92.5
22	19.8	60	149.3
23	21.0	70	233.7
24	22.4	80	355.1
25	23.8	90	525.8
26	25.2	100	760.0
27	26.7		

Table 6–2 lists the vapor pressure of pure water at various temperatures. The effect of dissolved particles on the vapor pressure will be discussed in the chapter dealing with solutions.

Bear in mind that the vapor pressure of water is not dependent on the *amount* of water, any more than barometric pressure depends on the diameter of the mercury column. Pressure is always described as force per unit area (Fig. 6–15).

Figure 6–15   Water vapor pressure at 29.0°C.

Example 6.5

What is the pressure of a dry gas if the volume collected over water at 26°C is measured at 742.0 torr?

1. Using Dalton's law:

$$P_{dry\ gas} = P_{total} - P_{water\ vapor}$$

2. Find the pressure due to water vapor at 26°C from Table 6–2:

$$\text{water vapor pressure @ } 26°C = 25.2 \text{ torr}$$

3. Solve for the dry gas pressure:

$$P_{dry\ gas} = 742.0 \text{ torr} - 25.2 \text{ torr}$$

$$\boxed{P_{dry\ gas} = 716.8 \text{ torr}}$$

Exercise 6.2

1. If a gas occupies 2.30 liters at 710 torr, what volume will it occupy at standard pressure?
2. At 0.82 atm, a gas occupies 172.0 mL. What volume will the gas occupy at 792.0 torr? (Remember that 1 atm = 760 torr.)
3. If a 1.4 liter container of a gas at 10.0 atm is altered so that the volume is increased to 4.2 liters, what is the final pressure in torr?
4. A 350.0 mL sample of gas is collected over water at 40°C and a barometric pressure of 756.7 torr. What is the pressure due to the dry gas at that temperature?

6.7 TEMPERATURE, VOLUME, AND CHARLES' LAW

Boyle's law was entirely adequate for the description of pressure-volume relationships at constant temperature. But what about temperature changes? Jacques Charles, in 1787, determined experimentally that the volume of a fixed mass of gas is *directly* proportional to the *absolute temperature.* He observed that the volume of a sample of gas at 0°C would expand or decrease by 1/273 of the 0°C volume for each degree of temperature rise or fall.

It is important to note that the volume varies with the *absolute* temperature, or degrees Kelvin, and not with the Celsius temperature. One reason for this is that the use of degrees below zero (−5°C, −15°C, −82°C, for example) would produce the fiction of *negative volumes* in calculations that relate volume and temperature. Another reason involves the definition of absolute zero, the temperature at which a gas

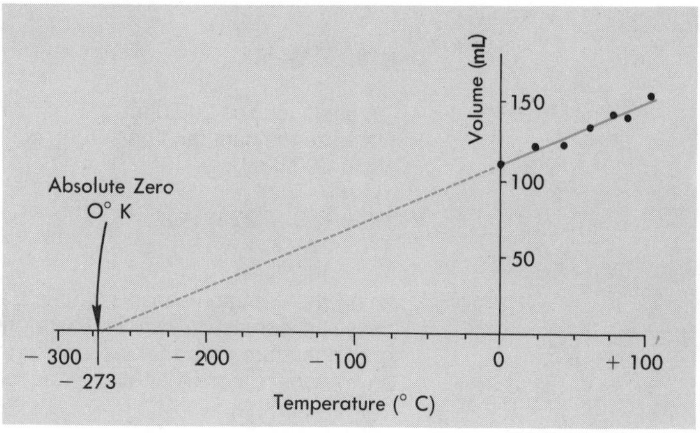

Figure 6–16 The extrapolation of volume-temperature data to absolute zero.

volume is *theoretically* reduced to zero.* This point can be determined by **extrapolating** from empirical data (Fig. 6–16). Extrapolation, as a graphing technique, means to extend a line beyond the points of experimental measurement.

Observe in Figure 6–16 that the volume of the gas decreases with the temperature. This was the essential observation that led Charles to state the direct proportionality of temperature and volume at constant pressure.

$$V \propto T$$

Mathematically stated:

$$V = kT$$

or

$$k = \frac{V}{T},$$

which means that the ratio of a gas volume to the absolute temperature at a fixed pressure is constant.

The value of the proportionality constant depends on the original volume and the atmospheric pressure. Thus, a doubling of the absolute temperature results in a doubling of the volume, and if the absolute temperature is reduced to 1/4 its original value, the volume is likewise reduced by the same fraction.

---

*Recall from Chapter 3 that there is in fact no substance existing in the gas state below 4°K.

6.7 TEMPERATURE, VOLUME, AND CHARLES' LAW

Example 6.6

A gas occupies 200.0 mL at 0°C. What volume will it occupy at 27°C?
1. Organize the data, and be sure to convert °C to °K:
   V_1 = 200.0 mL
   V_2 = ?
   °K = (°C + 273)
   T_1 = 273°K
   T_2 = 300°K
2. Using the common-sense approach, it is clear that a rise in temperature will cause an increase in volume. Therefore, multiply the original volume by the kind of temperature ratio that will produce a larger number as the answer. This is an *improper* (larger than unity) *fraction*.

$$V_2 = 200.0 \text{ mL} \times \frac{300 \text{ °K}}{273 \text{ °K}} \longleftarrow \text{improper fraction}$$

$$\boxed{V_2 = 220 \text{ mL}}$$

6.8 COMBINING BOYLE'S LAW AND CHARLES' LAW

Solving problems that include both temperature and pressure changes creates no difficulty. Recall that pressure and temperature changes are independent phenomena. Both P and T may work in concert or in opposition to either increase or decrease a volume of gas. The system may be manipulated so that P and T effectively cancel out each other. For example, we could decrease the pressure on the gas at the same time that the temperature is decreased so that the volume remains unchanged. The logical approach used in the common-sense method is to consider the effects of T and P separately and to make whatever calculations are appropriate.

Example 6.7

If 150 mL of gas is collected at 20°C and 700 torr, what volume will it occupy at STP?
1. Organize the data, paying careful attention to the proper units:
   V_1 = 150 mL
   V_2 = ?
   T_1 = 293°K
   T_2 = 273°K (standard temperature)
   P_1 = 700 torr
   P_2 = 760 torr (standard pressure)
2. Consider the temperature change. It is going *down* (from 293°K to 273°K). Therefore, the temperature ratio must be the kind of fraction that will *reduce* the volume (Charles' law). It is a *common fraction*.

THE BEHAVIOR OF GASES

3. Consider the pressure change. It is going *up* (from 700 to 760 torr). Therefore, the pressure ratio must be such that it too *reduces* the volume (Boyle's law).
4. Calculate the answer:

$$V_2 = 150 \text{ mL} \times \frac{273°K}{293°K} \times \frac{700 \text{ torr}}{760 \text{ torr}}$$

$$\boxed{V_2 = 129 \text{ mL}}$$

Example 6.8

If 67.0 mL of a gas is collected over water at a temperature of 15°C and a pressure of 1.20 atm, what volume will the dry gas occupy at 40°C and 770.0 torr?

1. Organize the data, paying special attention to the use of *compatible pressure units*, to the use of *Kelvin temperature*, and to *correcting the wet gas pressure to dry gas pressure*.
   (a) Change pressure units to torr:

$$1.20 \text{ atm} \times \frac{760 \text{ torr}}{1 \text{ atm}} = 912 \text{ torr}$$

   (b) Correct the total pressure, 912 torr, by subtracting the water vapor pressure at 15°C (see Table 6–2).

Because

$$P_{total} = P_{gas} + P_{water\ vapor}$$

then

$$P_{gas} = P_{total} - P_{water\ vapor}$$

$$P_{gas} = 912 \text{ torr} - 12.8 \text{ torr}$$
$$\text{(total)} \qquad \text{(water vapor pressure at 15°C)}$$

$$P_{dry\ gas} = 899 \text{ torr}$$

   (c) Convert the temperatures to °K:
   °K = °C + 273
   15°C = 288°K
   40°C = 313°K
   (d) Arrange the data:
   V_1 = 67.0 mL
   V_2 = ?
   T_1 = 288°K
   T_2 = 313°K
   P_1 = 899 torr
   P_2 = 770 torr

2. Consider the *individual* effects of temperature and pressure on the original volume of 67.0 mL.
   (a) The temperature is *rising*. Therefore, the volume should *expand* proportionately. Multiplication of 67.0 mL by an *improper fraction* of the temperatures is indicated.
   (b) The pressure on the gas is *decreasing*, causing an *expansion* of the gas volume. Again, an *improper fraction* achieves this effect.

3. Calculate the answer:

$$V_2 = 67.0 \text{ mL} \times \frac{313°K}{288°K} \times \frac{899 \text{ torr}}{770 \text{ torr}}$$

$$\boxed{V_2 = 85.0 \text{ mL}}$$

6.8 COMBINING BOYLE'S LAW AND CHARLES' LAW

Example 6.9

If 0.62 g of a gas occupies a 140 mL volume at 100°C and 0.98 atm, what is the molar mass of this gas? What is the density of this gas at STP?

1. In order to use the *mole concept* in the solution of this problem, it is necessary to find the *volume* of the gas at STP.

Caution: A common error made by students is to alter the *mass* of the gas by the combined gas law equation. This cannot be done. If temperature and pressure changes could alter mass, it would be a gross contradiction of the con-servation laws. Indeed, it would be a curious panacea for obese people if they could reduce their weight by sitting in a refrigerator. A method for incorporating mass into a gas law equation is made possible by another approach, which is introduced in the next section, dealing with the **equation of state.**

2. Organize the data, bearing in mind that the final conditions are at STP:

   V_1 = 140 mL
   V_2 = ?
   T_1 = 373°K
   T_2 = 273°K
   P_1 = 0.98 atm
   P_2 = 1.0 atm

3. Calculate the answer:

$$V_2 = 140 \text{ mL} \times \frac{273°\cancel{K}}{373°\cancel{K}} \times \frac{0.98 \cancel{\text{atm}}}{1.0 \cancel{\text{atm}}}$$

$$\boxed{V_2 = 100 \text{ mL at STP}}$$

4. Calculate the density from the equation. Change 100 mL to liters, since gas densities are reported in grams per liter.

$$d = \frac{m}{V}$$

$$d = \frac{0.62 \text{ g}}{0.10 \text{ L}}$$

$$\boxed{d = 6.2 \text{ g/L at STP}}$$

5. The mole concept states that 1 mole of any gas at STP occupies 22.4 liters. Therefore:

$$\text{molar mass} = \frac{6.2 \text{ g}}{1 \cancel{L}} \times \frac{22.4 \cancel{L}}{1 \text{ mole}}$$

$$\boxed{\text{molar mass} = 140 \frac{g}{\text{mole}}}$$

Exercise 6.3

1. If a gas occupies 1.3 liters at −20°C, what volume will it occupy at 30°C, assuming that pressure remains constant?
2. A gas occupies 220 mL at STP. What volume will this gas occupy at 80°C and 0.04 atm?

THE BEHAVIOR OF GASES

3. Correct a dry gas volume for STP if 75 mL is collected over water at 27°C and 705 torr.
4. If 240 mL of a gas collected at 60°C and 1.4 atm weigh 820.0 mg, calculate its density under the conditions of temperature and pressure described. Calculate also the molar mass of the gas.

6.9 THE IDEAL GAS LAW AND THE EQUATION OF STATE

We have seen how Charles' law and Boyle's law were conveniently combined in a common-sense approach to solving problems dealing with the behavior of gases. While this method permits consideration of pressure and temperature as individual effects on volume, these two factors are, nevertheless, related to each other and to the number of molecules in a container. The kinetic molecular theory states that the kinetic energy of particles is related to the absolute temperature because a higher temperature means greater molecular speed. Although the molecular mass remains constant, the increased velocity causes an increased kinetic energy.

$$KE = \frac{mv^2}{2}$$

While the product of pressure (force per unit area) and volume is not the same as kinetic energy, the PV product is proportional to the number of molecules present in a container and to their absolute temperature. Expressing the number of molecules as *moles*, n, we can write in shorthand fashion,

$$PV \propto nT$$

Mathematically expressed:

$$PV = R\,nT$$

proportionality constant (the **gas constant**)

The proportionality constant may be calculated from P, V, n, and T data for 1 mole of any ideal gas at STP.

$$P = 1 \text{ atm}$$
$$V = 22.4 \text{ liters}$$
$$n = 1 \text{ mole}$$
$$T = 273°K$$

Because $PV = Rnt$

$$R = \frac{PV}{nT}$$

Substituting the STP data:

$$R = \frac{(1 \text{ atm})(22.4 \text{ liters})}{(1 \text{ mole})(273°K)}$$

$$R = 0.0821 \frac{\text{L-atm}}{\text{mole °K}}$$

For problems in which the pressure is expressed in torr, it is more convenient to use an alternative value for the gas constant, 62.4 L = torr/mole °K.

$$R = \frac{760 \text{ torr}}{\text{atm}} \times \frac{0.0821 \text{ L-atm}}{\text{mole °K}}$$

$$R = 62.4 \frac{\text{L-torr}}{\text{mole °K}}$$

The equation, commonly known as the **equation of state** or the **ideal gas law equation**, is usually written:

$$PV = nRT$$

The equation of state is a very versatile tool, useful in solving many problems more quickly than by the application of the combined laws of Boyle and Charles. We do emphasize, however, that the equation of state represents an alternative that is definitely related to the fundamental observations of P, V, n, and T relationships. For example:

$V \propto T$ (volume is directly proportional to the absolute temperature)

and

$V \propto \dfrac{1}{P}$ (volume is inversely proportional to the pressure)

and

$V \propto n$ (volume is directly proportional to the number of moles)

Combining these proportionalities:

(M) Molar mass $= \dfrac{g}{mole}$

$$V \propto \frac{nT}{P}$$

$n (\# moles) = \dfrac{g}{M\left(\frac{g}{mol}\right)} = \dfrac{g}{n\left(mol\right)}$

Mathematically expressed:

$$V = R\frac{nT}{P}$$

and rearranging:

$$PV = nRT$$

Now let us consider a variety of problems in which the equation of state can be usefully employed.

Example 6.10

What volume will 0.200 mole of a gas occupy at 1.30 atm and a temperature of 27°C?
1. Organize the data:
    P = 1.30 atm
    V = ?
    n = 0.200 mole
    $R = \dfrac{0.0821 \text{ L-atm}}{\text{mole }°K}$
    T = 300°K
2. Use the equation of state to solve for volume:

$$P \quad \times V = \quad n \quad \times \quad R \quad \times \quad T$$

$$1.30 \text{ atm} \times V = 0.200 \text{ mole} \times \frac{0.0821 \text{ L-atm}}{\text{mole }°K} \times 300°K$$

$$V = \frac{0.200 \text{ mole} \times 0.0821 \text{ L-atm} \times 300°K}{1.30 \text{ atm} \times \text{mole }°K}$$

$$\boxed{V = 3.79 \text{ L}}$$

Example 6.11

If 0.660 g of a gas occupies 180 mL at 90°C and 740 torr, calculate the molar mass.
1. Express n in a more useful form. Remember that:

✳ $$n = \frac{\text{mass}}{\text{molar mass}} = \frac{g}{\dfrac{g}{\text{mole}}} = m$$

$$PV = \frac{g}{m}RT$$

Use M to represent the molar mass:

$$n = \frac{g}{M}$$

---

2. Substitute $\dfrac{g}{M}$ for n in the equation of state:

$$PV = \frac{gRT}{M} \text{ or, } PVM = gRT$$

3. Organize the data and substitute into the equation. Perform the dimensional analysis:

$$P = 740 \text{ torr}$$
$$V = 180 \text{ mL} = 0.180 \text{ L}$$
$$g = 0.660 \text{ g}$$
$$R = \frac{62.4 \text{ L-torr}}{\text{mole } °K}$$
$$T = 363°K$$
$$M = ?$$

P $\times$ V $\times$ M = g $\times$ R $\times$ T

$$740 \text{ torr} \times 0.180 \text{ L} \times M = 0.660 \text{ g} \times \frac{62.4 \text{ L-torr}}{\text{mole } °K} \times 363°K$$

$$M = \frac{0.660 \text{ g} \times 62.4 \text{ L-torr} \times 363°K}{740 \text{ torr} \times \text{mole } °K \times 0.180 \text{ L}}$$

$$\boxed{M = 112 \text{ g/mole}}$$

Example 6.12

What mass of carbon dioxide (CO_2) will be in a 4.00 liter container at 1.80 atm and a temperature of $-50°C$?

1. Organize the data:
P = 1.80 atm
V = 4.00 L
M = 44.0 g/mole (the formula mass)
$$R = \frac{0.0821 \text{ L-atm}}{\text{mole } °K}$$
T = 223°K
g = ?

2. Solve for mass of CO_2:

P $\times$ V $\times$ M = g $\times$ R $\times$ T

$$1.80 \text{ atm} \times 4.00 \text{ L} \times \frac{44.0 \text{ g}}{1 \text{ mole}} = g \times \frac{0.0821 \text{ L-atm}}{\text{mole } °K} \times 223°K$$

$$g = \frac{1.80 \text{ atm} \times 4.00 \text{ L} \times 44.0 \text{ g} \times \text{mole } °K}{\text{mole} \times 0.0821 \text{ L-atm} \times 223°K}$$

$$\boxed{\text{mass of } CO_2 = 17.3 \text{ g}}$$

THE BEHAVIOR OF GASES

Example 6.13

What is the density of nitrogen gas, N_2, at 30°C and 2.10 atm pressure?
1. Because density is mass divided by volume,

$$d = \frac{m}{V} = \frac{g}{V}$$

we can rearrange the equation of state to suit our convenience.

$$PVM = gRT$$

By proper use of dimensional analysis, the density of the gas at the stated conditions can be found.

2. Organize the data and solve for $\dfrac{mass}{volume}$

$P = 2.10$ atm
M of N_2 = 28.0 g/mole (from the formula mass)
$R = \dfrac{0.0821 \text{ L-atm}}{\text{mole °K}}$
$T = 303°K$

$$P \quad \times V \times \quad M \quad = g \times \quad R \quad \times \quad T$$

$$2.10 \text{ atm} \times V \times \frac{28.0 \text{ g}}{1 \text{ mole}} = g \times \frac{0.0821 \text{ L-atm}}{\text{mole °K}} \times 303°K$$

$$\frac{g}{V} = \frac{2.10 \text{ atm} \times 28.0 \text{ g} \times \cancel{\text{mole}} \, \cancel{°K}}{1 \, \cancel{\text{mole}} \times 0.0821 \text{ L-}\cancel{\text{atm}} \times 303 \cancel{°K}}$$

$$\boxed{d = 2.36 \text{ g/L}}$$

In Chapter 5 we dealt with stoichiometric problems that were characterized as moles-to-moles, moles-to-mass, mass-to-moles, and mass-to-mass. Now we are ready to investigate problems dealing with volumes of gases at standard and nonstandard conditions. We can label these problems as *mass-to-volume*, *volume-to-mass*, and *volume-to-volume* types.

Mass-to-Volume Problems

When a gas is involved in a reaction, it is often more convenient to deal with the volume of the gas rather than with its mass. A 0.01 mole sample of chlorine gas at STP is easier to measure as 224 mL rather than as 0.71 g. Once again, the mole concept forms the basis of calculations concerned with gases. This type of problem asks what volume of B (if substance B is produced in the gaseous state) will be obtained from a given mass of A. If the volume of B is measured at STP, the problem is relatively simple and straightforward. If the conditions are not standard, the equation of state is recommended as a method for the solution of such problems. Consider another general equation:

$$2 \text{ moles A(s)} \rightarrow 1 \text{ mole B(g)}$$

The sequence of steps would be:

$$\text{mass A} \rightarrow \text{moles A} \rightarrow \text{moles B} \rightarrow \text{volume B}$$

The only new step in this sequence is the change from moles of B to volume of B. This is done by multiplying the actual number of moles of B by the *molar volume*, 22.4 L/mole at STP:

$$\text{moles} \times \frac{22.4 \text{ L}}{\text{mole}} = \text{L}$$

Or you may use the equation of state to convert moles to liters at nonstandard conditions:

$$V = \frac{nRT}{P}$$

1. Change mass of A to moles of A.
2. Multiply moles of A by the molar ratio to obtain moles of B.
3. Multiply moles of B by 22.4 L/mole at STP, or use the equation of state.

4. In sequence:

$$\text{mass of A} \times \frac{1 \text{ mole A}}{\text{mass A}} \times \frac{1 \text{ mole B}}{2 \text{ moles A}} \times \frac{22.4 \text{ L B}}{1 \text{ mole B}}$$

$$= \text{liters of B at STP}$$

or

$$\text{mass of A} \times \frac{1 \text{ mole A}}{\text{mass A}} \times \frac{1 \text{ mole B}}{2 \text{ moles A}} = \text{moles of B}$$

non-standard

Finally,

$$V = \frac{nRT}{P}$$

Example 6.14

What volume of hydrogen gas can be collected at STP when 2.0 g of zinc reacts with sufficient hydrochloric acid?
1. Determine the stoichiometry by writing the balanced equation for the reaction:

$$Zn(s) + 2HCl(aq) \rightarrow ZnCl_2(aq) + H_2(g)$$

2. The stoichiometry indicates that the molar ratio is: one mole of zinc produces one mole of hydrogen.
3. Because the answer is to be expressed as a volume of hydrogen gas at STP, the molar volume of an ideal gas is required. This is 22.4 L/mole.
4. The molar mass of Zn is 65.4 g/mole.
5. Write the sequence of steps:

$$\text{mass Zn} \rightarrow \text{moles Zn} \rightarrow \text{moles H}_2 \rightarrow \text{volume H}_2$$

6. Perform the calculation:

$$V = 2.0 \text{ g Zn} \times \frac{1 \text{ mole Zn}}{65.4 \text{ g Zn}} \times \frac{1 \text{ mole H}_2}{1 \text{ mole Zn}} \times \frac{22.4 \text{ L}}{1 \text{ mole}}$$

$$\boxed{V = 0.68 \text{ L of H}_2 \text{ at STP}}$$

Volume-to-Mass Problems

A volume-to-mass problem might ask what mass of A would be required to produce a given volume of B measured at STP or some other conditions. The sequence of operations would be:

$$\text{volume B} \rightarrow \text{moles B} \rightarrow \text{moles A} \rightarrow \text{mass A}$$

---

1. The conversion of the volume of B to moles of B could be done as follows:

$$\text{liters of B} \times \frac{1 \text{ mole}}{22.4 \text{ liters}} = \text{moles of B at STP}$$

Otherwise, for moles of gas under *any* conditions:

$$n = \frac{PV}{RT}$$

2. The remainder of the sequence is routine:

$$\left(\begin{array}{c}\text{Actual number}\\ \text{of moles of B}\end{array}\right) \times \left(\frac{2 \text{ moles A}}{1 \text{ mole B}}\right) \times \left(\begin{array}{c}\text{molar mass}\\ \text{of A}\end{array}\right) = \begin{array}{c}\text{grams}\\ \text{of A}\end{array}$$

Example 6.15

What mass of $KClO_3$ is needed to produce 600 mL of oxygen gas measured at 27°C and 750 torr of pressure?

1. Write the balanced equation for the reaction:

$$KClO_3(s) \rightarrow KCl(s) + \tfrac{3}{2} O_2(g)$$

2. The molar ratio indicates that one mole of $KClO_3$ produces exactly 1.5 moles of O_2.
3. Organize the data:

$$P = 750 \text{ torr}$$

$$R = 62.4 \frac{\text{L-torr}}{\text{mole } °K}$$

$$T = 300 °K$$

$$V = 600.0 \text{ mL} = 0.6000 \text{ L}$$

4. Use the equation of state to find the number of moles of O_2 produced.

$$P \quad \times \quad V \quad = n \times \quad R \quad \times \quad T$$

$$750 \text{ torr} \times 0.600 \text{ L} = n \times \frac{62.4 \text{ L-torr}}{\text{mole } °K} \times 300°K$$

$$n = \frac{750 \text{ torr} \times 0.600 \text{ L} \times \text{mole } °K}{62.4 \text{ L-torr} \times 300°K}$$

$$\boxed{n = 0.0240 \text{ mole of } O_2}$$

5. Write the sequence of steps for the factor-label solution:

$$\text{moles } O_2 \rightarrow \text{moles } KClO_3 \rightarrow \text{mass } KClO_3$$

THE BEHAVIOR OF GASES

6. Calculate the mass of $KClO_3$:

$$\text{mass of } KClO_3 = 0.0240 \text{ mole } O_2 \times \frac{1 \text{ mole } KClO_3}{1.5 \text{ moles } O_2} \times \frac{122.5 \text{ g } KClO_3}{1 \text{ mole } KClO_3}$$

$$\boxed{\text{mass of } KClO_3 = 1.96 \text{ g}}$$

Example 6.16

If 4.2 grams of iron(III) oxide react with carbon monoxide to produce iron and carbon dioxide, what volume of carbon dioxide can be collected over water at 0.92 atm and 23°C?

1. Write the balanced equation:

$$Fe_2O_3(s) + 3 \ CO(g) \rightarrow 2 \ Fe(s) + 3 \ CO_2(g)$$

2. Note the molar ratio of iron(III) oxide to carbon dioxide: one mole of Fe_2O_3 is required to produce three moles of CO_2.
3. Calculate the number of moles of CO_2 produced from 4.2 g of Fe_2O_3, which has a molar mass of 159.7 g/mole.

$$\text{mass } Fe_2O_3 \rightarrow \text{moles } Fe_2O_3 \rightarrow \text{moles } CO_2$$

$$n = 4.2 \text{ g } Fe_2O_3 \times \frac{1 \text{ mole } Fe_2O_3}{159.7 \text{ g } Fe_2O_3} \times \frac{3 \text{ moles } CO_2}{1 \text{ mole } Fe_2O_3}$$

$$\boxed{n = 0.079 \text{ mole of } CO_2}$$

4. Organize the data and use the equation of state to find the volume of CO_2 gas:

$$n = 0.079 \text{ mole}$$

$$R = 0.0821 \ \frac{\text{L-atm}}{\text{mole } °K}$$

$$T = 296°K$$

$$P = P_{CO_2} - P_{H_2O} \ [\text{at } 23°C \text{ the } P_{H_2O} \text{ is } 21.0 \text{ torr (Table 6–2)}]$$

$$= 21.0 \text{ torr} \times \frac{1 \text{ atm}}{760 \text{ torr}} = 0.0276 \text{ atm}$$

$$P = 0.92 \text{ atm} - 0.0276 \text{ atm} = 0.89 \text{ atm}$$

$$P \quad \times V = \quad n \quad \times \quad R \quad \times \quad T$$

$$0.89 \text{ atm} \times V = 0.079 \text{ mole} \times \frac{0.082 \text{ L-atm}}{\text{mole } °K} \times 296°K$$

$$V = \frac{0.079 \text{ mole} \times 0.082 \text{ L-atm} \times 296°K}{\text{mole } °K \times 0.89 \text{ atm}}$$

$$\boxed{V = 2.2 \text{ L of } CO_2}$$

Volume-to-Volume Problems

When two or more reactants or products are in the gaseous state in a reaction at constant temperature and pressure, the volume ratio is the same as the molar ratio. Recall Avogadro's hypothesis, which states that equal numbers of moles of gases at the same temperature and pressure occupy equal volumes. A generalized equation, $2A(g) \rightarrow 5B(g)$, states in effect that just as 2 moles of gas A produce 5 moles of gas B, so will 2 volumes (expressed as liters or milliliters, for example) of gas A produce 5 volumes of gas B. Therefore, a volume-to-volume problem follows the sequence

$$\boxed{\text{volume of A} \times \text{volume ratio} = \text{volume of B}}$$

For the general equation above, assume that 300 mL of gas A is changed to gas B at constant temperature and pressure. What volume of gas B results?

$$(300 \text{ mL of A}) \times \left(\frac{5 \text{ mL B}}{2 \text{ mL A}}\right) = 750 \text{ mL of B}$$

Example 6.17

What volume of hydrogen gas is needed to react completely with 4.0 liters of oxygen gas in the formation of water? The temperature and pressure conditions are the same for both gases.

1. Balance the equation to obtain the stoichiometry:

$$H_2(g) + \tfrac{1}{2} O_2(g) \rightarrow H_2O(g)$$

moles:	1	$\frac{1}{2}$	1
volumes:	1	$\frac{1}{2}$	1
liters:	1	$\frac{1}{2}$	1

2. The $1:\frac{1}{2}$ ratio obviously means that 8 liters of hydrogen are needed to combine with 4.0 liters of oxygen.
3. Solving by the factor-label method, the one-step operation is as follows:

$$V = 4.0 \text{ L of } O_2 \times \frac{1 \text{ L } H_2}{0.5 \text{ L } O_2}$$

$$\boxed{V = 8.0 \text{ L of } H_2}$$

THE BEHAVIOR OF GASES

1. What volume of carbon dioxide can be collected at STP as a result of thermal decomposition of 1.7 g of ammonium carbonate? The equation is:

$$(NH_4)_2\,CO_3(s) \xrightarrow{\Delta} CO_2(g) + 2NH_3(g) + H_2O(l)$$

2. What mass of P_4 is needed to react completely with 450 mL of chlorine gas measured at 32.0°C and 710 torr of pressure? The skeleton equation is:

$$P_4(s) + Cl_2(g) \rightarrow PCl_3(s)$$

3. What mass of calcium is required to react with water in the production of 320 mL of hydrogen when collected over water at 29.0°C and 1.22 atm?

4. Given the following skeleton equation:

$$C_3H_8(g) + O_2(g) \rightarrow CO_2(g) + H_2O(l)$$

what volume of propane (C_3H_8) is needed to produce 0.70 L of CO_2 at STP?

QUESTIONS & PROBLEMS

6.1 What are the four parameters used to describe the behavior of gases?

6.2 Briefly summarize the main points of the kinetic molecular theory.

6.3 What is an ideal or perfect gas? Under what conditions will an ordinary gas behave most nearly like an ideal gas?

6.4 What does the Maxwell-Boltzmann distribution say about the relationship between temperature and the average kinetic energy of gas molecules?

6.5 What is Avogadro's principle?

6.6 Illustrate the law of combining volumes diagrammatically by showing how 2 liters of hydrogen can combine with 1 liter of oxygen to produce 2 liters of water vapor.

6.7   State Boyle's law in words and algebraically.

6.8   If 0.200 mole of nitrogen gas, 0.400 mole of oxygen gas, and 0.700 mole of carbon dioxide exert a total pressure of 750 torr, what is the partial pressure of the oxygen?

6.9   State Charles' law in words and algebraically.

6.10   What are two reasons for relating gas volumes to degrees Kelvin rather than degrees Celsius?

6.11   If 350 mL of a gas is collected over water at 50°C and 740 torr, what volume will the dry gas occupy at STP?

6.12   What is the molar mass of a gas if it has a density of 1.2 g/L at 20°C and 0.88 atm pressure?

6.13   What volume will be occupied by 0.025 mole of a gas at 27°C and 0.84 atm pressure?

6.14   Calculate the molar mass of a gas if 1.3 g occupy a volume of 460 mL at 10°C and 770 torr pressure.

6.15   What is the density of chlorine gas, Cl_2, at − 73°C and 4.0 atm pressure?

6.16   A gas collected over water has a volume of 980 mL at 20°C and 764 torr. What is the volume of dry gas at STP?

6.17   A 300 mL sample of gas at STP has a mass of 0.988 g. What is its molar mass?

6.18   At 24°C and 735 torr a sample of gas has a volume of 245 mL and a mass of 0.762 g. What is its molar mass?

6.19   A sample of a gas contains 1.67×10^{25} molecules. What is its volume in liters at 26°C and 740 torr?

6.20   What is the mass of 1.50 liters of $CO_2(g)$ measured at 27°C and 738 torr? How many molecules are present in this sample?

6.21   Calculate the molar mass of a gas if 0.646 g occupies a volume of 56.2 mL when measured over water at 27°C and 769 torr. If its empirical formula is CH_2, what is its molecular formula?

6.22   A gas is composed of 85.7 per cent carbon and 14.3 per cent hydrogen. A 0.147 g sample collected over water at 22.0°C and 732 torr has a volume of 88.0 ml. Calculate its empirical formula, molar mass, and molecular formula.

6.23   A gaseous hydrocarbon contains 7.75 per cent H and the remainder C. At a temperature of 273°C and a pressure of 1130 torr, 1.00 g of the gas occupies a volume of 0.578 liter. Calculate the molecular formula of the gas.

6.24   Ammonia is produced from the reaction of $N_2(g)$ and $H_2(g)$ at 300°C and 10 atm. How many liters of ammonia could be obtained from the reaction of 310 grams of hydrogen with excess nitrogen?

6.25   One gram of carbon is allowed to react with one liter of $O_2(g)$ at STP. How many liters of $CO_2(g)$ measured at STP will be produced? What is this volume if measured at 27°C and 730 torr?

6.26   What is the volume in liters of 1.20 grams of $SO_3(g)$ at 18°C and 755 torr?

6.27   A gas occupies a volume of 15.00 liters at 1.00 atm and 27°C. To what Kelvin temperature must it be cooled in order to fill a 5.00 liter flask at the same pressure?

6.28   Hydrogen gas (175 mL) was collected over water at a temperature of 25°C and an atmospheric pressure of 750 torr. What is the volume of the gas at STP? What is the mass of the gas?

6.29   Methane, oxygen, and sulfur dioxide gases are in a closed container at a total pressure of 770 torr. If 1.00 g of each gas is present, calculate the partial pressure of SO_2 in the mixture.

6.30   Calculate the density at STP of a gas if 0.810 g of the gas occupies 280 mL at 91°C and 570 torr. What is the molar mass of the gas?

6.31   Calculate the molar mass of a gas whose density is 1.97 g/L at 17°C and 76.0 torr.

CHAPTER SEVEN

LEARNING OBJECTIVES

At the completion of this chapter, you should be able to:

1. Explain why the Rutherford atomic model is not satisfactory.

2. Describe the photoelectric effect and state which model of light it supports.

3. Describe the Bohr model of the atom.

4. Write and define each term in the Bohr frequency rule equation.

5. Define the ground state of an electron in an atom.

6. Summarize the contributions of de Broglie and Schrödinger that led to a wave model of the atom.

7. State the uncertainty principle and explain its significance.

8. Calculate the maximum number of electrons that can occupy each principal energy level in an atom.

9. List the symbols that represent energy sublevels in atoms.

10. Write the maximum number of electrons that can occupy each sublevel.

11. Sketch the probability regions about an atomic nucleus that illustrate the first two sublevels.

12. State the Pauli Exclusion Principle.

13. Write the electronic configuration of an atom by using orbital notation.

14. List the four quantum numbers and their symbols.

15. Discuss the meaning and significance of the four quantum numbers.

16. Relate the values of the n, l, and m_l quantum numbers to the position of representative elements on the periodic table.

17. Specify values for the four quantum numbers assigned to the last electron added to any given representative element.

18. Illustrate the electronic configuration of atoms by pictorial representation.

19. Define and apply Hund's rule.

20. Describe the probability distribution of electrons in atoms in terms of the four quantum numbers.

21. Explain what is meant by the term *periodicity* as applied to the elements.

22. Sketch the arrangement of groups and periods on the periodic table.

23. Illustrate the position of related groups of elements on the periodic table.

24. Sketch and describe the generalized characteristics of elements on a periodic table.

25. Explain the significance of the *orbital blocks* of the periodic table.

ATOMIC STRUCTURE AND THE PERIODIC TABLE

ATOMIC STRUCTURE
AND THE PERIODIC TABLE

In 1911, Ernest Rutherford performed a series of dramatic experiments from which he derived an acceptable qualitative picture of the nuclear character of the atom. He suggested that an atom is composed of a tiny, dense, positively charged nucleus about which negatively charged electrons whirl in circular orbits. He postulated that the coulombic force of attraction between the oppositely charged particles is just enough to hold the electron in its orbital track despite the centrifugal force that constantly tends to pull it away (Fig. 7–1).

This was a beautiful model. Its logical simplicity was impressive, since it seemed to be perfectly analogous to the relationship between the sun and the planets of our solar system. However, the planets and the sun are not electrically charged particles, and gravitational interaction is not described by the same laws of classical physics as are oscillating electrical charges.

As early as 1879, James Clerk Maxwell demonstrated that light, magnetism, and electricity are related quantitatively, and his work indicated that oscillating electrical charges emit electromagnetic radiation as they lose energy. The amount of energy radiated, which ranges from as low as that of radio waves or as high as that of gamma radiation, depends on the frequency of the oscillations. Heinrich Hertz proved Maxwell's conclusions by using oscillating electrical charges to produce radio waves.

* see pg. 175

The fact that oscillating electrical charges must lose energy in the form of electromagnetic radiation destroyed Rutherford's theoretical basis for his planetary model of the atom. As the electron revolved about the nucleus it would have to emit energy. The only way in which the electron could compensate for this energy loss would be to make a smaller orbit; that is, a decrease in orbital radius would have to occur. The more energy lost, the tighter the circle would be-

Figure 7–1 The Rutherford planetary model of the atom.

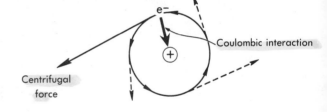

Figure 7–2 The spiral of atomic self-destruction.

come, until the electron spirals into the nucleus, resulting in the destruction of the atom (Fig. 7–2).

However, the atom, in a manner characteristic of all matter, does not tend toward self-destruction. The dilemma that Rutherford's model created was *how* could this most logical system be explained, since the electron does indeed behave like a particle oscillating about the nucleus.

7.2 THE NATURE OF LIGHT

A very significant aspect of energy relations is the production and measurement of light. The importance of light will become apparent as we get into the subject of atomic structure. Visible light is a small part of a huge range of types of **radiant energy** or **electromagnetic radiation.**

Electromagnetic radiation encompasses a range, or **spectrum,** from alternating current (low energy) to the cosmic radiation of stars (high energy). Electromagnetic radiation may be interpreted as the outward radiation of energy waves due to (1) the vibrations (oscillations) of electrically charged particles; (2) the radiation of energy that results from the alternation (a rapid shifting of positive and negative poles) of electrical current; (3) molecular vibrations; or (4) nuclear fission and fusion reactions. The word "electromagnetic" derives from the fact that the radiation we are talking about has electrical and magnetic components, which move at right angles to each other (Fig. 7–3). The part of the electromagnetic spectrum being radiated depends on the energy involved. Because radiant energy is classically considered to be a wave phenomenon, it is necessary to relate the characteristics of waves to the amounts of energy associated with those characteristics.

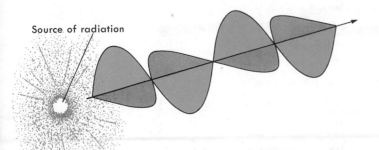

Source of radiation

Figure 7–3 A model of electromagnetic radiation, illustrating the magnetic and electrical components.

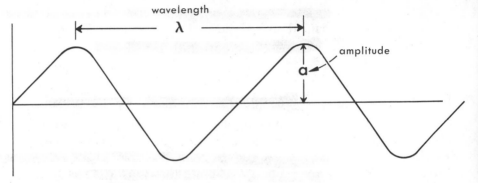

Figure 7–4   Wavelength and amplitude.

Wavelength

It helps to think of waves in terms of a stone that is dropped into a still pond. This old but durable analogy presents a picture of crests and troughs radiating outward from the point of disturbance. The essential difference between the water wave and the electromagnetic wave is that the electromagnetic wave apparently does not require matter in order to move. It can proceed through a vacuum. In fact, precise measurements indicate the speed of electromagnetic radiation in a vacuum to be 3.0×10^{10} cm/sec (186,000 miles per second). This universal constant is designated by the letter c (remember, $E = mc^2$).

The distance from the peak of one crest to the peak of the next is called the **wavelength,** and it is symbolized by the Greek letter lambda (λ). The other symbol (a) in Figure 7–4 represents the height of the wave, called the **amplitude.**

speed of
light = c
= 3.0×10^{10} cm

Frequency

Another essential characteristic of a wave is its **frequency,** symbolized by another Greek letter, nu (ν). Frequency means the number of waves, or crest-trough cycles, that pass an observation point in one second. The unit of measure for frequency is reciprocal seconds (per second, or $\dfrac{1}{\text{sec}}$, or sec^{-1}), since time is the only unit of measure. "Cy-

Figure 7–5   Frequency.

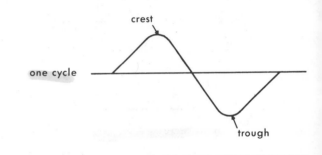

cles per second" has been replaced by a modern term, **hertz (Hz)**, named after Heinrich Hertz, who discovered electromagnetic waves. One cycle per second equals one hertz.

Wavelength, Frequency, and Energy

Comparing two "bits" of electromagnetic radiation as they pass an observation point moving at the same speed (the speed of light), we can see that the shorter the wavelength, the greater the number of cycles that will pass the observation point (Fig. 7–6).

From the idealized diagram (Fig. 7–6), we can also see that the wavelength is inversely proportional to the frequency:

$$\lambda \propto \frac{1}{\nu}$$

$$\lambda = k\frac{1}{\nu}$$

The proportionality constant (k) relating frequency and wavelength is the speed of light. Therefore, the equation may be rewritten:

$$\lambda = \frac{c}{\nu}$$

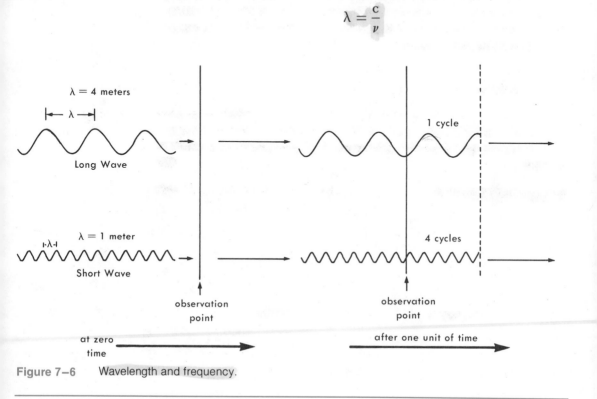

$\lambda = 4$ meters

$\vdash \lambda \dashv$

Long Wave

1 cycle

$\lambda = 1$ meter

Short Wave

4 cycles

observation
point

observation
point

at zero
time

after one unit of time

Figure 7–6   Wavelength and frequency.

ATOMIC STRUCTURE AND THE PERIODIC TABLE

The other relationships derived from this equation are as follows:

$$c = \lambda\nu \text{ (speed of light} = \text{the product of frequency and wavelength)}$$

and

$$\nu = \frac{c}{\lambda}\left(\text{frequency} = \frac{\text{speed of light}}{\text{wavelength}}\right)$$

The following example is an illustration of how we can perform a routine calculation involving frequency and wavelength.

Example 7.1

What is the frequency of radiation having a wavelength of 500 nm?
1. Organize the data and be sure the units are compatible.

$$c = 3 \times 10^{10} \frac{\text{cm}}{1 \text{ sec}}$$

$$\lambda = 500 \text{ nm} \times \frac{1 \text{ cm}}{10^7 \text{ nm}} = 5.00 \times 10^{-5} \text{ cm}$$

$$\nu = ?$$

$\nu = \frac{c}{\lambda}$

2. Solving for frequency:

$$\nu = \frac{3 \times 10^{10} \text{ cm}}{1 \text{ sec}} \times \frac{1}{5.00 \times 10^{-5} \text{ cm}} = \frac{0.6 \times 10^{15}}{1 \text{ sec}}$$

cycles/sec = hertz

$$\boxed{\nu = 6 \times 10^{14} \text{ Hz}}$$

---

higher energy → higher ν ∴
shorter λ

see
* pg. 171.

Many observations establish that higher energy radiation has a shorter wavelength and thus a higher frequency. Radio waves having wavelengths in the meter or kilometer range are much less energetic than X-rays of the same amplitude. The wavelengths of X-rays are in the angstrom region. This difference in energy is related to the source of the radiation (see Fig. 7–7).

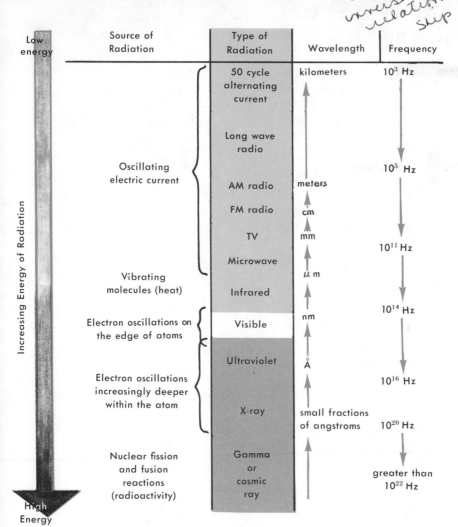

inverse relationship

Source of Radiation	Type of Radiation	Wavelength	Frequency
	50 cycle alternating current	kilometers	10^3 Hz
	Long wave radio		
Oscillating electric current		meters	10^5 Hz
	AM radio		
	FM radio	cm	
	TV	mm	10^{11} Hz
Vibrating molecules (heat)	Microwave	μ m	
	Infrared		10^{14} Hz
Electron oscillations on the edge of atoms	Visible	nm	
Electron oscillations increasingly deeper within the atom	Ultraviolet	Å	10^{16} Hz
	X-ray	small fractions of angstroms	10^{20} Hz
Nuclear fission and fusion reactions (radioactivity)	Gamma or cosmic ray		greater than 10^{22} Hz

Low energy

High Energy

Increasing Energy of Radiation

Figure 7–7   The electromagnetic spectrum.

Exercise 7.1

1. What is the frequency of electromagnetic radiation having a wavelength of 200 mm?
2. Calculate the wavelength of light if the frequency is 4.0×10^{12} Hz.
3. What is the speed of sound if a tuning fork vibrating at 120 cycles per second produces a wavelength of 2.76 meters?
4. What frequency of light has a wavelength of 210 Å?

Let us close by taking a brief look at the visible light region of the electromagnetic spectrum. Figure 7–8 illustrates the components of the visible part of the spectrum, which is a very small part of the whole electromagnetic spectrum. The violet part of the spectrum has the shortest wavelength and therefore the highest energy. All wavelengths

ATOMIC STRUCTURE AND THE PERIODIC TABLE

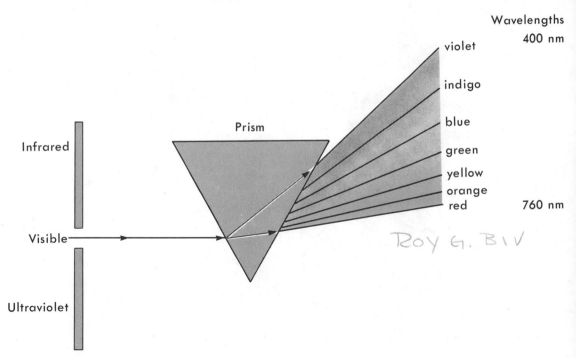

Wavelengths

400 nm

violet

indigo

blue

green

yellow

orange

red

760 nm

Infrared

Visible

Ultraviolet

Prism

ROY G. BIV

Figure 7—8   The colors of the visible range of the electromagnetic spectrum.

(© 1980 Sidney Harris.)

" I DON'T USE IT ANY MORE, SINCE
I GOT MY MICROWAVE OVEN."

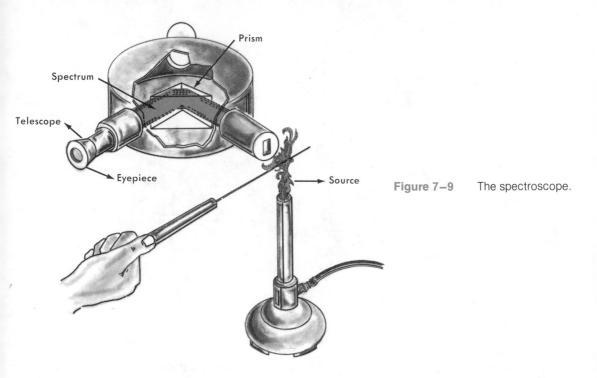

Prism

Spectrum

Telescope

Eyepiece

Source

Figure 7–9   The spectroscope.

of light traveling through a vacuum have the same speed (c), but through a **quartz prism,** the violet is slowed more than the red and as a result is bent more sharply. The speed increases with the color as the colors move toward red.

A spectroscope, illustrated in Figure 7–9, is an optical instrument that reveals the frequency composition of light. The light enters a narrow slit and is focused into a beam by the lens. The beam is then passed through a prism, which refracts (bends) all the light. Different frequencies of light are bent through different angles. This separation of light into its component frequencies forms a pattern on the photographic plate that is known as a **spectrum.** The dispersion of sunlight or white light of an incandescent bulb gives a continuous or "rainbow" spectrum containing all visible wavelengths (Table 7–1). By visible light we mean that which we can see with our eyes (approximately 3800 Å to 7600 Å).

TABLE 7–1 Range of Wavelengths for Visible Light

Color	λ in Å	
Infrared	>7500	low energy
Red	6100–7500	
Orange	5900–6100	
Yellow	5700–5900	
Green	5000–5700	
Blue	4500–5000	
Violet	4000–4500	
Ultraviolet	<4000	high energy

ATOMIC STRUCTURE AND THE PERIODIC TABLE

The dispersion of light emitted by excited atoms of an element produces an **emission** or **line spectrum.** When an element is heated in a flame or excited electrically in a gas discharge tube it gives off a characteristic color. You have seen the red glow of neon gas. Sodium vapor produces a yellow light, and mercury gives off a greenish glow. When the light emitted by excited atoms is examined with a spectroscope, a discontinuous spectrum containing colored lines of specific wavelengths only is observed. This line spectrum of an element is unique; no two elements have the same spectrum. Figure 7–11 shows the line spectrum for the hydrogen atom.

Different colors correspond to different frequencies. Blue light has a frequency of about 7.5×10^{14} Hz, and red light has a lower frequency, 4.3×10^{13} Hz. Frequencies lower than the frequency of red light are called infrared (*infra* = below) frequencies; those greater than violet frequencies are known as ultraviolet (*ultra* = beyond).

7.3 THE ORIGIN OF THE QUANTUM THEORY

It was Max Planck, in 1901, who daringly proposed that the accepted physical laws of Maxwell, Newton, and others do not necessarily apply to radiaton. Planck assumed that radiant energy is not a continuous ill-defined blur like a rainbow of color but rather is a stream of distinct, or discrete, units of energy called **quanta.** The word, **quantum,** suggests that each discrete packet of energy has a very specific energy content that is proportional to its frequency (rate of oscillation).

Planck's **quantum theory** of electromagnetic radiation effectively resolved a dilemma of his day that had arisen when scientists attempted to explain light in terms of the Maxwell-Boltzmann distribution for the energies of gas molecules at various temperatures. The average kinetic energy of a gas molecule is a statistical average at a given temperature. The evening-out of the available energy to all the molecules present due to continued collisions of the molecules, accepting the fact that many molecules will have more or less than the average, is called the **equipartition principle.** While the principle of equipartition of heat energy among gas molecules works nicely, it must be remembered that there are a finite number of gas molecules in any container. The evening-out of energy can go just so far. However, light is another matter. The possible number of frequencies over which radiant energy can be spread, if the equipartition principle is applied, is infinite. The equipartition principle simply cannot be applied to electromagnetic radiation. If it were true, our friendly reading lamp would bathe us in a small portion of deadly gamma radiation. Planck's quantization of light resolved this problem. A single quantum of light is restricted to a definite amount of energy that depends only upon its frequency.

When light of sufficient frequency strikes a metallic surface, elec-

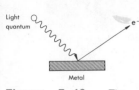

Light
quantum

e⁻

Metal

Figure 7–10 The photoelectric effect.

trons are ejected from some of its atoms. This phenomenon, called the **photoelectric effect,** was observed by Heinrich Hertz in 1887 and explained by Albert Einstein in 1905. Hertz noticed that electric sparks jump between two electrodes much more readily when the electrodes are illuminated by ultraviolet light.

Einstein suggested that when light of sufficient energy and specific frequency impinges on a metallic surface, many electrons obtain enough energy to be emitted, and the metal becomes positively charged. He explained that the energy needed to remove electrons from a particular metal consists of photons, or quanta, which are absorbed in the process and that the energy of light quanta absorbed must be equal to the energy required to detach the electron plus the kinetic energy contained in the ejected electron (Fig. 7–10). The basis of the well-known photoelectric cell is the fact that electrons are emitted from very active metals such as sodium or cesium when they are exposed to visible light.

By 1916 the validity of the Einsteinian photoelectric equation had been fully confirmed by Robert Millikan. It enabled us to interpret other photoelectric phenomena, led to the acceptance of light quanta, and established the duality of the character of light. By duality we mean that in order to explain some properties of light, such as diffraction, we need the wave theory; to explain other properties, such as photoemission and atomic spectra, we use the particle or photon theory.

7.4 THE BOHR ATOM

In his search for a theoretical justification for the Rutherford planetary model, the Danish physicist Niels Bohr brilliantly wedded Planck's quantum theory of electromagnetic radiation to the studies of the hydrogen spectrum made by Johann Balmer in 1885.

When an electric discharge is passed through a tube of hydrogen gas at low pressure, a peculiar light is emitted. When viewed through a prism or diffraction grating, the visible light is observed to be made of several distinct lines of color. This **line spectrum** produced by the gas discharge tube must be the result of hydrogen atoms interacting with the high-energy electrons moving through the tube (Fig. 7–11).

As the high-energy electrons strike them, the H_2 molecules are split into atoms. The electrons of the hydrogen atoms seem to absorb some of the available energy, as they exhibit new patterns of oscillation that result in the emission of electromagnetic radiation. The frequencies of these emitted photons must be very specific in order to produce the fine lines of color that characterize the hydrogen spectrum. The wavelength and therefore the frequency ($\nu = c/\lambda$) can be determined precisely with the aid of a spectrometer. Johann Balmer, in his efforts to find a simple formula that could relate these observed wavelengths mathematically, was ultimately rewarded for his perseverance.

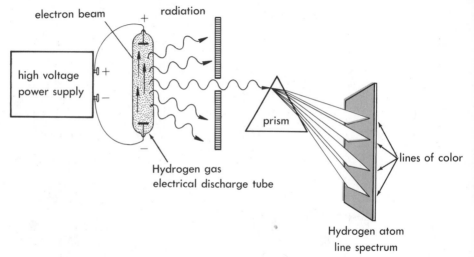

Figure 7–11   Production of the hydrogen spectrum.

Niels Bohr explained the Balmer lines of the hydrogen spectrum by postulating that the single electron of the hydrogen atom is able to have only very distinct energy values. In effect, he *quantized* the mechanical energy of the electron "particle" as Planck had previously quantized the energy of the photon "particle."

Like Rutherford, Bohr viewed the atom as a miniature solar system, with the electron revolving about the nucleus much as planets revolve about the sun. He postulated that the electron of a hydrogen atom is restricted to specific **energy levels.** Electrons in these energy levels revolve about the nucleus in fixed orbits, called **stationary states,** without radiating any energy. Each of the several energy levels represents a definite but different amount of energy, and these values of energy can be used to calculate the precise energies absorbed when atoms are excited to higher levels.

When an excited electron makes a **quantum jump** to a higher energy level and then "falls" back to a lower one, the excitation energy is emitted as a *specific light quantum.* This energy is exactly equal to the difference between the energy values of the orbits involved.

Bohr's energy-level concept is somewhat analogous to a man on a ladder that has irregularly spaced rungs. Each rung may represent an energy level. The lowest rung, where the man has the least potential energy, is like the orbit of the normal or unexcited electron. It is the orbit nearest the nucleus. The lowest energy level to which an electron can "fall" is called the **ground state** of that electron. As the man moves up and down the ladder, he must move from rung to rung, since standing between the rungs is physically impossible or "forbidden." Bohr's energy levels also have their "forbidden" regions, which is to say that the electron can move only from one energy level to another—all other

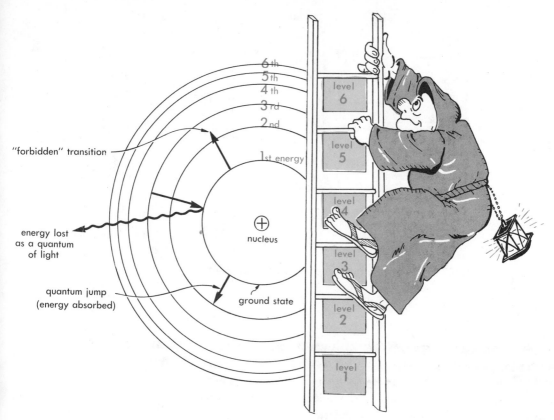

Figure 7–12   Hydrogen atom energy levels and a ladder analogy.

orbital radii are forbidden. Figure 7–12 represents the energy levels of the Bohr atom and a corresponding ladder analogy.

The energy of the emitted photon produced by an electron as it makes a nearly instantaneous all-or-nothing shift from a higher to a lower energy level is expressed by an equation known as the **Bohr frequency rule:**

$$\epsilon_2 - \epsilon_1 = h\nu$$

where

ϵ_2 = the energy of an electron in the higher energy level
ϵ_1 = the energy of an electron in the lower energy level
h = Planck's constant
ν = frequency of the emitted photon

Bohr was able to calculate the energy of the electron for each energy level by equating the mechanical energy of the orbiting electron to the balancing force of attraction of the positive nucleus. He found that the energy of each level is equal to the total energy of the electron divided by the square of the small whole number assigned to each level. This number, called the **principal quantum number,** is symbolized by the letter n. For the ground state, n = 1, and then n successively equals 2, 3, 4, 5, . . . The equation is:

$$E = -\frac{\text{total energy}}{n^2}$$

The negative sign indicates an increasing loss in potential energy as the electron moves to orbits closer to the nucleus. The total energy of the hydrogen electron calculated by Bohr and experimentally verified by measuring the amount of energy needed to remove the electron from the atom is -313.6 kcal/mole, or -21.8×10^{-12} erg per atom.* The energy levels, called **stationary states**† by Bohr, are listed in Table 7–2.

The data from Table 7–2 are pictorially represented in Figure 7–13, which shows the relationship between the Bohr frequency rule and the spectral lines produced by the hydrogen electron quantum jumps between energy levels. (See also Fig. 7–14.)

* $\dfrac{-313.6 \text{ kcal}}{6.02 \times 10^{23} \text{ atom}} \times \dfrac{4184 \text{ J}}{\text{kcal}} \times \dfrac{10^7 \text{ erg}}{\text{J}} = -21.8 \times 10^{-12} \text{ erg/atom}$

†The name "stationary state" refers to the absence of radiation while the electron remains in a particular orbit.

TABLE 7–2 **The Energy Levels of Hydrogen Atoms**

n (the energy level)	**Energy $= -313.6/n^2$** (kcal/mole)			**Energy $= -1312/n^2$** (kJ/mole)
1 (ground state)	$\dfrac{-313.6}{1^2} = -313.6$		$=$	-1312
2	$\dfrac{-313.6}{2^2} = -78.4$		$=$	$- 328$
3	$\dfrac{-313.6}{3^2} = -34.8$		$=$	$- 146$
4 ("excited states")	$\dfrac{-313.6}{4^2} = -19.6$		$=$	$- 82.0$
5	$\dfrac{-313.6}{5^2} = -12.5$		$=$	$- 52.5$
6	$\dfrac{-313.6}{6^2} = -8.7$		$=$	$- 36.4$

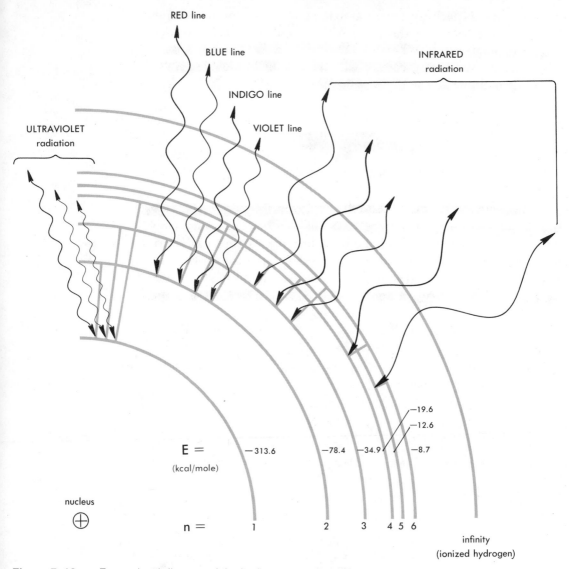

Figure 7–13 Energy level diagram of the hydrogen atom.

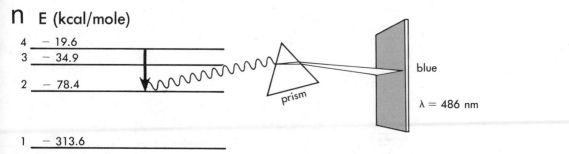

Figure 7–14 Electron transition of an electron of hydrogen.

Taken as a whole, Bohr's attempt to explain the structure of the hydrogen atom in terms of Planck's theory of the quantization of energy, the supporting evidence of spectrum analysis, and the emerging atomic model were very exciting in 1913. Although Bohr's model, with the added refinement of elliptical orbits, was found to be inadequate when applied to many-electron atoms, his principal quantum numbers remain as a permanent fixture in the modern atomic model. His concept of the orbit also persists in the word **orbital,** although the meaning is different.

In the early 1920s, improved spectroscopic instruments and observation of the effects of external magnets on gas discharge tubes caused the atomic scientists of that day to begin moving in a new direction. It was a movement toward a consideration of the electron as having a dual nature, both particle and wave.

Louis de Broglie, in 1924, made a very startling proposal. He reasoned that since the quantum theory so beautifully permits the wave phenomenon of light to be endowed with the properties of particles, variously called "discrete packets," "corpuscles," "quanta," and "photons," why not consider electrons (generally accepted as particles of matter) in terms of a wave model? In other words, de Broglie suggested that the electron might actually have a wavelength that would be inversely proportional to its **momentum.** Momentum, symbolized by the Greek letter rho (ρ), is a property of matter. It is the product of a particle's mass and velocity:

$$\rho = mv$$

As a result of manipulating the energy, mass, and wavelength relationships demonstrated by Einstein, de Broglie showed that the proportionality constant relating the electron's wavelength and momentum is Planck's "quantum of action." This is called Planck's constant.

$$\lambda_{e-} = \frac{h}{mv}$$

λ_{e-} = the wavelength of the electron

where

h = Planck's constant = 6.63×10^{-34} J · sec
$\left.\begin{array}{l} m = \text{mass} \\ v = \text{velocity} \end{array}\right\}$ $m \times v = \rho$

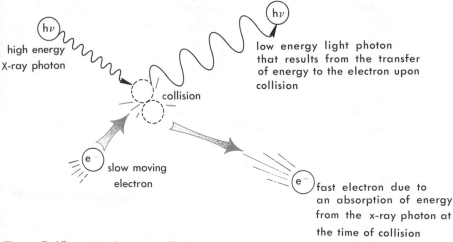

Figure 7–15   The Compton effect.

While there was abundant evidence supporting the dual (wave-particle) model for light, de Broglie's interesting hypothesis did not have empirical origins. The photoelectric effect, in addition to the experiment of Arthur Compton, clearly showed that photons of light behave like particles when they transfer kinetic energy upon colliding with electrons. Compton's observations were made when high-energy X-ray photons transferred kinetic energy to electrons (Fig. 7–15).

However, just three short years after de Broglie's proposal, experimental evidence was provided in support of the wave properties of electrons.

Of course, de Broglie's equation could not be $E = hc/\lambda$, because matter is not electromagnetic radiation. As particles of matter oscillate at various frequencies, radiation can be emitted. Although "matter waves" possess the typical properties of waves (ν, λ, amplitude), they cannot move at the speed of light and should not be confused with light. The de Broglie relation between wavelength and momentum is illustrated in Figure 7–16.

7.6   WAVE MODEL OF THE ATOM

As a direct consequence of the experimentally supported wave character of electrons, Erwin Schrödinger, in 1926, described the dis-

Figure 7–16   An aspect of the de Broglie hypothesis.

ATOMIC STRUCTURE AND THE PERIODIC TABLE

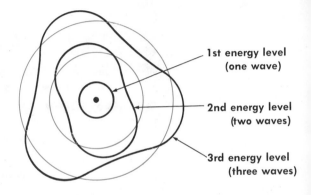

Figure 7–17   The de Broglie electron waves.

1st energy level
(one wave)

2nd energy level
(two waves)

3rd energy level
(three waves)

tribution of electrons in atoms by application of the mathematics used to describe waves. Whereas de Broglie fitted his wave concept of electron motion to Bohr's orbits, saying in effect that the number of waves would be equal to the principal quantum number (i.e., the number of the energy level) as illustrated in Figure 7–17, Schrödinger departed from the notion of circular and elliptical electron orbits and suggested three-dimensional wave forms.

Schrödinger's **wave mechanics** (mathematical description of wave motion) quantitatively described the electron waves so that they agree with the energy characteristics of Bohr's frequency rule for the hydrogen atom. Schrödinger's equations did indeed produce the correct values for the frequencies and intensities of the observed spectral lines. However, to some extent, Schrödinger's results worked for many-electron atoms, whereas the Bohr model did not.

A significant contribution that Bohr did make at this time was his **principle of complementarity,** which helped to resolve the argument that was raging over the nature of the electron. Bohr suggested that the model of the electron that probably described its nature most accurately is a duality concept. In other words, rather than having the particle model stand in conflict with the wave model, they should be accepted as mutually complementary concepts. Therefore, the electron, as in the case of light, was generally considered to have the properties of *both* matter and waves.

7.7 THE UNCERTAINTY PRINCIPLE

Shortly after Schrödinger's wave-mechanical explanation of electron behavior in atoms, Werner Heisenberg* submitted a mathematical

*An interesting historical note is that while the new quantum theory is commonly linked to Schrödinger's name, Heisenberg simultaneously announced the same results. However, his mathematical format was entirely different from Schrödinger's.

answer to the problem of a still imperfect description of the atom. In his **uncertainty principle** he stated that it is impossible to measure simultaneously the path (orbital trajectory) and the momentum of an orbiting electron. This uncertainty is due to the fact that the physical act of observing and measuring an object will alter, to some degree, those properties that are being measured. We are able to see a baseball in flight because photons of visible light strike the ball and are reflected from its surface to our eyes. Even though the ball absorbs some of the radiant energy reaching it, the photons have such small mass that we would hardly expect them to alter the momentum of the ball. On the other hand, photons of very short wavelength (high energy) would be required to detect a particle the size of an electron.

When the electron is struck by high energy photons, it is instantly moved to a new position, and its original velocity is changed by the massive input of energy received from the photon. This is analogous to trying to determine the position of a ping-pong ball in flight by intercepting its movement with your hand. The interaction of your hand with the ping-pong ball will change the position and velocity of the ball.

The effect of acceptance of the uncertainty principle was to abandon any attempt to develop a perfect planetary model of the atom. The original model of Bohr, in which electrons described clearly defined orbits about the nucleus, had served its purpose. A new direction was needed.

7.8 THE NEW MODEL OF THE ATOM

In 1928, Max Born suggested that Schrödinger's wave equations should be interpreted as describing the regions around a nucleus where electrons could be found in terms of *probability* (i.e., the most probable location of the electrons). The uncertainty principle had clearly demonstrated the futility of attempting to gain a precise knowledge of electron behavior.

It was also during this period that improved spectroscopic apparatus (as well as studies of magnetic effect on electron behavior) indicated that the original energy levels are, in fact, subdivided. What Bohr saw as a single line of color in the line spectrum was usually found to be made of several lines of color very close together. An external magnet often served to fractionate these lines even further (Fig. 7–18).

As a result of the fine-line observations, it was generally con-

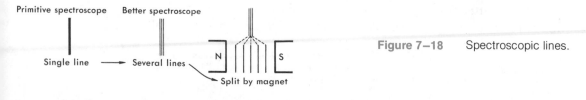

Primitive spectroscope   Better spectroscope

Single line ⟶ Several lines   Split by magnet

Figure 7–18   Spectroscopic lines.

TABLE 7–3 Spectrum Terminology

Spectrum Term	Symbol	Maximum e⁻ population
sharp	s	2 (1 pair)
principal	p	6 (3 pair)
diffuse	d	10 (5 pair)
fundamental	f	14 (7 pair)

cluded that the principal energy levels would have to be subdivided into secondary levels or **sublevels** (often called **subshells**), occupied by electrons of differing total energy. The energy sublevels are symbolized by letters taken from spectrum terminology outlined in Table 7–3.

It was the discovery of sublevels that 10 years earlier had led Bohr and Arnold Sommerfeld to suggest elliptical orbits. Schrödinger, however, was able to incorporate the sublevel distinctions and magnetic field interpretations into a refined method of describing the probability distributions of electrons in atoms. The shapes of the probability regions for s, p and d orbitals are described by simulating electron density clouds around three-dimensional axes called **Cartesian coordinates** (Fig. 7–19A, B, C).

The f region is much more difficult to represent diagrammatically. Seven probability regions are required to accommodate a maximum of 14 electrons. The Bohr energy-level diagram can be put to good use in illustrating the progressive subdivisions of the principal energy levels (Fig. 7–16).

The following should be observed from Figure 7–16:

1. The number of sublevels is equal to the number assigned to the energy level (the principal quantum number). The first energy level has one subdivision, the s probability region. The second energy level has two subdivisions, the s and p probability regions.
2. Generally there are not more than four sublevels described in any case. The s, p, d, and f probability regions are the practical limits.
3. The writing of a principal energy level number plus the s, p, d, or f symbol, describes an **orbital.** * For example:
   A 1s orbital signifies a region of spherical probability close to the nucleus.
   A 3s orbital signifies a region of spherical probability farther from the nucleus (Fig. 7–17).
   A 2p orbital signifies a distribution of electron probability along the x, y, or z axis of the Cartesian coordinates. A specific x-axis orientation could be written $2p_x$.
4. When the maximum electron population of any orbital is related to the number of probability regions, it is observed that there is *one*

---

*An orbital is a region in space around the nucleus where the probability of finding an electron is greatest.

Figure 7–19   A,   An s probability region—always a spherical region of probability around the nucleus.

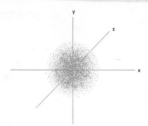

Figure 7–19   B,   The three p probability or probability regions optionally called p_x, p_y, and p_z to indicate difference in directional orientation as well as slight difference in a magnetic field.

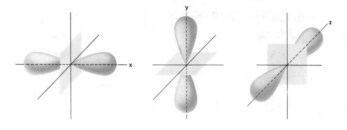

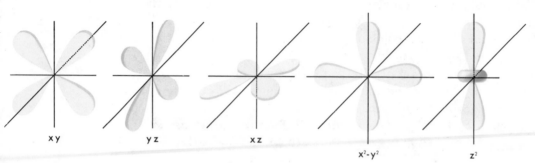

xy   yz   xz

x²-y²   z²

Figure 7–19   C,   The five d probability regions that may optionally be distinguished by the subscripts indicated in the diagram.

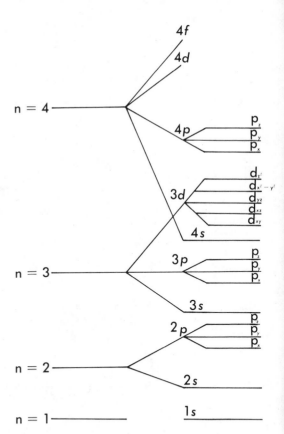

Figure 7–20   The subdivisions of energy levels.

probability region for every *two* electrons. For example, a *p* sublevel has three orbitals, which can accommodate a maximum of six electrons. Other orbitals are illustrated in Table 7–4.

Figure 7–21   Orbital notation.

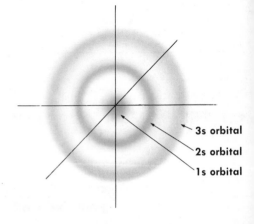

3s orbital
2s orbital
1s orbital

TABLE 7–4

Orbital	Number of probability regions	Maximum number of electrons
s	1	2
p	3	6
d	5	10
f	7 probability regions	14

7.9 THE PAULI EXCLUSION PRINCIPLE

The possibility of having two electrons in a single probability region was explained by Wolfgang Pauli in 1929, and it helped to rectify some remaining problems in the new atomic model. Pauli stated in his **exclusion principle** that no more than two electrons may

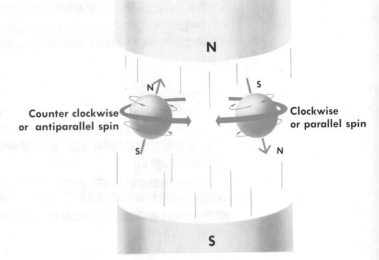

Figure 7–22   Idealized electron spin in a magnetic field.

occupy any single probability region (orbital)—and then only if they have opposite "spin." A model of **electron spin** suggests that the electron be visualized as a spherical particle rotating about an axis. The direction of the spin may be described as clockwise or counterclockwise.

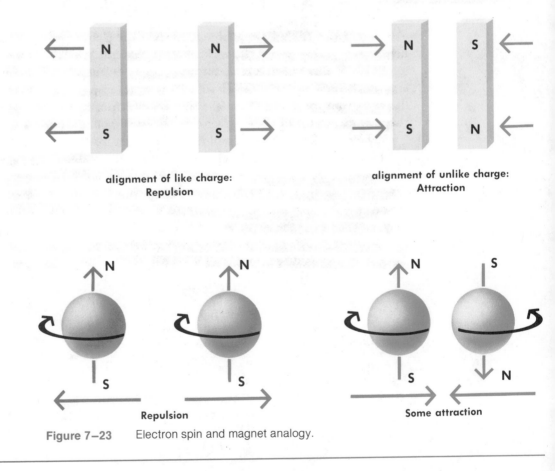

Figure 7–23   Electron spin and magnet analogy.

Since a rotating electrically charged particle behaves like a magnet that has a north and south pole, the type of spin is alternately described as being *parallel* or *antiparallel* to the direction of the field of an external magnet (Fig. 7–22).

The total energies of the electrons differ slightly, depending on their parallel or antiparallel orientation to the external magnet. It is the opposite polarization of the electrons that makes the model so attractive. By having their N and S (north and south) poles reversed, the electrons can approach each other more closely as they occupy a single region (Fig. 7–23).

It should be emphasized, however, that the concept of electron spin is only a model, used to explain the slight energy differences between two electrons in the same probability region. The electron is not a ball, and it does not rotate about an axis. However, the rotating-ball model does enable the chemist to cope with the mysterious electron in a meaningful way, so long as the model is not confused with reality.

7.10 DISTRIBUTION OF ELECTRONS IN ATOMS (ELECTRONIC CONFIGURATION)

The chemical and physical properties of the elements depend on how many electrons are in the various energy levels (classically referred to as **shells**). The determination of the number of electrons in the *outermost* shell, or the highest energy level, is of primary importance, because these electrons are involved in the sharing and transferring activities that take place during the formation of chemical bonds among atoms.

The order for mentally "building" the structure of atoms according to the distribution of the electrons follows the **aufbau principle.** The aufbau principle says, in effect, that electrons are arranged by assigning the first ones to the lowest energy levels and then moving toward higher energy levels in known order until all the electrons are distributed among the orbitals (Table 7–5).

In point of fact, the total energy required for an electron to occupy a *d* or *f* orbital of a lower principal energy level may be *higher* than the

TABLE 7–5

Energy Level n	$2n^2$	Maximum Electron Population
1	$2(1)^2$	2
2	$2(2)^2$	8
3	$2(3)^2$	18
4	$2(4)^2$	32
5	$2(5)^2$	50

ATOMIC STRUCTURE AND THE PERIODIC TABLE

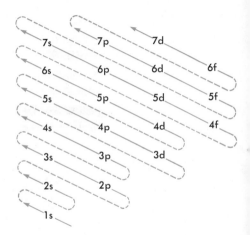

Figure 7–24 The sequence of filling atomic orbitals in atom "building."

energy of an electron in the s or p orbital of the principal energy level above it. For example, referring to Figure 7–24, you can see that the total energy of a $3d$ electron is *greater* than that of a $4s$ electron. Figure 7–24 illustrates a kind of "road map" for distributing electrons according to the aufbau principle. The arrows indicate the proper order of electron filling.

The method used to describe the electronic configuration of atoms is to write the principal energy level number (n), then the orbital letter designation (s, p, d, or f), and finally a superscript—not to be confused with an exponent—to indicate the number of electrons in that orbital. For example, $3d^7$ means third energy level, with the d orbitals occupied by 7 electrons.

The first ten elements of the periodic table are listed in Table 7–6.

A very helpful device for visualizing the distribution of electrons in greater detail is to make a **pictorial representation** (or orbital diagram). This method uses a *platform* _____, *box* ☐, *or circle* ○ to represent each orbital, and the electrons are illustrated by *arrows* pointing up and down, $\updownarrow$, to represent the "spin" direction. A common alternative is the use of diagonal lines to represent unpaired, ⊘, or paired, ⊗, electrons.

TABLE 7–6 Electronic Configurations of The First Ten Elements

Symbol and Atomic Number	Element	Electronic Configuration
$_1$H	hydrogen	$1s^1$
$_2$He	helium	$1s^2$
$_3$Li	lithium	$1s^2 2s^1$
$_4$Be	beryllium	$1s^2 2s^2$
$_5$B	boron	$1s^2 2s^2 2p^1$
$_6$C	carbon	$1s^2 2s^2 2p^2$
$_7$N	nitrogen	$1s^2 2s^2 2p^3$
$_8$O	oxygen	$1s^2 2s^2 2p^4$
$_9$F	fluorine	$1s^2 2s^2 2p^5$
$_{10}$Ne	neon	$1s^2 2s^2 2p^6$

An important experimental observation governing pictorial representation is expressed in terms of **Hund's rule,** or the **principle of maximum multiplicity,** which states that in the *filling of the p, d, and f orbitals, there must be at least one electron in each orbital before pairing begins for atoms in the ground state.* This means that in regions of equivalent probability, electrons require less energy to occupy orbitals singly than they do to pair off with those of opposite "spin."

Consider nitrogen as an example of pictorial representation and Hund's rule.

$$_7N \quad 1s^2 \quad 2s^2 \quad 2p^3 \qquad \text{electronic configuration}$$

$$\underline{\uparrow\downarrow} \quad \underline{\uparrow\downarrow} \quad \underline{\uparrow} \quad \underline{\uparrow} \quad \underline{\uparrow}$$

$$\text{x axis} \quad \text{y axis} \quad \text{z axis}$$

correct pictorial representation

It would be wrong, according to Hund's rule, to write:

$$\underline{\uparrow\downarrow} \quad \underline{\uparrow\downarrow} \quad \underline{\uparrow\downarrow} \quad \underline{\uparrow} \quad \underline{}$$

$$\text{x axis} \quad \text{y axis} \quad \text{z axis}$$

Since electrons enter separate orbitals with parallel spin, the pairing of electrons in the p_x orbital should not occur until there is at least one electron on each of the three equivalent p orbitals.

Consider the electronic configuration and pictorial representation of a larger atom, iron. The atomic number, 26, indicates the distribution of 26 electrons according to the aufbau principle. It may be useful to employ the orbital diagram, Figure 7–24.

$$_{26}Fe \quad 1s^2 \quad 2s^2 \quad 2p^6 \quad 3s^2 \quad 3p^6 \quad 4s^2 \quad 3d^6$$

$$\underline{\uparrow\downarrow} \quad \underline{\uparrow\downarrow} \quad \underline{\uparrow\downarrow} \quad \underline{\uparrow\downarrow} \quad \underline{\uparrow\downarrow} \quad \underline{\uparrow\downarrow} \quad \underline{\uparrow\downarrow}$$

$$\underline{\uparrow\downarrow} \qquad\qquad \underline{\uparrow\downarrow} \qquad\qquad \underline{\uparrow}$$

$$\underline{\uparrow\downarrow} \qquad\qquad \underline{\uparrow\downarrow} \qquad\qquad \underline{\uparrow}$$

$$\underline{\uparrow}$$

$$\underline{\uparrow}$$

Notice the application of Hund's rule in the $3d$ orbital.

In order to avoid drawing the many arrows representing the larger atoms, a commonly used shortcut involves the symbols for the **noble gases,** which are found at the far right of each period on the periodic table of the elements (see Fig. 7–27). A period is a horizontal row of elements. Iron, for example, is in the fourth period. The noble gas that marks the end of the previous period is argon. Therefore, the first 18 electrons of iron, which are the total number of electrons for argon, are represented by the notation [Ar] meaning "argon core." The electronic configuration and pictorial representation for iron may thus be written,

$_{26}$Fe $1s^2$ $2s^2$ $2p^6$ $3s^2$ $3p^6$ $4s^2$ $3d^6$

$\underbrace{}_{\text{[Ar]}}$ $\underline{\uparrow\downarrow}$ $\underline{\uparrow\downarrow}$ $\underline{\uparrow}$ $\underline{\uparrow}$ $\underline{\uparrow}$ $\underline{\uparrow}$

Example 7.2

Write the electronic configuration and pictorial representation for sulfur.

$_{16}$S $1s^2$ $2s^2$ $2p^6$ $3s^2$ $3p^4$

$\underbrace{}_{\text{[Ne]}}$ $\underline{\uparrow\downarrow}$ $\underline{\uparrow\downarrow}$ $\underline{\uparrow}$ $\underline{\uparrow}$

Example 7.3

Write the electronic configuration and pictorial representation for cadmium.

$_{48}$Cd $1s^2$ $2s^2$ $2p^6$ $3s^2$ $3p^6$ $4s^2$ $3d^{10}$ $4p^6$ $5s^2$ $4d^{10}$

$\underbrace{\phantom{1s^2 \quad 2s^2 \quad 2p^6 \quad 3s^2 \quad 3p^6 \quad 4s^2 \quad 3d^{10} \quad 4p^6 \quad 5s^2}}_{\text{[Kr]}}$ $\underline{\uparrow\downarrow}$ $\underline{\uparrow\downarrow}$ $\underline{\uparrow\downarrow}$ $\underline{\uparrow\downarrow}$ $\underline{\uparrow\downarrow}$ $\underline{\uparrow\downarrow}$

Exercise 7.2

Write the electrohic configuration and pictorial representation for the following species:

(a) oxygen atom
(b) scandium atom
(c) barium atom
(d) bismuth atom
(e) iodine atom

(f) lead atom
(g) calcium atom
(h) fluorine atom
(i) manganese atom
(j) arsenic atom

7.11 THE QUANTUM NUMBERS

Up to this point we have described the electronic configuration of atoms in terms of **spectroscopic notation.** In other words, the probability distribution of electrons in atoms has been represented by $1s^2$, $2s^2$, $2p^6$ . . . and so on. We have made use of the spectroscopic notations in the aufbau principle. There is, however, an alternative method that comes to us as a legacy of the work of Bohr, Schrödinger, and others,

---

which describes the probable whereabouts and energy of electrons in atoms in terms of four **quantum numbers.**

The first quantum number, called the **principal quantum number,** n, comes from Niels Bohr and was described in Section 7.4. These numbers relate to the distance between the nucleus and the principal energy levels. When n = 1 for the hydrogen atom, the electron is in the ground state. As the total energy of the electron increases by the definite amounts described in Bohr's theory, the value for n successively equals 2, 3, 4, and so on to infinity. A principal energy level of infinity signifies the point at which an atom becomes ionized by the loss of at least one electron.

The concept of energy sublevels as interpreted by Sommerfeld in 1916, when improved spectrometers showed that Bohr's spectral lines were actually made up of groups of fine lines (see Figure 7–18), led to the second quantum number, called the **orbital,** or **azimuthal quantum number** (*l*). The orbital quantum number, therefore, relates to the shape of the electron cloud. It corresponds to energy sublevels, which are symbolized by *s, p, d,* and *f* in spectroscopic notation.

The number of sublevels is equal to the principal quantum number (n) to which it is related. This means that the first energy level (n = 1) has only one sublevel, the second energy level (n = 2) has two sublevels, and so on. The numbers actually employed (from Schrödinger) are: *l* = 0, 1, 2, 3, which corresponds to the *s, p, d,* and *f* sublevels respectively. Table 7–7 shows the relationship between the spectroscopic notation and the principal and orbital quantum numbers.

TABLE 7–7 **Spectroscopic Notation and Related Principal and Orbital Quantum Numbers**

Spectroscopic Notation	*n* (principal quantum number)	*l* (orbital quantum number)
1s	1	0
2s	2	0
2p	2	1
3s	3	0
3p	3	1
3d	3	2

Notice that the *l* (orbital quantum number) value corresponding to an *s* sublevel is always "zero." For a *p* sublevel, *l* = 1; for a *d* sublevel, *l* = 2; and if an *f* sublevel were included, *l* would be equal to 3. Theoretically, *l* values could go higher than 3, just as *s, p, d, f,* can continue alphabetically to *g, h, i,* . . . , but as long as we are dealing with the "unexcited" states of atoms (ground states) of the known elements, we have no practical use for *l* values above 3.

The third quantum number, called the **magnetic quantum number,** m_l, becomes apparent when an external magnetic field is applied to gas discharge tubes during observation of the line spectra. The effective splitting of the spectral lines in a magnetic field (called the Zeeman

ATOMIC STRUCTURE AND THE PERIODIC TABLE

Spectroscopic Notation	l	m_l
s	0	0
p_x	1	+1
p_y	1	0
p_z	1	−1
d_{xy}	2	+2
d_{xz}	2	+1
d_{yz}	2	0
$d_{x^2-y^2}$	2	−1
d_{z^2}	2	−2

effect) suggests slightly different energy values for those electrons occupying p, d, and f energy sublevels. In other words, *orbital electrons* ($l = 0$), which occupy a spherical, and therefore symmetrical, proba-bility region about the nucleus, would not be affected by the magnetic field; but the p ($l = 1$), d ($l = 2$), and f ($l = 3$) orbital electrons, which have nonsymmetrical distributions, are affected. This further subdivi-sion, which occurs as long as the magnetic field is applied, has been notated by the x, y, and z designation we have used in relation to the p orbital electrons and the xy, xz, yz, $x^2 - y^2$, and z^2 descriptions applied to the d orbital electrons. See Table 7–4 for a review of orbital shapes.

The magnetic quantum number, which is related to and limited

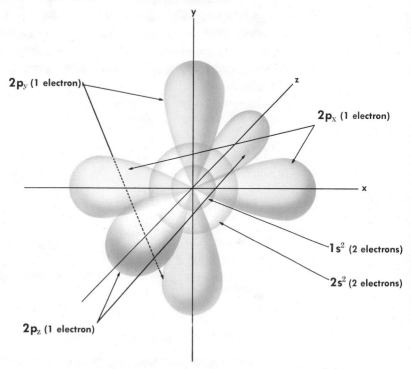

2p$_y$ (1 electron)

2p$_x$ (1 electron)

1s^2 (2 electrons)

2s^2 (2 electrons)

2p$_z$ (1 electron)

Figure 7–25   Three-dimensional model of electron distribution in nitrogen.

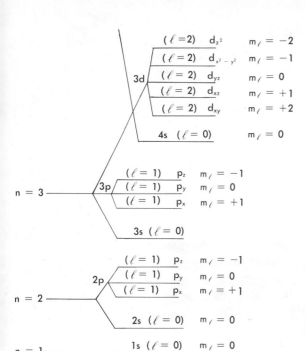

$(\ell=2)$ d_{z^2} $m_\ell = -2$
$(\ell=2)$ $d_{x^2-y^2}$ $m_\ell = -1$
3d $(\ell=2)$ d_{yz} $m_\ell = 0$
$(\ell=2)$ d_{xz} $m_\ell = +1$
$(\ell=2)$ d_{xy} $m_\ell = +2$

4s $(\ell=0)$ $m_\ell = 0$

$(\ell=1)$ p_z $m_\ell = -1$
3p $(\ell=1)$ p_y $m_\ell = 0$
$(\ell=1)$ p_x $m_\ell = +1$

3s $(\ell=0)$

$(\ell=1)$ p_z $m_\ell = -1$
2p $(\ell=1)$ p_y $m_\ell = 0$
$(\ell=1)$ p_x $m_\ell = +1$

2s $(\ell=0)$ $m_\ell = 0$

1s $(\ell=0)$ $m_\ell = 0$

n = 3

n = 2

n = 1

Figure 7–26   Relationships among n, ℓ, and m_ℓ quantum numbers.

by the ℓ value, has the same numerical value as m_ℓ but is differentiated by + and = signs. See Table 7–8 and Figures 7–25 and 7–26 for the relationship between ℓ and m_ℓ values.

The fourth quantum number is called the **spin quantum number**, m_s. There is only one numerical value for the spin quantum number, which is $\frac{1}{2}$. The opposite spin of two electrons in a single probability region, stated in the Pauli exclusion principle, is indicated by labeling one electron with a $+\frac{1}{2}$ spin value and the other electron with a $-\frac{1}{2}$ spin number. Figure 7–22 illustrates the model. The exclusion principle can be restated in terms of quantum numbers as follows: *No two electrons in an atom can have the same set of four quantum numbers.*

Summarizing the quantum-number relationships, we should bear in mind that each solution to the Schrödinger wave equation is characterized by a set of definite values for four numbers, called quantum numbers, commonly represented by the letters n, ℓ, m_ℓ and m_s. These numbers can assume the following values:

$$n = 1, 2, 3, 4, 5, \ldots \text{ (any integer)}$$
$$\ell = 0, 1, 2, 3, 4, \ldots (n-1)$$
$$m_\ell = +\ell, (\ell-1), (\ell-2), \ldots, 0 \ldots, (-\ell+2), (-\ell+1), -\ell$$
$$m_s = \pm\frac{1}{2}$$

Summary of Quantum Numbers

n = *Principal* quantum number and relates to the average *distance* of the electron from the nucleus. In general, the greater the value of

ATOMIC STRUCTURE AND THE PERIODIC TABLE

n, the more energy that an electron possesses; however, there is some overlapping of levels.

l = *Orbital* quantum number and defines the *shape* of the electron cloud. It corresponds to sublevels or subshells within the principal energy level.

Example: If n = 1, l = 0, and the orbital is spherical. If l = 1 the electrons are called p electrons; if l = 2, they are called d electrons. An electron in the n = 4 level, where l = 1, is called a $4p$ electron.

m_l = *Magnetic* quantum number and represents the *direction* of maximum extension in space of the electron cloud (relative to an applied magnetic field) in which an electron would have the highest probability of being found.

Example: m_l has $(2l + 1)$ values.

$$\text{if } n = 1, l = 0, \text{ and } m_l = 0$$
$$\text{if } n = 2, l = 1, \text{ and } m_l = -1, 0, +1$$
$$\text{if } n = 3, l = 2, \text{ and } m_l = -2, -1, 0, +1, +2$$

m_s = *Spin* quantum number and represents the magnetic effect related to the electron spin. Values are $+\frac{1}{2}, -\frac{1}{2}$.

Combinations of Quantum Numbers

n	l	m_l	m_s	Number of e⁻	Spectroscopic Notation	Total Number of e⁻
1	0	0	$\pm\frac{1}{2}$	2	$1s^2$	2
2	0	0	$\pm\frac{1}{2}$	2	$2s^2$	
2	1	+1	$\pm\frac{1}{2}$			8
2	1	0	$\pm\frac{1}{2}$	6	$2p^6$	
2	1	−1	$\pm\frac{1}{2}$			
3	0	0	$\pm\frac{1}{2}$	2	$3s^2$	
3	1	+1	$\pm\frac{1}{2}$			
3	1	0	$\pm\frac{1}{2}$	6	$3p^6$	
3	1	−1	$\pm\frac{1}{2}$			18
3	2	+2	$\pm\frac{1}{2}$			
3	2	+1	$\pm\frac{1}{2}$			
3	2	0	$\pm\frac{1}{2}$	10	$3d^{10}$	
3	2	−1	$\pm\frac{1}{2}$			
3	2	−2	$\pm\frac{1}{2}$			

Consider phosphorus as an example illustrating the relationship between the spectroscopic system of notation and the quantum number assignment method.

Spectroscopic notation: $_{15}P$ $1s^2\ 2s^2\ 2p^6\ 3s^2\ 3p^3$

Pictorial representation:

$$\underset{1s}{\uparrow\downarrow}\ \underset{2s}{\uparrow\downarrow}\ \underset{\underbrace{\uparrow\downarrow\ \uparrow\downarrow\ \uparrow\downarrow}_{2p}}{}\ \underset{3s}{\uparrow\downarrow}\ \underset{\underbrace{\uparrow\ \uparrow\ \uparrow}_{3p}}{}$$

Quantum number method

	n	ℓ	m_ℓ	m_s
1s	1	0	0	$+\frac{1}{2}$ and $-\frac{1}{2}$
2s	2	0	0	$+\frac{1}{2}$ and $-\frac{1}{2}$
2p	$\begin{cases} 2 \\ 2 \\ 2 \end{cases}$	1 1 1	+1 0 −1	$+\frac{1}{2}$ and $-\frac{1}{2}$ $+\frac{1}{2}$ and $-\frac{1}{2}$ $+\frac{1}{2}$ and $-\frac{1}{2}$
3s	3	0	0	$+\frac{1}{2}$ and $-\frac{1}{2}$
3p	$\begin{cases} 3 \\ 3 \\ 3 \end{cases}$	1 1 1	+1 0 −1	$+\frac{1}{2}$ $+\frac{1}{2}$ $+\frac{1}{2}$

Notice that the application of Hund's rule dictates the singular distribution of the 3p electrons in each of the three separate orbitals.

Example 7.4

Write the set of four quantum numbers for each electron in any fully occupied 4d orbital.

n	ℓ	m_ℓ	m_s
4	2	+2	$+\frac{1}{2}$ and $-\frac{1}{2}$
4	2	+1	$+\frac{1}{2}$ and $-\frac{1}{2}$
4	2	0	$+\frac{1}{2}$ and $-\frac{1}{2}$
4	2	−1	$+\frac{1}{2}$ and $-\frac{1}{2}$
4	2	−2	$+\frac{1}{2}$ and $-\frac{1}{2}$

Example 7.5

Write the set of four quantum numbers for each electron of an oxygen atom.
1. Write the electronic configuration and pictorial representation via the aufbau principle as an aid to the assignment of the quantum numbers.

$_8O$ $1s^2$ $2s^2$ $2p^4$

      ⇅   ⇅   ⇅   ↑   ↑

2. Complete the quantum number table.

	n	ℓ	m_ℓ	m_s
$1s^2$	1	0	0	$+\frac{1}{2}$ and $-\frac{1}{2}$
$2s^2$	2	0	0	$+\frac{1}{2}$ and $-\frac{1}{2}$
$2p^2$	$\begin{cases} 2 \\ 2 \\ 2 \end{cases}$	1 1 1	+1 0 −1	$+\frac{1}{2}$ and $-\frac{1}{2}$ $+\frac{1}{2}$ $+\frac{1}{2}$

ATOMIC STRUCTURE AND THE PERIODIC TABLE

One of the most significant observations that can be made from the electronic configuration of atoms is that the s and p orbitals of each successive energy level are filled to the maximum before electrons begin occupying the d and f orbitals of higher energy levels. The total number of electrons in the s and p orbitals is *eight*. While the "octet" portends no special magic, it is nevertheless characteristic of the highest energy levels of the remarkably stable noble gases, except for helium. Since changes in matter that lead toward increased stability (and lower energy) appear as a fundamental driving force behind chemical change in general, the "octet" configuration of the noble gases will provide a focal point for our discussion of the nature of the chemical bond (Chap. 8).

Just as the "octet" arrangement of the highest energy level characterizes the noble gases (helium forms a "duet"), the electron distributions in the highest energy levels of other related groups of atoms bear directly on their physical and chemical properties. One of the most useful tools available to chemists as an aid in seeing the orderly relationship among the atoms and, in a larger sense, among the elements they compose, is the periodic table.

7.12 CHEMICAL PERIODICITY

The modern periodic table (see Fig. 7–28) consists of an arrangement of elements in three broad categories—metals, nonmetals, and metalloids. The metals are located on the left side of the table and the nonmetals on the right side, with the metalloids between them (Fig. 7–27). Most of the elements are metals, and all the metals (except Hg) are solids at room temperature.

The light metals on the far left of the table are soft, easily cut, and chemically reactive. Some of the other metals are soft too, but most are quite hard. Metals tend to lose electrons in chemical reactions.

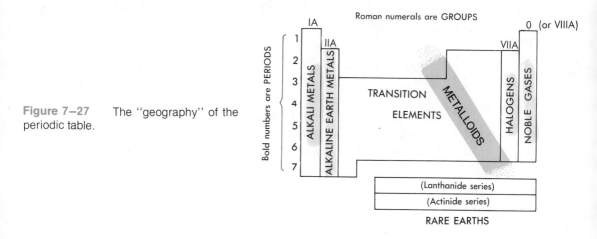

Figure 7–27   The "geography" of the periodic table.

A high proportion of the nonmetals are gaseous at room temperature; one (bromine) is a liquid, and a few are solids. Except for carbon in the form of the diamond, the solid nonmetals are soft and brittle. All the diatomic elements that we have considered to this point (H_2, O_2, and N_2, and the halogens) are nonmetals. In chemical reactions the nonmetals tend to gain or share electrons.

There is a gradual transition from metallic to nonmetallic properties across the periodic table from left to right. The metalloids, or semimetals, lie between the metals and nonmetals. They behave like both metals and nonmetals, depending upon the substances with which they react and the conditions of the reaction. There is disagreement about the number of elements that should be classified as metalloids, but they generally include the elements B, Si, Ge, As, Sb, Te, and Po. Actually, the division between metals, nonmetals, and metalloids is an arbitrary one, made for our convenience in discussing the properties and activities of elements. Later we will look at the degree of metallic or nonmetallic character, a much more realistic approach to the behavior of the elements.

The **periodic law** states that properties of the elements are periodic functions of their atomic numbers. In the periodic table the elements are arranged in order of their increasing atomic numbers into 7 horizontal rows called periods and 8 vertical columns called groups. The periods are numbered in arabic numerals from 1 to 7 on the left edge of the table. The groups are designated by Roman numerals, and the noble gases are usually designated as Group O. Each group identified by a Roman numeral is further subdivided into subgroups A and B. Elements in the A subgroups and in Group O are known as the representative elements;* Group IA elements are called the alkali metals, group IIA are the alkaline earth metals, and Group VIIA to the right of the table are called halogens. Hydrogen is found in two places on the table. It appears in Group IA and also in Group VIIA, but it is neither an alkali metal nor a halogen. The unique properties of hydrogen will be discussed later.

If you examine the periodic table closely you will find that the elements in a period (horizontal row) all have the same number of main energy levels (Fig. 7–28). The elements hydrogen and helium constitute the shortest period, having only two elements in a single energy level. Period 2 (Li to Ne) and period 3 (Na to Ar) contain 8 elements each, followed by periods 4, 5, and 6, containing 18, 18, and 32 elements, respectively. The seventh period includes 14 elements but is incomplete. Most periodic tables abstract elements 57 to 71 of period 6 and place these 14 elements (Lanthanide series) with period 7 (Actinide series) in horizontal rows below the table and detached from it.

---

*Some periodic tables exclude the noble gases from the category of representative elements.

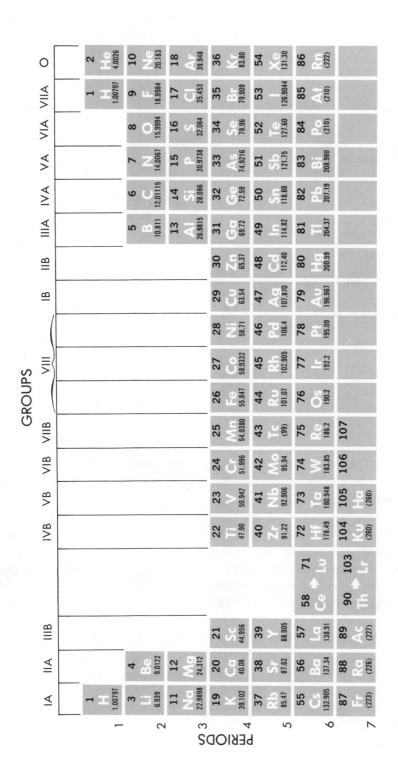

Figure 7-28   The periodic table.

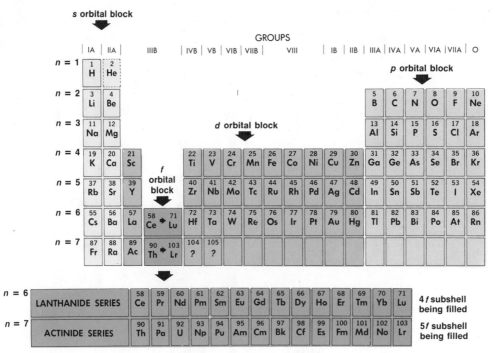

Figure 7–29   The *s, p, d,* and *f* blocks of the periodic table.

The elements in the long periods between Groups IIA and IIIA are the transition metals, which include the B subgroups and Group VIII. Transition metals are generally harder and stronger than the A subgroup metals and have many properties that are quite different. Group VIII includes horizontal blocks of three elements having the same characteristics. For example, Fe, Co, and Ni are strongly magnetic.

From the aspect of electron configuration, the periodic table is divided into four major regions or "blocks" of elements showing the highest occupied orbital in an atom. Figure 7–29 shows such a division into the *s, p, d,* and *f* orbital blocks. The *s* orbital block is 2 atoms wide and includes Groups IA and IIA. Here the last electron is added to the *s* orbital. The *p* orbital block is 6 elements wide and includes Groups IIIA through O. The *d* block is 10 elements wide, and the *f* block is 14 elements wide. Table 7–9 gives the relationship between the number of groups in each block and the number of electrons that

TABLE 7–9 Relationship between the Number of Groups in an Orbital "Block" and the Maximum Electron Population for the Orbital

Orbital	Maximum Number of Electrons	Number of Groups in the Orbital Block
s	2	2
p	6	6
d	10	10
f	14	14

can occupy the orbitals. The element helium is placed in Group IIA in Figure 7–29 because the last electron added is an s electron, similar to the Group IIA elements. On periodic tables, however, helium is placed in Group O because its highest energy level is completely filled with two electrons and its properties resemble those of the other noble gases.

The arrangement of the elements in the periodic table has some irregularities and inconsistencies. You have already noticed that hydrogen appears in two places, above Group IA and also above Group VIIA. However, its properties and activities differ from those of both groups. It is placed in Group IA because it has one electron in its (only) energy level; but hydrogen has little similarity to other members of Group IA. It is definitely not an alkali metal, but it fits even less satisfactorily anywhere else. It is not a halogen but is placed in Group VIIA because, like the halogens, it lacks one electron to complete its outer energy level.

The transition elements are another special case. The properties of the regular transition elements (atomic numbers 21 to 30, 39 to 48, and 72 to 80) are not well related vertically. The highest energy level patterns of these elements are similar to one another, and transition metal reactions involve both outer and inner electrons. The Lanthanide series (57 to 71) elements are so much alike that they all seem to fall into one space on the table, Group IIIB of the sixth period. Similarly, the Actinides fit into Group IIIB of the seventh period. Finally, the properties of the first member of any group are not really characteristic of the group. Beryllium is quite different in its reactions from the other Group IIA elements, and fluorine is unlike the other halogens. Despite such inconsistencies, the marvel is that the periodic table relates so many properties so well.

An understanding of the fundamentals of modern atomic structure and the systematic arrangement of the elements according to the properties that stem from the varieties of electronic configuration will form the basis for probing the nature of the chemical bond in Chapter 8.

7.1 What was wrong with the Rutherford planetary model of the atom?

7.2 What model of light evolved from the photoelectric effect?

7.3 What equaton did Einstein derive that relates energy and frequency of electromagnetic radiation?

7.4 State the Bohr frequency rule.

7.5 What is a "stationary state" in an atom?

7.6 Calculate the energy, in kcal/mole, of an electron transition in hydrogen atoms where the electrons shift from the eighth to the second energy level.

7.7 Rearrange the de Broglie equation so that it is ideally set up to solve for the mass of an electron.

7.8 What is the Compton effect?

7.9 How did the Heisenberg uncertainty principle serve to modify Schrödinger's contribution?

7.10 What conflict was resolved to some extent by the principle of complementarity?

7.11 What is the Pauli exclusion principle?

7.12 Write the electronic configuration and pictorial representation of cobalt.

7.13 Write electronic configurations for F, Cl, Br, and I. How many electrons are associated with the highest principal (n) quantum numbers in these elements? (These are *valence electrons;* see Chapt. 8.)

7.14 Write the set of four quantum numbers for each electron of an iron atom.

7.15 Describe the probability distribution of 10 electrons in a $4f$ orbital of an atom in terms of the four quantum numbers.

7.16 Draw a rough outline of a periodic table. Label the groups and periods. Shade in the area containing the elements that have the lowest ionization energies.

7.17 Classify the following elements as alkali, noble gas, alkaline earth, halogen, metalloid, transition, or rare earth: Ca, Pb, U, Xe, K, Co, Rb.

7.18 What is the fundamental difference between the Group A and B elements?

7.19 Explain the meaning of the notation:
(a) $4p^4$
(b) $2s^1$
(c) $3p^1$

7.20 What is the maximum number of electrons allowed in the $n = 3$ energy level?

7.21 What values of m_ℓ are allowed for $n = 2$?

7.22   How many p electrons are allowed for n = 4?

7.23   Tabulate the four quantum numbers for all electrons in the n = 3 energy level.

7.24   Which atom has the electronic configuration:
(a) $1s^2\ 2s^2\ 2p^2$

(b) $1s^2\ 2s^2\ 2p^6\ 3s^1$
(c) $1s^2\ 2s^2\ 2p^6\ 3s^2\ 3p^5$

7.25   List all values for the quantum numbers for the last electron added to these atoms:
(a) F
(b) Se
(c) Ba
(d) N

CHAPTER EIGHT

LEARNING OBJECTIVES

At the completion of this chapter, you should be able to:

1. Describe chemical bonding by the atomic orbital method.

2. Write electron-dot formulas for atoms and chemically combined atoms.

3. Define the term *ionization energy*.

4. Explain the relationship between ionization energy and metallic character.

5. Describe ionic bonds.

6. Define the term *electronegativity*.

7. Define and give illustrative examples of the following: oxidation number, oxidation, reduction.

8. State the rules governing the formation of ionic bonds.

9. Describe covalent bonds.

10. State the octet principle and cite exceptions to it.

11. Distinguish between polar molecules and polar bonds.

12. Demonstrate the relationship between electronegativity and per cent of ionic character.

13. Predict the degree of polarity of molecules on the basis of symmetry or asymmetry of covalent bonds.

14. Describe the special characteristics of metallic, network, and hydrogen bonds.

15. Predict the geometry of selected molecules.

16. Identify the type of orbital hybridization associated with the structure of special molecules.

17. Sketch a periodic chart so that contrasting regions of electronegativity and ionization energy are illustrated.

CHEMICAL BONDS

CHEMICAL BONDS

So far, we have been talking about compounds, calculating formulas for compounds, and naming compounds, all without discussing what sort of forces hold the atoms together. These forces are called chemical bonds. There are several different types of chemical bonds.

A study of the properties of different compounds shows that they fall roughly into two groups. One group is *solid* at room temperature, melting only at very high temperatures. These compounds do not look shiny and metallic. If you hit a lump of such a substance with a hammer, it will crumble into distinct granules. One example of this group is table salt.

A second group of compounds are *gases* or *liquids* at room temperature, or solids that melt at fairly low temperatures or blacken and burn when heated. Examples are water, gasoline, plastics, and sugar.

The differences in properties are due to differences in the kinds of chemical bonds present; that is, differences in the forces that hold the units together. All chemical bonds involve the *valence electrons* of the atoms. **Valence electrons** are those associated with the highest **n** quantum number. Bonds differ in the way that these electrons are associated with the atomic nuclei present. Another property common to the formation of most chemical bonds is that a structure is created that has greater stability and lower energy than do the isolated atoms.

If two atoms collide (this is rarely the way in which chemical bonds actually are formed, but it is a good way to imagine what happens), the valence electrons of each are momentarily in a position to be attracted by the positive charge on the nucleus of both atoms at the same time. The result may be favorable enough that the unit stays together, with the electrons shared by the two atoms. This sharing holds the two atoms together in a unit, and it is called a **covalent bond.**

On the other hand, one nucleus may attract the electrons much more strongly than the other. The result is that one or more electrons are transferred from one atom to another. This results in charged particles called **ions.** An atom having "too many" electrons, or a net negative charge is called an **anion.** One left with "too few" electrons, or a net positive charge is called a **cation.** The attraction between ions of opposite charge is called an ionic bond.

Of all substances in nature, only the noble gases exist in the form of individual uncombined atoms. The majority of matter is in the form of compounds, or mixtures of compounds, rather than free elements. It is a fundamental rule of nature that the most probable state of matter corresponds to the state of lowest energy. Since elements are more often found in the combined state than in the free (uncombined) state, it is a reasonable conclusion that the combined state must be the state of lower energy than the free state. We conclude that elements such as the noble gases, which occur naturally as free elements, must have atomic characteristics that correspond to a comparatively low state of energy. As we discuss the aspects of chemical bonding, we will find that many elements tend to combine by assuming the most stable electronic configuration possible; in many cases, the configuration of the noble gases.

The electronic configurations of the noble gases, except for helium, correspond to a higher energy level containing 8 valence electrons (Fig. 8–1). In each horizontal row of the periodic table, the noble gas has the highest first ionization energy (Section 8.4), suggesting the great stability in an octet (or duet in the case of He) of valence electrons. The implication is that atoms of elements may attain stability by becoming **isoelectronic** (Greek *isos* = equal) with a noble gas.

The apparent stability of the noble gases lies in the fact that the *s* and *p* orbitals composing their outer energy levels are filled. The drive toward the *octet* configuration may involve losing, gaining, or sharing of electrons. The specific mechanism used to describe a particular chemical bond depends on the types of atoms involved. The periodic table of the elements is an invaluable aid in the chemist's effort to categorize the elements with regard to the types of bonds they tend to form.

$_2$He $\qquad\qquad\qquad\qquad\qquad\qquad\qquad\qquad\qquad\qquad\qquad 1s^2$

$_{10}$Ne $\qquad\qquad\qquad\qquad\qquad\qquad\qquad\qquad\qquad 1s^2 \quad 2s^2\,2p^6$

$_{18}$Ar $\qquad\qquad\qquad\qquad\qquad\qquad 1s^2\;2s^2\;\;2p^6 \quad 3s^2\,3p^6$

$_{36}$Kr $\qquad\qquad\qquad 1s^2\;2s^2\;\;2p^6\;\;3s^2\,3p^6\;3d^{10}\;4s^2\,4p^6$

$_{54}$Xe $\qquad 1s^2\;2s^2\;\;2p^6\;3s^2\,3p^6\;3d^{10}\;4s^2\;4p^6\;4d^{10}\;5s^2\,5p^6$

$_{86}$Rn $\;1s^2\;2s^2\;2p^6\;3s^2\,3p^6\;3d^{10}\;4s^2\;\;4p^6\;4d^{10}\;4f^{14}\;5s^2\;5p^6\;5d^{10}\;6s^2\,6p^6$

Figure 8–1   The electronic configuration of the noble gases.

CHEMICAL BONDS

It is a convenience to use a symbolism known as the Lewis electron dot structure to show the formation of bonds between atoms. The
Lewis structure shows the symbol of the element and the number of
outer electrons present in the element (Fig. 8–2). In the notation, the
symbol represents the nucleus of the atom of an element plus all of the
electrons except the valence electrons. The valence electrons are shown
as dots (or circles or x's).

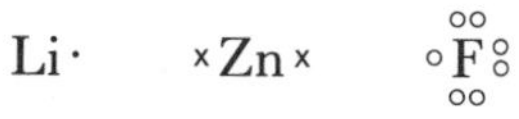

$$\text{Li} \cdot \qquad \times \text{Zn} \times \qquad {}^{\circ\circ}_{\circ\circ}\text{F}{}^{\circ}_{\circ}$$

The use of dots, x's, and circles is merely a convenience that we
use to keep track of where electrons came from and where they go in
the transferring or sharing process; they do not represent different
types of electrons.

Figure 8–2 shows Lewis diagrams for the representative elements.
Note that all elements in a group have the same number of valence
electrons, and this number corresponds with the group number. For
example, the VIIA elements have 7 dots; phosphorus, in Group VA,
has the same dot diagram as nitrogen; the Group IA elements have a
single valence electron; and so forth.

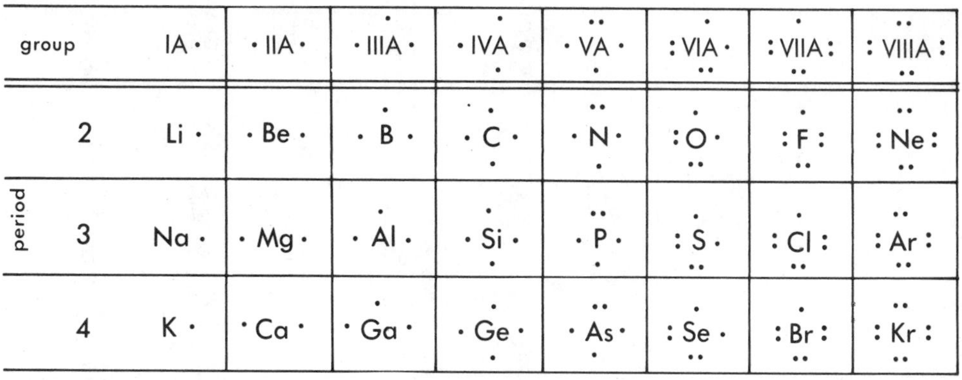

group	IA ·	· IIA ·	· IIIA ·	· IVA ·	· VA ·	: VIA ·	: VIIA:	: VIIIA :
2	Li ·	· Be ·	· B ·	· C ·	· N ·	:O·	:F:	:Ne:
3	Na ·	· Mg ·	· Al ·	· Si ·	· P·	:S·	:Cl:	:Ar:
4	K ·	·Ca·	·Ga·	· Ge ·	· As·	:Se·	:Br:	:Kr:

Figure 8–2   Electron-dot formulas of selected elements in groups IA through VIIA.

The number of dots can be determined from the electronic configuration:

		number of valence electrons	electron dot formula
$_{11}$Na	[Ne] $3s^1$	1	Na·
$_{20}$Ca	[Ar] $4s^2$	2	·Ca·
$_{13}$Al	[Ne] $3s^2\ 3p^1$	3	·Al·
$_6$C	[He] $2s^2\ 2p^2$	4	·Ċ·
$_{15}$P	[Ne] $3s^2\ 3p^3$	5	·P̈·
$_{16}$S	[Ne] $3s^2\ 3p^4$	6	·S̈:
$_{17}$Cl	[Ne] $3s^2\ 3p^5$	7	·C̈l:

8.4 IONIZATION ENERGY

In order to understand the concept of ionic bonding (Section 8.5), and covalent bonding (Section 8.6), it is necessary that we explore the meaning and significance of ionization energy.

The energy necessary to remove an electron from an atom in the gaseous state is known as **ionization energy.** Ionization energies are determined experimentally and usually are expressed in units of kcal or electron volts.* For example, consider the following general equation, in which we use M to represent an element:

$$M(g) + E_1 \rightarrow M^+(g) + e^-$$

*One electron volt (abbreviated eV) is the energy acquired by an electron upon passing through a potential difference of one volt; $1\ eV = 3.8 \times 10^{-20}$ calorie.

TABLE 8–1 Ionization Energies of Gaseous Atoms (kcal/mole and eV)

Atomic Number	Element	First Electron eV	First Electron kcal	Second Electron eV	Second Electron kcal	Third Electron eV	Third Electron kcal
1	H	13.6	314				
2	He	24.6	567	54.4	1254		
3	Li	5.4	124	75.6	1744	122	2823
4	Be	9.3	215	18.2	420	154	3548
5	B	8.3	191	25.2	580	37.9	875
6	C	11.3	260	24.4	562	47.9	1104
7	N	14.5	335	29.6	683	47.4	1094
8	O	13.6	314	35.2	811	54.9	1267
9	F	17.4	402	35.0	807	62.7	1445
10	Ne	21.6	497	41.1	947	64.0	1500
11	Na	5.1	118	47.3	1091	71.7	1652
12	Mg	7.6	176	15.0	347	80.1	1848
13	Al	6.0	138	18.8	434	28.4	656
14	Si	8.2	188	16.3	377	33.5	772
15	P	11.0	254	19.7	453	30.2	696
16	S	10.4	239	23.4	540	35.0	807
17	Cl	13.0	300	23.8	549	39.9	920
18	Ar	15.8	363	27.6	637	40.9	943
19	K	4.3	99	32.2	734	47.8	1100
20	Ca	6.1	141	11.8	274	50.9	1181

E_1 is called the first ionization energy. It is the minimum energy required to remove the first electron from gaseous element M to form gaseous ion M^+. The second ionization energy, E_2, is the amount required to remove the second electron:

$$M^+(g) + E_2 \rightarrow M^{2+}(g) + e^-$$

The third ionization energy applies to the removal of the third electron.

Table 8–1 lists ionization energies for the first 20 elements in kcal and in electron volts, when one mole of electrons is removed from a mole of atoms in the gaseous state. We find that the first ionization energy of Na is 118 kcal:

$$Na(g) + 118 \text{ kcal} \rightarrow Na^+(g) + e^-$$

The second ionization energy for Na is 1091 kcal, showing that a tenfold increase in energy is required to remove the second electron. As a result, the Na^{2+} ion would not be observed under ordinary conditions.

The first ionization energy for Mg is 176 kcal:

$$Mg\ (g) + 176 \text{ kcal} \rightarrow Mg^+\ (g) + e^-$$

To remove the second electron requires 347 kcal:

$$Mg^+ + 347 \text{ kcal} \rightarrow Mg^{2+}\ (g) + e^-$$

The third ionization energy is a whopping 1848 kcal!

$$Mg^{2+}\ (g) + 1848 \text{ kcal} \rightarrow Mg^{3+}\ (g) + e^-$$

This large amount of energy is not available in ordinary chemical reactions. Consequently, magnesium always loses two electrons in its chemical reactions to become a Mg^{2+} ion.

Experiments show that it is always easier to remove the first electron from an atom than it is the second, and the removal of the third one is even more difficult. There are several explanations for this observation. One is that the removal of an electron increases the net positive charge so that the remaining electron cloud is drawn in closer to the nucleus. Another is that the removal of the electron reduces the total electron repulsion, so that the ion formed becomes smaller in size and the electrons are held more tightly. An interesting observation is the marked difference between the first and second ionization energies for Li, Na, and K, and the similar gaps that exist between the second and third ionization energies for Be, Mg, and Ca. These gaps are shown in Table 8–1 by the stepped lines.

Such gaps of energy are characteristic between the second and third ionization energies of Group IIA elements just as between the

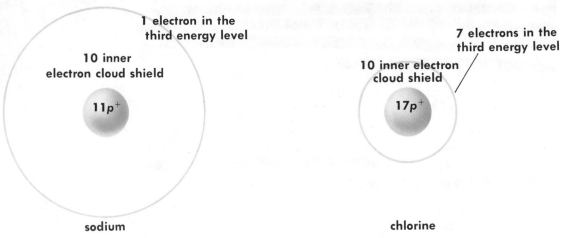

Figure 8–3   A model representing the shielding effect of 10 inner electrons in both sodium and chlorine.

first and second ionization energies for sodium and the other Group IA elements. Notice in Table 8–1 the great difference between the ionization energies of Group IA and Group VIIA elements. A classic example is found in the comparision between sodium (Group IA metals) and chlorine (Group VIIA nonmetals).

Both sodium and chlorine have electrons occupying three energy levels. However, sodium has only an 11-proton positive charge holding the single electron in the third energy level against the shielding effect of the 10 electrons that occupy the two inner levels. Compare this to chlorine, which has a 17-proton attractive force for the third energy level electrons against the same 10 inner electron shield. These structural differences are summarized pictorially in Figure 8–3.

We can get a good idea of the applications and limitations of the periodic table by studying the properties of some of the elements. Let us examine closely the characteristics of elements in Groups IA and VIIA and study their differences (Table 8–2).

The Group IA alkali metals are very much alike in physical appearance and chemical behavior. The key to this similarity lies in the fact that they all contain one electron in their outer energy level.

TABLE 8–2 The Principal Energy Level Distribution of Electrons in the Alkali Metals

Period	Element	Energy Level Electron Distribution
2	lithium	$_3$Li $1s^2 2s^1$
3	sodium	$_{11}$Na $1s^2 2s^2 2p^6 3s^1$
4	potassium	$_{19}$K $1s^2 2s^2 2p^6 3s^2 3p^6 4s^1$
5	rubidium	$_{37}$Rb $1s^2 2s^2 2p^6 3s^2 3p^6 3d^{10} 4s^2 4p^6 5s^1$
6	cesium	$_{55}$Cs $1s^2 2s^2 2p^6 3s^2 3p^6 3d^{10} 4s^2 4p^6 4d^{10} 5s^2 5p^6 6s^1$
7	francium	$_{87}$Fr $1s^2 2s^2 2p^6 3s^2 3p^6 3d^{10} 4s^2 4p^6 4d^{10} 4f^{14} 5s^2 5p^6$ $5d^{10} 6s^2 6p^6 7s^1$

We also find that each alkali metal immediately *follows* a noble gas. Chemically, the group IA elements are the most active metals known. They react with virtually every nonmetallic element to form ionic compounds in which the Group IA ion formed has a charge of $1+$, the same as its group number.

Table 8–3 shows some properties of the alkali metals. We find that the atomic radius and density increase with increase of atomic

TABLE 8–3 Some Properties of the Alkali Metals

Period	2	3	4	5	6
	Li	Na	K	Rb	Cs
Atomic no.	3	11	19	37	55
Atomic radius (Å)	1.52	1.86	2.27	2.48	2.65
Atomic mass (amu)	6.94	22.990	39.102	85.468	132.906
m.p. (°C)	179	97.8	63.2	39.0	28.6
Density (g/ml)	0.534	0.97	0.86	1.52	1.87
Ionization energy (kcal)	124	118	99	98	90

number, and the ionization energy decreases. All metals have relatively low ionization energies.

Table 8–4 shows the electron distribution in the halogens. The Group number, VII, gives the number of electrons in the highest

TABLE 8–4 Principal Energy Level Electron Distribution of the Halogens (Valence Electrons are Encircled)

Period	Element	Energy Level Electron Distribution
2	fluorine $_9$F	$1s^2$ $2s^2$ $2p^5$
3	chlorine $_{17}$Cl	$1s^2$ $2s^2$ $2p^6$ $3s^2$ $3p^5$
4	bromine $_{35}$Br	$1s^2$ $2s^2$ $2p^6$ $3s^2$ $3p^6$ $3d^{10}$ $4s^2$ $4p^5$
5	iodine $_{53}$I	$1s^2$ $2s^2$ $2p^6$ $3s^2$ $3p^6$ $3d^{10}$ $4s^2$ $4p^6$ $4d^{10}$ $5s^2$ $5p^5$
6	astatine $_{85}$At	$1s^2$ $2s^2$ $2p^6$ $3s^2$ $3p^6$ $3d^{10}$ $4s^2$ $4p^6$ $4d^{10}$ $4f^{14}$ $5s^2$ $5p^6$ $5d^{10}$ $6s^2$ $6p^5$

energy level. Each halogen has 7 electrons in the highest level and appears immediately *before* a noble gas in the periodic table. When a halogen gains an electron it forms a halide ion with a charge of $1-$. Table 8–5 lists properties of halogens.

TABLE 8–5 Some Properties of the Halogens

Period	2	3	4	5
	F	Cl	Br	I
Atomic no.	9	17	35	53
Atomic rad (Å)	0.72	1.00	1.14	1.35
Atomic mass (amu)	18.99	35.45	79.90	126.9
m.p. (°C)	−220	−101	−7.3	114
Density (g/ml)	1.11	1.57	3.14	4.94
Ionization energy (kcal)	402	300	275	242

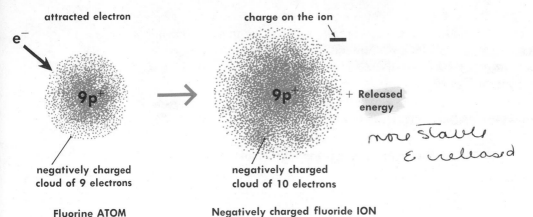

attracted electron

charge on the ion

e⁻

9p⁺

9p⁺ + Released energy

more stable & released

negatively charged cloud of 9 electrons

negatively charged cloud of 10 electrons

Fluorine ATOM

Negatively charged fluoride ION

Figure 8–4 The attraction of fluorine for an electron.

8.5 IONIC BONDS

Reviewing an atom (or a group of atoms), with a net electrical charge is called an **ion.** In chemical notation, an ion is represented by the symbol of the atom, with its charge shown as a right-hand superscript. Thus the sodium ion (cation) is written Na^+, and the chloride ion (anion) is Cl^-.

Ionic bonds result from the force of attraction between oppositely charged ions. The process of ionic bond formation comes about through the interaction between metallic atoms and nonmetallic atoms or groups of nonmetallic atoms. The metallic atoms, having low ionization energies, tend to *transfer* electrons to nonmetals that have a relatively strong attraction for additional electron(s). Fluorine, which has the strongest attraction for an electron, is illustrated in Figure 8–4. The removal of an electron (or electrons) from a metallic atom reduces the repulsion between the remaining electrons, so that the size of the electron cloud gets smaller. The reverse occurs when nonmetallic atoms accept electrons, in which case the electron cloud increases in size. Consequently, positive ions always have a smaller radius than the

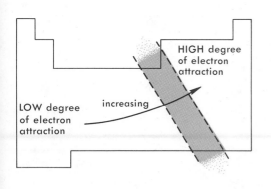

HIGH degree of electron attraction

LOW degree of electron attraction

increasing

Figure 8–5 The periodicity of electron attraction.

CHEMICAL BONDS

atom from which they are formed, and negative ions have a larger radius than the atoms from which they are formed. See page 467 of Appendix II for the illustration of relative atomic and ionic sizes.

The tendency of elements to retain electrons is summarized on the periodic chart (Fig. 8–5). The periodicity of the attraction of electrons by elements is shown in Figure 8–5.

When a metallic atom having a low ionization energy transfers one or more electrons (depending on the number of valence electrons) to a nonmetallic atom, strongly attracting ions result. A typical example of the formation of an ionic bond may be investigated in the production of sodium chloride, NaCl, which is common table salt:

$$1s^2\ 2s^2\ 2p^6\ 3s^1 \quad \xrightarrow[\text{energy required}]{\text{low ionization}} \quad \overbrace{1s^2\ 2s^2\ 2p^6}^{10\ e^-} + e^-$$

sodium atom $\qquad\qquad$ sodium ion $\quad$ "octet"

$$_{11}\text{Na} \quad \longrightarrow \quad _{11}\text{Na}^+ \quad + \quad e^-$$

$$1s^2\ 2s^2\ 2p^6\ 3s^2\ 3p^5 + e^- \rightarrow \quad \overbrace{1s^2\ 2s^2\ 2p^6\ 3s^2\ 3p^6}^{18\ e^-}$$

chlorine atom $\quad$ (high attraction $\quad$ chloride ion $\quad$ "octet"
$\qquad\qquad\quad$ for an electron)

$$_{17}\text{Cl} \quad \longrightarrow \quad _{17}\text{Cl}^-$$

Figure 8–6 illustrates the formation of sodium chloride more diagrammatically.

The interaction between a sodium atom and a chlorine atom may be illustrated by the electron-dot representation:

$$\text{Na} \cdot \quad + \quad \overset{\text{transfer}}{\ddot{\underset{\circ\circ}{\text{Cl}}}} \rightarrow \text{Na}^+, \quad [\ddot{\underset{\circ\circ}{\text{Cl}}}]^-$$

The resulting pair of ions (Na^+, Cl^-) is not meant to suggest the formation of a molecule (although NaCl(g) does exist as a diatomic molecule).

The anions and cations that are formed attract each other to produce an infinite three-dimensional crystalline array of alternating rows of positive and negative ions (Fig. 8–6). Each chloride ion is attracted to six sodium ions that are its closest neighbors and, to a weaker degree, to other sodium ions farther away. Similarly, each

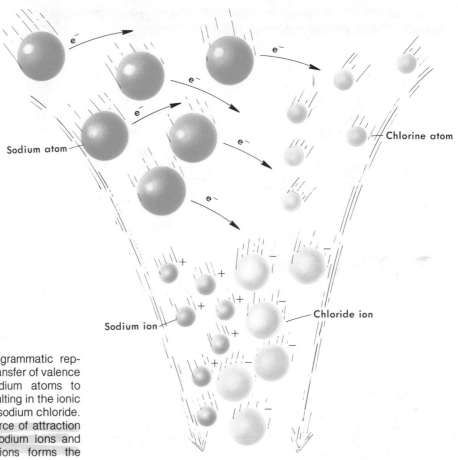

Figure 8–6 Diagrammatic representation of the transfer of valence electrons from sodium atoms to chlorine atoms, resulting in the ionic crystal structure of sodium chloride. The electrostatic force of attraction between positive sodium ions and negative chloride ions forms the crystal.

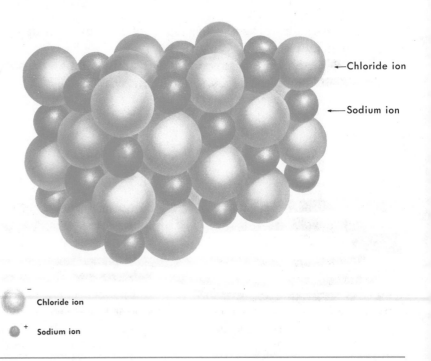

CHEMICAL BONDS

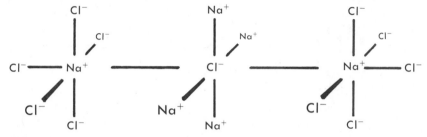

Figure 8–7   The NaCl crystal lattice structure.

sodium ion is attracted to an equal number of chloride ions. Thus, when we speak of an ionic bond, we are referring to the type of force that holds ions together and not of a bond between two specific ions. The formula NaCl merely represents the simplest ratio of ions in the crystal structure.

The three-dimensional aspect can also be represented by another diagram (Fig. 8–7).

The structure of the crystal lattice for sodium chloride serves as a model for many other ionic compounds, even though the ratios of ions involved and the basic geometry may vary. For purposes of clarity, the examples selected will be very typical members of their type. Bearing in mind that there are many exceptions, our guiding principle will be that *ionic bonds are most commonly formed between metals exhibiting common oxidation numbers of +1 or +2 and nonmetals.* Some selected examples of ionic bonding are illustrated in Table 8–6.

TABLE 8–6 **Selected Examples of Ionic Bonding**

Low Ionization Energy Metal		Electron-attracting Nonmetal	Electron Dot Formula
K×	+	·F̈:	K⁺, [×F̈:]⁻
potassium	+	fluorine	potassium fluoride
Li× Li×	+	·S̈:	2Li⁺, [×S̈:×]²⁻
lithium	+	sulfur	lithium sulfide
×Ca×	+	·Ï: ·Ï:	Ca²⁺, 2[×Ï:]⁻
calcium	+	iodine	calcium iodide
×Mg×	+	·B̈r: ·B̈r:	Mg²⁺, 2[×B̈r:]⁻
magnesium	+	bromine	magnesium bromide
×Sr×	+	·S̈:	Sr²⁺, [×S̈:×]²⁻
strontium	+	sulfur	strontium sulfide

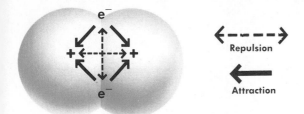

Repulsion

Attraction

Figure 8–8 The forces of attraction and repulsion in tension between the two atoms of the hydrogen molecule.

Notice the net effect of electrical neutrality in each example. Recall that *the sum of the positive and negative charges in an ionic compound must equal zero.*

8.6 COVALENT BONDS

Metallic elements that have relatively low ionization energies tend to lose electrons in chemical reactions, and atoms of nonmetals have a tendency to gain electrons. But when atoms of nonmetals combine with each other, they tend to *share* one or more electrons to form a molecule. Bonds composed of shared electrons are called *covalent bonds*.

Lithium, sodium, and the other Group IA elements each have one valence electron. We have seen that IA elements tend to lose their single valence electron to become positive ions with a charge 1+. In so doing, the IA atom attains a noble gas structure. Lithium attains the duet structure of helium, and the other IA atoms attain octet structures when they become 1+ ions. Now let us examine how two

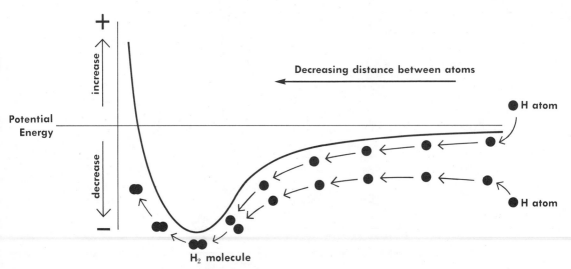

Figure 8–9 A graphical representation of developing energy change as H atoms form a bond.

hydrogen atoms may attain noble gas configurations by sharing a pair of electrons.

The most probable distance between two bonded atoms is such that the attractive forces between the positively charged nuclei and the electron pair are greater than the forces of repulsion (1) between the electrons and (2) between the nuclei. Figure 8–8 illustrates the conflict in forces between two hydrogen atoms in a molecule of hydrogen gas.

The ideal structure of the hydrogen molecule evolves from the loss of energy as two isolated hydrogen atoms approach each other in the drive toward greater stability. Figure 8–9 illustrates the loss of energy as the two atoms move closer and closer together until they stop at an equilibrium distance where the attraction is balanced by a sharply increasing force of repulsion.

While the hydrogen molecule illustrates the fundamental nature of the chemical bond, it represents only one method by which atoms become bonded.

The hydrogen atom has a single $1s$ electron. It can attain a duet structure by *attracting* one electron to form a hydride ion, H^-. But the hydrogen atom can also form a duet in another way. It can *share* its electrons with another nonmetallic atom to form a covalent bond. Two hydrogen atoms can satisfy their tendencies toward formation of a noble gas structure by sharing their electrons by overlapping their s orbitals to form a diatomic hydrogen molecule. This may be illustrated as follows:

$$H \quad + \quad H \quad \rightarrow \quad H_2$$
$$H\cdot \quad + \quad H^\times \quad \rightarrow \quad H^\times_\cdot H$$

Each hydrogen atom forms a duet by sharing $1s$ electrons.

The formation of octets will be illustrated by showing the covalent bonding in the diatomic molecules of fluorine and nitrogen. Two fluorine (Group VIIA element) atoms can attain a noble gas structure (Ne) by sharing their p electrons.

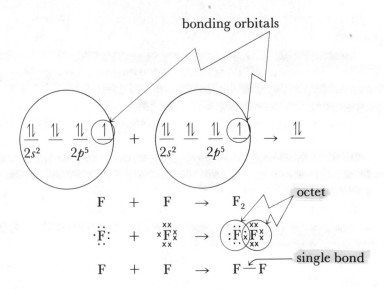

$$F + F \rightarrow F_2$$

octet

$$\cdot \ddot{F}{:} + {}^{x}_{x}{\overset{xx}{F}}{}^{x}_{x} \rightarrow {:}\ddot{F}{\overset{xx}{\underset{xx}{x}}}F{}^{x}_{x}$$

$$F + F \rightarrow F\!-\!F$$

single bond

Two nitrogen atoms will share six p electrons (i.e., three pairs) to form a triple-bonded diatomic molecule:

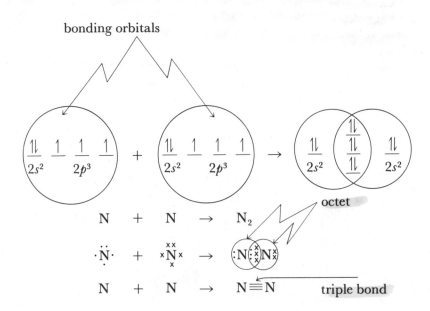

$$N + N \rightarrow N_2$$

octet

$$\cdot \ddot{N}{\cdot} + {}^{x}_{\ }{\overset{xx}{N}}{}^{x}_{x} \rightarrow {:}N{\overset{x}{\underset{x}{\overset{x}{x}}}}N{}^{x}_{x}$$

$$N + N \rightarrow N\!\equiv\!N$$

triple bond

Now let's consider bonding in the hydrogen fluoride molecule. The hydrogen atom ($1s^1$) and the fluorine atom ($1s^2 2s^2 2p^5$) each need one electron in order to attain a noble gas structure. The noble gas structure is realized when the hydrogen and fluorine atoms share a pair of electrons.

CHEMICAL BONDS

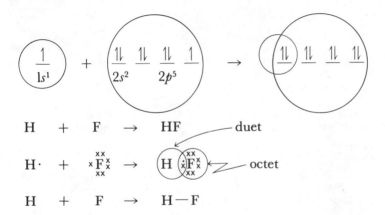

$$\text{H} \quad + \quad \text{F} \quad \rightarrow \quad \text{HF} \qquad \text{— duet}$$

$$\text{H·} \quad + \quad \overset{\times\times}{\underset{\times\times}{\times\text{F}\times}} \quad \rightarrow \quad \text{H}\,\overset{\times\times}{\underset{\times\times}{\times\text{F}\times}} \qquad \text{— octet}$$

$$\text{H} \quad + \quad \text{F} \quad \rightarrow \quad \text{H—F}$$

Molecules composed of more than two atoms can be handled in the same manner. Covalent bonding in water and ammonia further illustrates the application of the tendency toward octet and duet structures.

$$2\,\text{H·} \quad + \quad \overset{\times\times}{\underset{\times}{\times\text{O}\times}} \quad \rightarrow \quad \text{H}\overset{\times\times}{\underset{\underset{\text{H}}{\times\cdot}}{\times\text{O}\times}}$$

$$3\,\text{H·} \quad + \quad \overset{\times\times}{\underset{\times}{\times\text{N}\times}} \quad \rightarrow \quad \text{H}\overset{\times\times}{\underset{\underset{\text{H}}{\cdot\times}}{\times\text{N}\times}}$$

8.7 POLAR AND NONPOLAR COVALENT BONDS

When certain molecules are placed in an electrical field, they act as though they have a positive and a negative end. Molecules such as water and ammonia orient themselves toward the electrically charged plates of a condenser (Fig. 8–10). Such molecules that are sensitive to an electrical field are known as **polar molecules.** The partial charges are due to the unequal displacement of the covalently bonded electrons between atoms.

The concept of polarity can be grasped quickly if a bar magnet or our planet Earth is visualized. These examples, illustrated in Figure 8–11, have ends that are described as positive or negative, north or south.

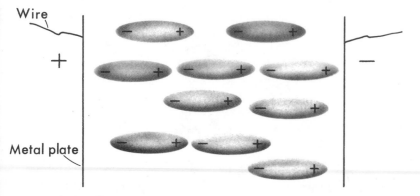

Figure 8–10   The orientation of polar molecules in an electrical field.

Bar magnet

Planet Earth

Figure 8–11 The concept of magnetic polarity.

Polarity is a descriptive term that is usefully applied to molecular structures and covalent bonds. A molecule can and often does have a physical separation between a region of distinct negative charge and one of relative positive charge. Such a molecule is commonly called a **dipole** (Fig. 8–12). It must be noted that the concept of polarity is meaningful only when applied to molecules and not to ionic crystals. An ionic substance, being composed of many distinctly charged particles (ions), cannot be thought of as having oppositely charged *ends*.

8.8 DEGREE OF POLARITY

When dissimilar atoms bond, the bonding electrons will be closer to the atom displaying the greater force of attraction for the electrons. In the resulting molecule, the atom having the greater attraction for the electrons acquires a partial negative charge and the second atom an equal and opposite partial positive charge. The bond is said to be polar, or more properly, to exhibit polar character. Polarity is indicated by

Figure 8–12 A model of an electric dipole structure.

Few electrons   Many electrons

$\delta+$ $\delta-$

Figure 8–13   Partial ionic character in a diatomic molecule.

placing a lowercase Greek letter delta (δ) at each end of the dipole (Fig. 8–13). Most covalent bonds have polar character. The only truly nonpolar bonds are those between identical atoms, as in H_2, Cl_2, Br_2, and the like. In the latter cases both atoms have the same attraction for electrons, and the charge in the molecule is symmetrically distributed.

We can visualize the gradual change in structure observed in diatomic compounds of varying degrees of polarity by the aid of models illustrated in Figure 8–14, in which the range of bonds is provided for comparison.

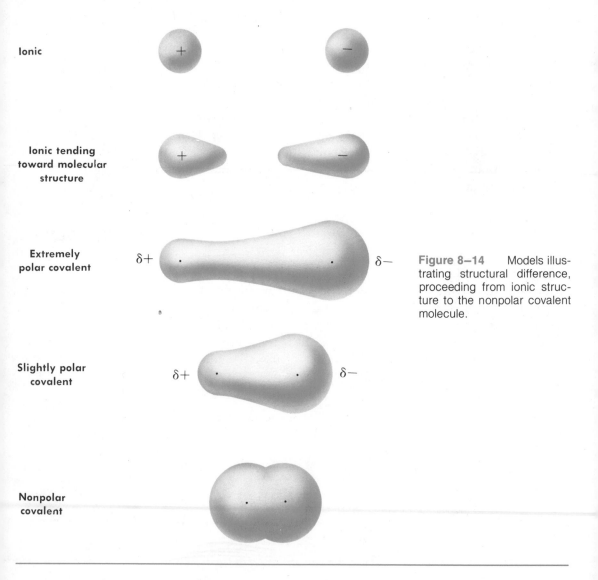

Ionic

Ionic tending toward molecular structure

Extremely polar covalent $\delta+$ $\delta-$

Slightly polar covalent $\delta+$ $\delta-$

Nonpolar covalent

Figure 8–14   Models illustrating structural difference, proceeding from ionic structure to the nonpolar covalent molecule.

Polar covalent bonds arise when two atoms have different electron-attracting ability but not so widely different that an ionic bond is formed. The relative value of the electron-attracting ability of atoms is called its **electronegativity.** Electronegativity is related to nuclear charge and atomic size.

We would correctly expect electronegativity to increase with increased nuclear charge and to decrease with an increase of an atom's radius. The radii of the representative elements (the Group IA elements) decrease slightly as you go from left to right across a horizontal row of the periodic table and increase greatly as you go down a vertical group. Thus electronegativity is a periodic property of atoms that generally increases in value as we proceed from left to right in a horizontal row and increases also as we go from the bottom to the top in a group. These observations are consistent with a decreasing atomic radius and increasing nuclear charge of atoms in a given horizontal row of the periodic table and with an increase of the size of atoms as you go down a group. Nonmetals have a relatively high electronegativity and metals a relatively low electronegativity. Observe the comparison between fluorine and lithium in Figure 8–15.

Both lithium and fluorine are second-period elements. The fluorine atom has a very small radius, so its nucleus can exert a strong pull on a pair of electrons in a chemical bond. By contrast, the attraction for an electron pair exhibited by a metallic atom such as lithium (with its relatively puny nuclear charge and wide radius) is negligible.

The distribution of electronegativity values on the periodic table is generalized in Figure 8–16. A more precise listing of the values, as developed by Linus Pauling, is illustrated in Figure 8–17. The rule of thumb that is most usefully derived from the table of electronegativity values is as follows: *the greater the difference between the electronegativity values of the two atoms sharing a pair of electrons, the more ionic is the character of the bond.* Some examples are illustrated in Figure 8–18.

A short summary of the relationship between electronegativity difference and degree of ionic character is seen in Table 8–7.

From the information in Table 8–7 we can formulate a useful guideline in making a tentative classification of compounds as ionic or covalent. Hence: if the electronegativity difference (symbolized Δen)

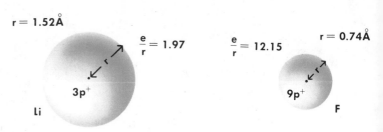

Figure 8–15 Comparison between the charge (e) to radius (r) ratios of Li and F, two atoms in the second period of the periodic table.

$r = 1.52\text{Å}$

$\dfrac{e}{r} = 1.97$

$\dfrac{e}{r} = 12.15$

$r = 0.74\text{Å}$

$3p^+$

$9p^+$

Li

F

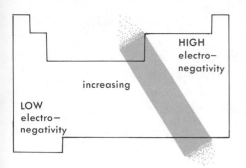

Figure 8–16   Electronegativity trend on the periodic table.

is 1.8 or more, the compound is described as being ionically bound. Some specific examples:

Compound	Δen	% Ionic Character
CsF	3.3	94
NaCl	2.1	67
HCl	0.9	19
HI	0.4	4

H																	He
2.1																	2.1
Li	Be											B	C	N	O	F	
1.0	1.5											2.0	2.5	3.0	3.5	4.0	
Na	Mg											Al	Si	P	S	Cl	
0.9	1.2											1.5	1.8	2.1	2.5	3.0	
K	Ca	Sc	Ti	V	Cr	Mn	Fe	Co	Ni	Cu	Zn	Ga	Ge	As	Se	Br	
0.8	1.0	1.3	1.5	1.6	1.5	1.8	1.8	1.8	1.8	1.9	1.6	1.6	1.8	2.0	2.4	2.8	
Rb	Sr	Y	Zr	Nb	Mo	Tc	Ru	Rh	Pd	Ag	Cd	In	Sn	Sb	Te	I	
0.8	1.0	1.2	1.4	1.6	1.8	1.9	2.2	2.2	2.2	2.4	1.7	1.7	1.8	1.9	2.1	2.5	
Cs	Ba	La	Hf	Ta	W	Re	Os	Ir	Pt	Au	Hg	Tl	Pb	Bi	Po	At	
0.7	0.9	1.1	1.3	1.5	1.7	1.9	2.2	2.2	2.2	2.4	1.9	1.8	1.8	1.9	2.0	2.2	

Figure 8–17   Electronegativity values of the elements. (After L. Pauling.)

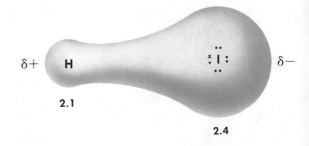

Electronegativity difference is 0.3. HI is very slightly polar and therefore has very little ionic character.

Figure 8–18   The relationship between electro-negativity and the degree of ionic character.

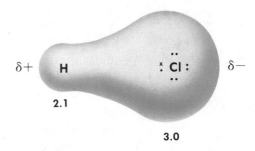

Electronegativity difference is 0.9. HCl is moderately polar.

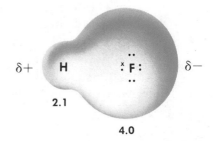

TABLE 8–7

Difference in Electronegativity	Per Cent of Ionic Character
0.2 to 1.0	1 to 22
1.2 to 1.8	30 to 55
2.0 to 3.2	63 to 92

8.10 POLARITY IN MULTI-ATOM MOLECULES

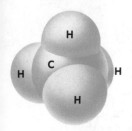

Figure 8–19 The structure of the methane molecule.

The fact that molecules composed of more than two atoms exhibit significant polarity can be explained in terms of the **symmetrical** or **asymmetrical** distribution of all the bonds between pairs of atoms in a polyatomic molecule. For example, methane is composed of 1 carbon atom bonded to 4 hydrogen atoms, and the hydrogen atoms are distributed at the four points of a regular tetrahedron, with the carbon atom in the center (Figure 8–19). Carbon has an electronegativity value that is slightly higher than that of hydrogen (H = 2.1, C = 2.5). This means that the pair of electrons constituting the shared pair bond between a carbon atom and a hydrogen atom will be closer to the carbon atom. Each of the four C—H bonds will be just the same. The result is that this carbon atom will be a center of negative charge and that each of the hydrogen atoms will constitute a region of equally strong positive charge. However, the *center* of positive charge will be equidistant from each hydrogen atom, which places this center exactly at the carbon atom. It may be said that in a *molecule* where there is a symmetrical (geometrically balanced) distribution of polar bonds, the centers of positive and negative charge *coincide*. Any molecule fitting such a description is described as being *nonpolar*. See Figure 8–20 for a visualization of the methane molecule symmetry. The single-plane model is not meant to suggest that molecules are "flat," any more than a wall map means that the world is flat. It is simply easier to illustrate polarity in this way.

The water molecule is known to have an angle of about 105° between the hydrogen atoms. This information, expressed in the format of an electron-dot formula, serves as the basis for the explanation of why water is a decidedly polar molecule. Notice that the electronegativity differences (indicated in Figure 8–21) are related to the unequal sharing of the electron pairs that cause the oxygen atoms to

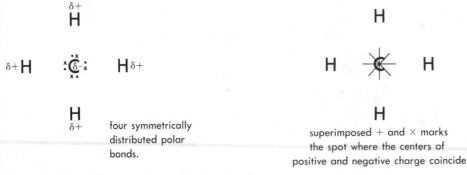

four symmetrically distributed polar bonds.

superimposed + and × marks the spot where the centers of positive and negative charge coincide

Figure 8–20 The nonpolar methane molecule.

CHEMICAL BONDS

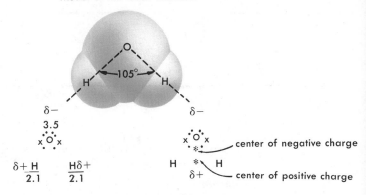

Figure 8–21 The polarity of the water molecule.

be the center of negative charge. The resulting polarity of the water molecule is clearly indicated in Figure 8–21. The asterisks (*) in the figure represent the separation between the centers of positive and negative charge.

The asymmetry (geometrical imbalance) of the polar bond distribution leads to a *separation* of the centers of charge—an absolute requirement for any molecule that has a dipole moment.

Exercise 8.2

Use the electronegativity values in Figure 8–17 to draw illustrative electron-dot formulas for the following molecules. Label each molecule as polar or nonpolar:
(a) carbon dioxide (CO_2), (a linear molecule)
(b) ammonia (NH_3)
(c) silicon tetrafluoride (SiF_4), (like CH_4)
(d) nitrogen (N_2)

8.11 OTHER TYPES OF BONDS

While the majority of compounds may be described in terms of ionic and covalent bonds, other classes of substances demonstrate bonding characteristics that are sufficiently unique that chemists assign them to special categories.

Some examples of bonds other than ionic and covalent types are described in the following sections.

Metallic Bond

This is the bonding among atoms in metals. One rather simplified description is to picture a metal as being composed of positive

---

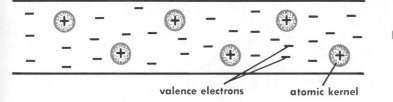

Figure 8–22   Metallic bonding.

valence electrons   atomic kernel

atomic kernels (nucleus plus lower energy level electrons) in a "sea" of valence electrons. Figure 8–22 illustrates metallic bonding diagrammatically.

This model explains the electrical conductivity of metals and their strength and heat conductivity. The metallic property of good electrical conductivity may be explained in part if we focus our attention on the "sea" of valence electrons. In metallic bonding, these valence electrons are not restricted to their usual atomic orbitals. While each atomic kernel can vibrate about a fixed position only (as long as the metal is in the solid state), the valence electrons have great freedom of mobility. We say that they are **delocalized,** which means that they are quite separated from their normal atomic orbitals.

When an electric current (a flow of electrons) is induced into a metal conductor, the free mobility of the valence electrons permits their movement in the direction of the positive pole. Figure 8–23 shows a comparison of the motion of valence electrons in a piece of metal that is part of an active electrical circuit and a piece of metal that stands alone.

Heat conduction is another property of metals that is related to the metal's distinctive type of bonding. The metals that are good electrical conductors are also good heat conductors. As the temperature is raised, the vibrations of the atomic kernels increase, and so does the speed of the randomly moving electrons. In fact, the heating of metals can result in a "boiling out" of electrons by a process called **thermionic emission.** At any rate, the increased kinetic energy of the valence electrons and atomic kernels causes a sharp increase in the collisions between them and results in the transfer of heat through the metal to its environment.

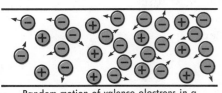

Random motion of valence electrons in a "free" piece of metal

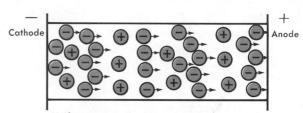

Uniform motion of valence electrons in a metal which is part of an active electrical circuit

Figure 8–23   The motion of valence electrons in metal.

CHEMICAL BONDS

It should also be noted that when a circuit is "overloaded" by drawing electrons through the metal wire at a rate in excess of the wire's capacity for transport, there is often a dangerous buildup of heat. This condition comes about because of a great increase in collisions between the electrons and the atomic kernels, resulting in heat buildup as the vibrating kernels impede the free flow of the electrons. This process is called **resistance.** The cooling of metal conductors lowers their resistance. In fact, it has been observed that the lowering of temperature to near absolute zero enables the metal conductors to function as **super conductors,** owing to the absence of resistance.

Network Bond

The network bond is a case of complex sharing of electron pairs. The covalent bonding is so extensive that huge crystals result rather than molecules. A typical example of network bonding is in the hardest material known, the diamond. In the diamond, each carbon atom is

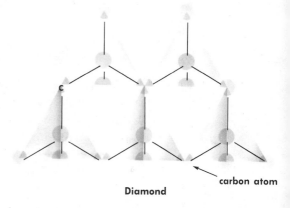

Figure 8–24   Examples of network bonding in diamond and quartz.

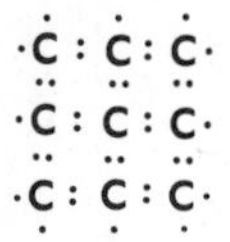

·C̈:C̈:C̈·
·C̈:C̈:C̈·
·C̈:C̈:C̈·

Figure 8–25 A model of a small fragment of a diamond crystal to illustrate multiple covalent bonding in a network crystal.

centrally located and covalently bonded to four other carbon atoms at the four corners of a tetrahedron, as illustrated in Figure 8–24.

Another example of network bonding is silicon carbide, commonly called carborundum, a compound that is nearly as hard as a diamond and is used as an industrial abrasive. Hardness, insolubility in water, and very high melting and boiling points are properties of these network-bonded compounds. The great strength of network bonds, demonstrated by the need for extraordinary energy requirements to break the bonds between the atoms, is directly due to the multiple covalent bonding in which the valence electrons are decidedly localized. See Figure 8–25 for a model of multiple covalent bonding in the diamond.

Silicon carbide is produced by roasting silicon dioxide (sand) with carbon in a very hot electric furnace.

$$SiO_2(s) + 2\ C(s) \xrightarrow{\Delta} SiC(s) + 2\ CO(g)$$

Since the number of atoms in these network-bonded substances is variable, they are often described as **macromolecules** or "giant" molecules.

Hydrogen Bond

This type of bond describes the tendency of small nonmetallic atoms of high electronegativity (notably *fluorine, oxygen,* and *nitrogen*) to "share" hydrogen atoms in compounds. Hydrogen bonding is a close dipole–dipole interaction involving polar molecules containing hydrogen bonded to a small, highly electronegative atom.

The empirical formula for hydrogen fluoride is HF. In reality, solid HF forms a zig-zag chain because of hydrogen bonding. Its actual structure might be H_6F_6, or some other number of atoms in a 1:1 ratio, written $(HF)_n$. Figure 8–26 illustrates hydrogen bonding in hydrogen fluoride.

Ice offers another example of hydrogen bonding. The fact that ice

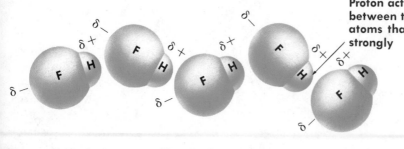

Proton acts as bridge between two fluorine atoms that attract it strongly

Figure 8–26 Ice offers another example of hydrogen bonding. The fact that ice floats in water is due to a reduction in density resulting from "hole" formation as hydrogen bonding spreads the molecules. See also Figure 8–27.

CHEMICAL BONDS

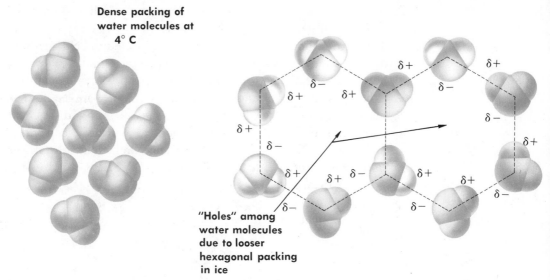

Dense packing of water molecules at 4° C

"Holes" among water molecules due to looser hexagonal packing in ice

Figure 8–27   Hydrogen bonding among water molecules.

floats in water is due to a reduction in density resulting from "hole" formation as hydrogen bonding spreads the molecules. Figure 8–27 represents the effect of hydrogen bonding on the arrangement of water molecules.

8.12   THE GEOMETRY OF SELECTED MOLECULES

It has been pointed out that the polarity of molecules is related to their physical structure. In a larger sense, it may be inferred that many of the physical and chemical properties of molecules depend upon the geometrical arrangement of valence electrons and the participating atoms.

The actual shapes of many molecules have been determined by measurements made in the laboratory with modern instruments. What we shall attempt to do in this section is to present a rational explanation of why the molecule is shaped as it is.

Experience permits the chemist to make predictions about the shapes of molecules on the basis of some guiding principles that work most of the time. However, these predicted shapes must be regarded merely as attractive hypotheses until supporting empirical evidence is available.

The guiding principles regarding the prediction of molecular shapes may be summarized as follows:

1. The central atom in a molecule will most often have the full complement of electron pairs (four) resulting from covalent bonding.

2. A central atom may have more or less than four bonds because the octet principle is not an inviolate law.
3. The pairs of bonding electrons will be expected to be distributed at maximum distances from each other, since forces of repulsion between pairs of electrons will tend toward this arrangement.
4. When the measured angles between atoms in a molecule deviate from what is expected, the explanation may be found in the electron-pair repulsion effect, owing to unbonded pairs of electrons. This interpretation is often termed the **valence shell electron-pair repulsion theory,** represented by the acronym **VSEPR** and called the "vesper" theory.

The application of these guidelines is illustrated by the following examples.

Water

The measured angle between the hydrogen atoms in the water molecule is about 105°, which is contrary to the 90° angle suggested by the electron-dot formula,

$$
\begin{array}{c}
\overset{\text{xx}}{\underset{\text{x}}{\text{x}}} \overset{\text{x}}{\text{O}} \overset{\text{x}}{\text{·}} \text{H} \\
\text{H}
\end{array}
$$

An examination of the electron configuration of oxygen, $1s^2\ 2s^2\ 2p^4$, indicates a total of six valence electrons. The two unpaired electrons can form covalent bonds with hydrogen, but the remaining two pairs of electrons are unbonded. The maximum distance between the electron pairs, both bonded and unbonded, is obtained by distributing them at the four corners of a regular tetrahedron as illustrated in Figure 8–28.

The incorrect model in Figure 8–28 is rectified by application of the VSEPR theory, which states that the two unbonded pairs of

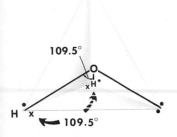

Figure 8–28   A model of the water molecule that does not agree with its actual shape.

CHEMICAL BONDS

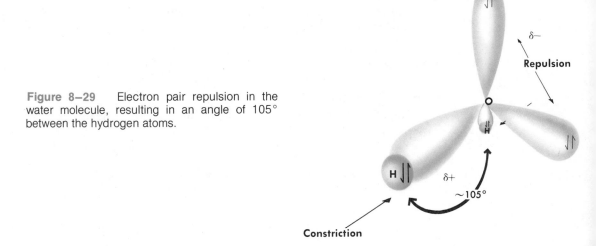

Figure 8–29   Electron pair repulsion in the water molecule, resulting in an angle of 105° between the hydrogen atoms.

electrons in the water molecule would repel each other to the point of distorting the normal tetrahedral structure which has a bond angle of 109.5° (Fig. 8–29).

8.13   ORBITAL HYBRIDIZATION

The formation of the tetrahedral arrangement of the valence electrons in oxygen is logically explained by the concept of **orbital hybridization.** Orbital hybridization means that a net probability distribution of electrons may result from the absorption of energy that allows the electronic configuration of an atom to achieve an

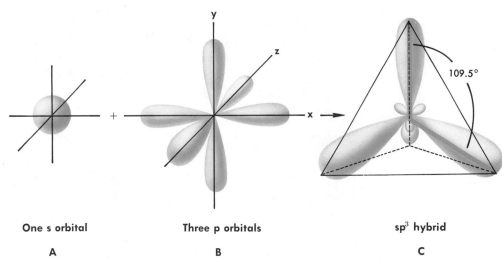

One s orbital	Three p orbitals	sp³ hybrid
A	B	C

Figure 8–30   Formation of an sp^3 hybrid orbital from one s orbital and three p orbitals.

excited state. As we discussed earlier, the normal configuration is called the **ground state.** In the case of oxygen, the tetrahedral distribution of electrons results from the $2s$ and $2p$ electrons forming what is called an sp^3 **hybrid.** An sp^3 hybrid orbital is a combination of one s orbital electron and three p orbital electrons in the excited state. The sp^3 hybrid is *not* a simple composite of an s and three p orbitals; it is an entirely different structure (Fig. 8–30).

The sp^3 hybridization can be illustrated by altering the ground state pictorial representation of oxygen:

ground state $_8$O $\quad$ 1s^2 $\quad$ $2s^2$ $\quad$ $2p_x^2$ $2p_y^1$ $2p_z^1$

| $\uparrow\downarrow$ | $\uparrow\downarrow$ | $\uparrow\downarrow$ | $\uparrow$ | $\uparrow$ |

excited state

| $\uparrow\downarrow$ | $\uparrow\downarrow$ | $\uparrow\downarrow$ | $\uparrow$ | $\uparrow$ |

$\quad\quad\quad$ $2sp^3$ $\quad$ $2sp^3$ $2sp^3$ $2sp^3$

A useful generalization that can be made at this point is that *sp^3 orbital hybridization is always equated with a tetrahedral structure.*

Ammonia

Ammonia is another example of sp^3 orbital hybridization. The electron-dot formula,

$$H \overset{\times\times}{\underset{\underset{H}{\cdot\times}}{\times\!N\!\times}} H$$

indicates four pairs of electrons (three bonded and one unbonded), which deviate slightly from the regular tetrahedral arrangement, about 107° instead of 109.5°, because of the electron pair repulsion. The electronic configuration and pictorial representation of nitrogen in the ground and excited states as illustrated in Figure 8–31 demonstrate the orbital hybridization.

ground state $_8$O $\quad$ 1s^2 $\quad$ $2s^2$ $\quad$ $2p_x^1$ $2p_y^1$ $2p_z^1$

| $\uparrow\downarrow$ | $\uparrow\downarrow$ | $\uparrow\downarrow$ | $\uparrow$ | $\uparrow$ |

excited state

| $\uparrow\downarrow$ | $\uparrow\downarrow$ | $\uparrow\downarrow$ | $\uparrow$ | $\uparrow$ |

$\quad\quad\quad$ $2sp^3$ $\quad$ $2sp^3$ $2sp^3$ $2sp^3$

CHEMICAL BONDS

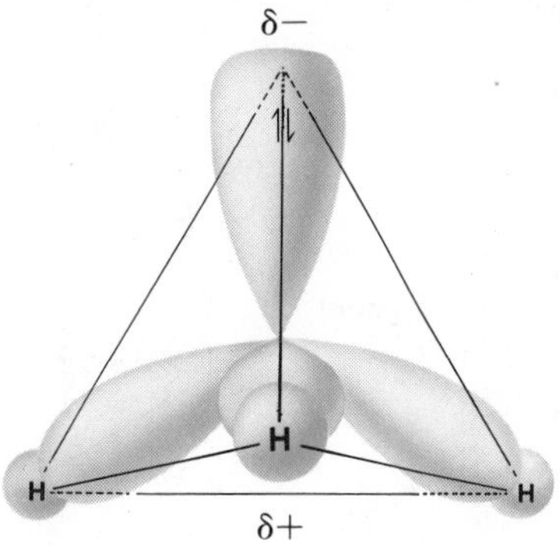

Figure 8–31   The pyramidal structure of the NH_3 molecule due to sp^3 hybridization of the orbitals.

Hence, covalent bonds are formed between nitrogen and hydrogen as the sp^3 orbitals of nitrogen overlap with the $1s$ orbital of each hydrogen atom.

Boron Trifluoride

An explanation of the fact that boron can form three bonds is found in the hybridization of the orbitals in the excited state. The electronic configuration and pictorial representation of boron are as follows:

ground state $_5B$ $1s^2$ $2s^2$ $2p_x^1$ $2p_y^0$ $2p_z^0$

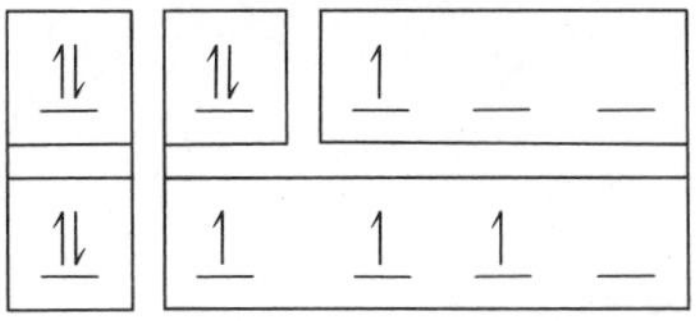

| ↑↓ | ↑↓ | ↑ | — | — |

excited state

| ↑↓ | ↑ | ↑ | ↑ | — |

sp^2 sp^2 sp^2

Notice that the energy absorption related to the excited state permits an **uncoupling** of the $2s$ electrons. The three unpaired electrons that constitute the sp^2 (one s orbital and two p orbital electrons) hybrid will be arranged at the three points of a triangle, each bond being 120° apart. *The sp^2 hybrid always has a trigonal planar distribution* (Fig. 8–32).

The fluorine atom has the following electronic configuration and pictorial representation:

$_9F$ $1s^2$ $2s^2$ $2p_x^2$ $2p_y^2$ $2p_z^1$

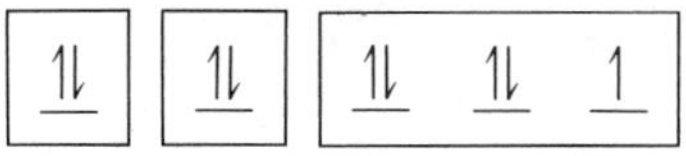

| ↑↓ | ↑↓ | ↑↓ | ↑↓ | ↑ |

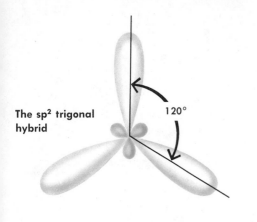

The sp² trigonal hybrid

120°

Figure 8–32   The sp² trigonal hybrid.

The single p orbital electron will form the covalent bond with boron. Remembering that a p orbital probability distribution resembles the shape shown in Figure 8–33, the "distorted" shape can be said to be due to the much higher probability that the fluorine electron will be at the site of the covalent bond.

The combined diagrammatic representation of the boron and fluorine atoms is illustrated in Figure 8–34. Further proof for the trigonal planar structure of BF_3 comes from experimental evidence that the molecule is indeed flat and nonpolar.

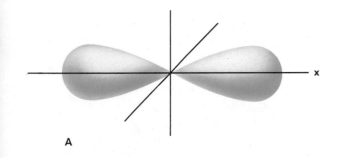

A

x

Figure 8–33   Normal and bonded p orbital electron in the fluorine atom. A, A normal p orbital distribution. B, "Distorted" p orbital due to bonding.

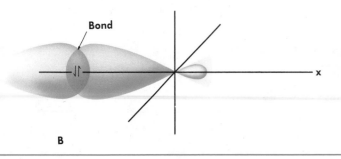

Bond

x

B

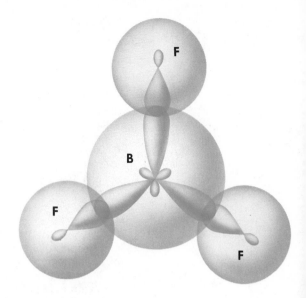

Figure 8–34 A model of sp^2 orbital hybridization in the BF_3 molecule.

It should be mentioned at this point that there are other types of molecular geometry related to other modes of orbital hybridization. But a discussion of these other geometrical forms is really beyond the scope of this text. As you continue your study of chemistry, the sub-

TABLE 8–8 **Summary of Some Common Molecular Shapes**

Number of Bonds to Central Atom	Type of Geometry around a Central Atom	Possible Orbital Notation	Model	Example
2	Linear	sp		CO_2
3	Trigonal planar	sp^2		BF_3
4	Tetrahedral	sp^3		CH_4

jects of complex ions and molecules and organic chemistry will be examined, which will give you the foundation for a meaningful investigation of the other geometrical forms and some alternative theories that have been advanced to help our understanding.

A summary of a few of the common molecular shapes is presented in Table 8–8.

Exercise 8.3

Draw the probable geometrical shapes of the following, and label the type of orbital hybridization.
1. CS_2, carbon disulfide
2. PI_3, phosphorus triiodide
3. NH_4^+, ammonium ion
4. SO_4^{2-}, sulfate ion
5. CCl_4, carbon tetrachloride

QUESTIONS & PROBLEMS

8.1   Define a chemical bond.

8.2   What is the essential difference between an ionic bond and a covalent bond?

8.3   What is meant by a polar bond? Explain and illustrate.

8.4   What is meant by a nonpolar molecule? Explain and illustrate.

8.5   Is it possible to have a nonpolar molecule containing polar bonds? Explain.

8.6   What is the difference between a polar bond and a polar molecule?

8.7   What does a calculated electronegativity difference (Δen) value tell you about a compound?

8.8   Does an atom having a low ionization energy also have a low electronegativity?

8.9   Define the term *ionization energy*.

8.10   Explain the relationship between ionization energy and metallic character.

8.11   What is hydrogen bonding?

8.12   Describe the molecules CH_4 (methane) and water in terms of polarity, symmetry, and separation of the centers of charge.

8.13   Write the electron-dot formulas for the following:
(a) sodium bromide
(b) potassium hydride
(c) calcium iodide
(d) silicon tetrafluoride
(e) hydrogen bromide

8.14   What is the significance of a high charge-to-radius ratio in an atom with regard to ionization energy?

8.15   Predict the most probable shapes for the following molecules.
(a) CH_3F
(b) PCl_3
(c) BH_3
(d) CCl_4

CHEMICAL BONDS

8.16 Write electron dot formulas for each of the following molecules and ions:
(a) sulfur dioxide (SO_2), a polar molecule
(b) sulfur trioxide (SO_3), a nonpolar molecule
(c) the ammonium ion ($NH_4{}^+$), a tetrahedral ion
(d) the hydroxide ion (OH^-), a linear ion

8.17 Predict which atom will have the greater first ionization energy:
(a) Na, Ar
(b) Na, Rb
(c) Li, F
(d) K, Cu

CHAPTER NINE

At the completion of this chapter, you should be able to:

1. Describe the essential characteristics of a solution.

2. Draw sketches to illustrate the dissociation of ionic compounds and polar molecules in water.

3. Contrast the following terms: *strong electrolyte, weak electrolyte, non-electrolyte.*

4. Use a table of solubilities to predict the solubility or insolubility of the products of a reaction.

5. Write and balance net ionic equations.

6. Identify spectator ions so that net ionic equations can be written.

7. List rules for writing ionic equations.

8. Define *molarity* and calculate molar concentrations of solutions from data provided.

9. Calculate final molarities or volumes in dilution problems.

10. Define *normality, equivalents,* and *equivalent mass,* and calculate normal concentrations of solutions from given data.

11. Calculate the number of equivalents from a given mass of a compound and its formula.

12. Describe the colligative effect of solute particles on the vapor pressure of a solution.

13. Explain the effect that the lowering of vapor pressure has on the boiling and freezing points of solvents.

14. Define *mole fraction.* Describe the effect of the mole fraction in terms of Raoult's law.

15. Define *molality* and calculate the molality of solution from available data.

16. Calculate the actual changes in the boiling and freezing points of solvents for any given molal concentration of a nonelectrolyte.

17. Use tables of boiling and freezing point data in the solution of related problems.

18. Find true molecular formulas of compounds from their empirical formulas and information on their freezing or boiling points.

SOLUTIONS

It is convenient to define a solution as a combination of a dissolved substance, called a **solute,** and a medium in which the solute is homogeneously dispersed, called the **solvent.** While a solvent could be a liquid, solid, or gas, the most common solvent is water. The symbol (aq) following the chemical formula of a compound designates a water solution of that compound.

Solute particles move about randomly in a solvent and are of ionic or molecular dimensions. When a mass of solid material is mixed with a solvent, there is an interaction between the unit structures (ions or molecules) of the solute and the solvent molecules. The forces of attraction in the solid structure of the solute are overcome, as are the attraction forces of the solvent, and the particles of solid dissolve. If the size of the dissociated or dispersed particles is large enough to reflect visible light, the mixture is called a **colloidal suspension** rather than a solution.

In a true solution the dissolved solute particles will neither settle to the bottom of a container nor can they be filtered out by usual laboratory methods of filtration, and they will not reflect light. The process of dissolving is commonly described as a physical change, because the solute and solvent can be recovered in their original forms by the simplest evaporation of the solvent. For example, the crystals of $CuSO_4 \cdot 5H_2O$ form a beautiful blue solution when dissolved in liquid water. Complete evaporation of the water restores the original components of the solution.

$$CuSO_4 \text{ (aq)} \xrightarrow{\text{evaporation}} \underset{\text{blue solid}}{CuSO_4 \cdot 5H_2O \text{ (s)}} + H_2O \text{ }(\ell)$$

9.2 THE SOLUTION PROCESS

Ionic Solutes and Water

When NaCl(s) dissolves in water, it is reasonable to assume that there is an attractive force between the positively and negatively charged ions and the dipolar water molecules. The salt dissolves because of the attractive forces between the anions, the cations, and the

---

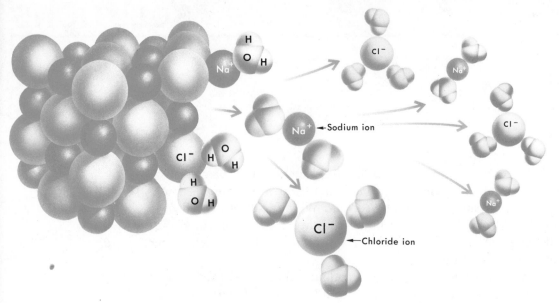

Figure 9–1   The dissociation of the sodium and chloride ions from the crystal structure because of ion-dipole interaction.

water dipoles. Such forces between positive ions and water (or negative ions and water) are described as **ion-dipole interactions.** The question that arises is why the ion-dipole interactions between NaCl and H_2O are apparently stronger than the ion-ion attractions within the ionic crystal structure of NaCl. The answer seems to be related to the neutralizing property of the "cage" of water molecules surrounding the surface ions of the crystal, which partially overcomes the electrostatic forces of attraction between the ions. Figure 9–1 illustrates the dynamic equilibrium of the dissolving process (also called **solvation**) resulting from the ion-dipole interaction between the sodium ions, chloride ions, and water molecules.

A second factor bearing on the solution process is the tendency of any system to move toward a more random distribution (entropy). The ionic lattice structure of crystalline NaCl is highly ordered. The interaction of the ions with water provides a means of disrupting the crystal structure to produce a random solution system.

The fact that many compounds increase in solubility when heated relates to the tendency toward randomness. Most solutions, when heated, increase their tendency toward increased randomness because of the energy that can be absorbed. For this reason, the application of heat usually permits more of the solid to dissociate.

At a given temperature, a particular compound has a specific solubility. This means that just so many grams can be dissolved in a stated mass of solvent to make a saturated solution. Solubility curves (Fig. 9–2) describe the solubility of a substance in terms of grams of solute per 100 g of solvent. These data are related to the temperature.

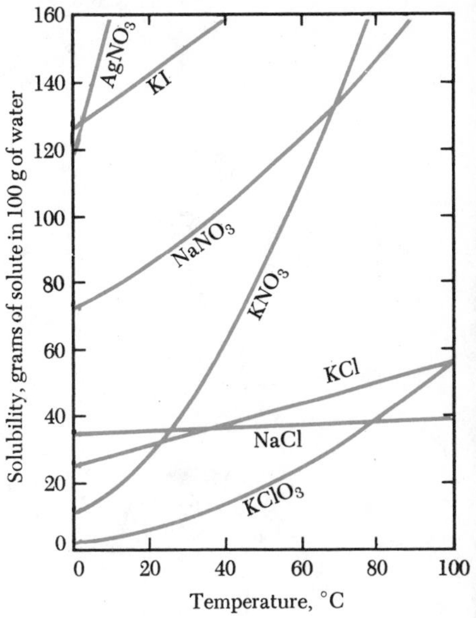

Figure 9–2 Solubility curves of salts in water.

For example, the information given for a compound may state the solubility as 40 g per 100 g of water at 0°C, and 80 g per 100 g of water at 100°C. Usually, when the temperature is raised (that is, when more heat energy is provided), more solid solute can dissolve in water. If the temperature is then lowered, the extra solute will usually crystallize out of solution. Occasionally, a special phenomenon is observed, in which the excess mass of solute remains in solution even though the temperature is lowered. A solution of this type is said to be **supersaturated,** and it is likely to be so unstable that an added crystal of solute or a sharp jolt will cause the crystallization of the extra mass of solute.

Heat is the one common factor that can affect the degree of the solubility of solid and liquid compounds. Grinding of large crystals into powder and vigorous stirring of the solvent during the dissolving process speeds up the solution process but does not increase the solubility of a substance. These techniques allow more surface of the solute to come in contact with the water dipoles within a given time. If the solubility of a substance is described as 84 g per 100 g of water at room temperature, stirring and grinding will not allow more than 84 g per 100 g of water to dissolve, but 84 g of powdered compound will probably dissolve in a fraction of the time required to dissolve the same mass of large crystals.

Polar Covalent Solutes and Water

A typical example of a polar covalently bonded compound is hydrogen chloride gas. When HCl(g) is mixed with water, it dissolves very rapidly as it forms the solution called hydrochloric acid (Fig. 9–3).

The interaction between the HCl(g) dipoles and the water dipoles is appropriately described as being of a **dipole-dipole** type. Interest-

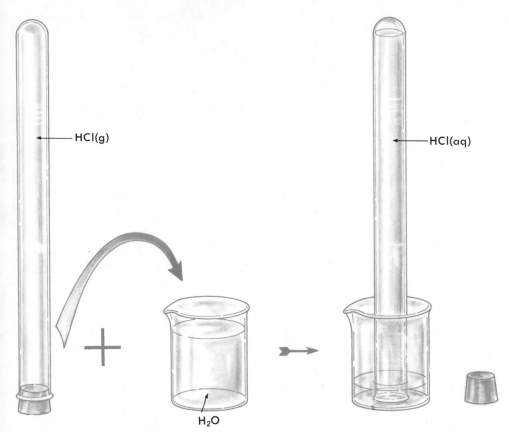

HCl(g)

H₂O

HCl(aq)

Figure 9–3   The solubility of HCl(g) in water.

ingly, the forces of attraction operating between the dipoles actually increase the degree of polarity of each dipole. These are described as **enhanced** dipoles. The hydrogen-to-chlorine bond polarity is enhanced to the point that there is a physical separation of the HCl molecule and a resulting formation of hydrated chloride and hydrogen ions. Figure 9–4 illustrates the enhancement of polarity that may occur in the process of dipole-dipole interaction.

The HCl molecules separate in water to become hydrated ions in a process known as **ionization.** The resulting hydrated hydrogen ion, or

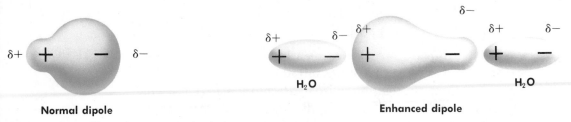

Normal dipole

Enhanced dipole

Figure 9–4   The enhancement of polarity.

SOLUTIONS

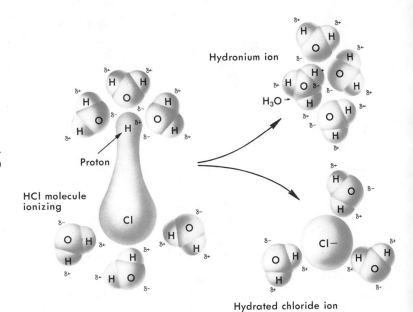

Figure 9–5 The dipole-dipole interaction between H_2O and HCl, leading to ionization.

Hydronium ion

H_3O^+

Proton

HCl molecule ionizing

Hydrated chloride ion

hydrated proton, is also known by the name **hydronium ion** (H_3O^+) (Fig. 9–5).

9.3 ELECTROLYTES AND NONELECTROLYTES

Although acids will be discussed more fully in the next chapter, it is useful to note at this time that the formation of hydrated H^+ ions in solution is a characteristic of acids. In fact, HCl(aq) is described as a strong acid because of the degree to which it undergoes ionization to form hydrated H^+ ions.

A simple test may be used to predict whether a solution process contains ions in solution. While an ionic crystal needs only to dissociate in order to produce ions in solution, a molecular compound may or may not undergo ionization. For example, hydrogen chloride ionizes almost completely in dilute solution to form ions (a characteristic of strong acids), but compounds such as sugar and alcohol do not ionize appreciably when they dissolve in water. Substances known as weak acids ionize slightly owing to strong bonds that are not easily overcome by dipole-dipole interaction. Soluble ionic substances and others which ionize extensively are known as strong **electrolytes.** Weak acids and slightly soluble ionic substances are termed weak electrolytes. Most organic (carbon) compounds and others that do not ionize appreciably are appropriately labeled **nonelectrolytes.** Table 9–1 presents a brief summary of some typical electrolytes and nonelectrolytes. An electrolyte may be defined as any substance which when dissolved in a

TABLE 9–1 Some Electrolytes and Nonelectrolytes

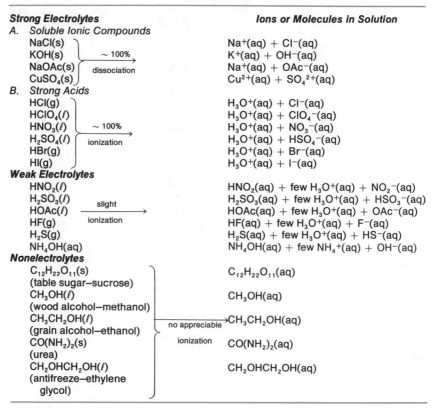

Strong Electrolytes | **Ions or Molecules in Solution**

A. *Soluble Ionic Compounds*

NaCl(s) ⎫	Na$^+$(aq) + Cl$^-$(aq)
KOH(s) ⎬ ~ 100%	K$^+$(aq) + OH$^-$(aq)
NaOAc(s) ⎬ dissociation	Na$^+$(aq) + OAc$^-$(aq)
CuSO$_4$(s) ⎭	Cu^{2+}(aq) + SO$_4^{2+}$(aq)

B. *Strong Acids*

HCl(g) ⎫	H$_3$O$^+$(aq) + Cl$^-$(aq)
HClO$_4$(l) ⎬ ~ 100%	H$_3$O$^+$(aq) + ClO$_4^-$(aq)
HNO$_3$(l) ⎬	H$_3$O$^+$(aq) + NO$_3^-$(aq)
H$_2$SO$_4$(l) ⎬ ionization	H$_3$O$^+$(aq) + HSO$_4^-$(aq)
HBr(g) ⎬	H$_3$O$^+$(aq) + Br$^-$(aq)
HI(g) ⎭	H$_3$O$^+$(aq) + I$^-$(aq)

Weak Electrolytes

HNO$_2$(l)	HNO$_2$(aq) + few H$_3$O$^+$(aq) + NO$_2^-$(aq)
H$_2$SO$_3$(l) ⟶ slight	H$_2$SO$_3$(aq) + few H$_3$O$^+$(aq) + HSO$_3^-$(aq)
HOAc(l) ⟶	HOAc(aq) + few H$_3$O$^+$(aq) + OAc$^-$(aq)
HF(g) ionization	HF(aq) + few H$_3$O$^+$(aq) + F$^-$(aq)
H$_2$S(g)	H$_2$S(aq) + few H$_3$O$^+$(aq) + HS$^-$(aq)
NH$_4$OH(aq)	NH$_4$OH(aq) + few NH$_4^+$(aq) + OH$^-$(aq)

Nonelectrolytes

C$_{12}$H$_{22}$O$_{11}$(s) ⎫	C$_{12}$H$_{22}$O$_{11}$(aq)
(table sugar–sucrose)	
CH$_3$OH(l)	CH$_3$OH(aq)
(wood alcohol–methanol)	
CH$_3$CH$_2$OH(l) ⟶ no appreciable	CH$_3$CH$_2$OH(aq)
(grain alcohol–ethanol)	
CO(NH$_2$)$_2$(s) ionization	CO(NH$_2$)$_2$(aq)
(urea)	
CH$_2$OHCH$_2$OH(l)	CH$_2$OHCH$_2$OH(aq)
(antifreeze–ethylene ⎭	
glycol)	

suitable liquid, or when melted, conducts an electrical current. The name electrolyte means "carrier of electricity."

One test for ions in solution is accomplished by a conductivity device called an electrolyte tester. A simple conductivity device consists of a light bulb connected in series with a pair of electrodes so that the circuit can be completed only when ions move between the two electrodes. Figure 9–6 illustrates an electrolyte tester in use.

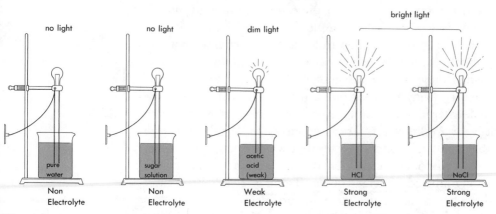

Figure 9–6   A simplified method for testing relative electrolyte strengths.

Nonpolar Covalent Solutes and Water

Nonpolar compounds are not very soluble in water. For example, iodine crystals dissolve slightly (0.029g/100g H_2O) to give water a faint yellow-brown tinge. Carbon dioxide is somewhat more soluble (0.145g/100g H_2O), accounting for the slight acidity of water that is exposed to the CO_2 in the air.

$$CO_2(g) + 2H_2O(\ell) \rightleftharpoons H_3O^+(aq) + HCO_3{}^-(aq)$$

Ordinary soft drinks contain CO_2 in solution under pressure; this gives the drink its "fizz." However, when the cap is removed from a warm bottle, the wet bubbles of carbon dioxide give the drink its layer of foam, or "head."

From another point of view, applicable to both I_2 and CO_2, the slight degree of solubility is an example of the phenomenon of **induced polarity.** The induction of polarity in molecules that are nonpolar comes about as a result of the influence of the water dipole on the charge distribution in the nonpolar molecule. Normally, the centers of positive and negative charge coincide as they do in the typical examples of I_2 and CO_2. However, when the polar water molecule interacts with the nonpolar molecules, a slight separation of the centers of charge occurs (Fig. 9–7), and the I_2 and CO_2 molecules temporarily become slightly polar.

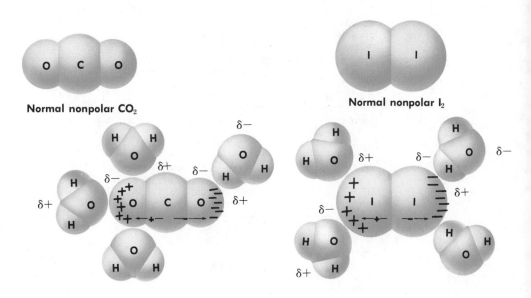

Normal nonpolar CO_2

Normal nonpolar I_2

Induced polarity in CO_2

Induced polarity in I_2

Figure 9–7   The slight water solubility of many nonpolar compounds is due to induced polarity.

**INSOLUBLE AND IMMISCIBLE
COMPOUNDS AND WATER**

The terms "insoluble" and "immiscible" are relative. All substances dissolve in water to some degree. However, for practical purposes, any substance added to water that does not appear to be dissolving or one that forms a solid (precipitate) product in a chemical reaction is described as being insoluble. The term "insoluble" is often qualified by saying "slightly" or "very." Or, for example, $PbCl_2$ is described as being insoluble in cold water but soluble in hot water. For practical purposes we say that a compound is soluble if we can dissolve 0.1 mole of solute per liter of solution and insoluble if not more than 0.001 mole can be dissolved per liter of solution.

Many ionic compounds are soluble. In those instances where an ionic compound deviates from this generalization, the explanation is found in the exceedingly strong electrostatic force of attraction between the cation and anion. For example, MgF_2 is insoluble because the ion-dipole interaction in water is not sufficient to overcome the great attraction of the Mg^{2+} and F^- in the crystal lattice. Additional examples of insoluble compounds include other fluorides, $AgCl$, $PbSO_4$, $Ca(OH)_2$, and many phosphates, carbonates, chromates, sulfides, and hydroxides.

TABLE 9–2 **Simplified Table of the Solubility of Common Salts in Water
at 20°C**

Anion	Cation	Solubility
acetate chlorate nitrate	nearly all	soluble
chloride bromide	lead(II), silver, mercury(I)	insoluble
iodide	all others	soluble
hydroxide	Group IA metals, barium, strontium	soluble
	all others*	insoluble

sulfate	mercury(I) and (II), calcium,
	    barium, strontium, silver,
	    lead(II)
		insoluble

	all others	soluble
carbonate	Group IA metals,** ammonium	soluble
phosphate chromate	all others	insoluble
sulfide	Group IA metals, ammonium, magnesium, calcium, barium	soluble
	all others	insoluble

*$Ca(OH)_2$ slightly soluble
**Li_3PO_4 insoluble

SOLUTIONS

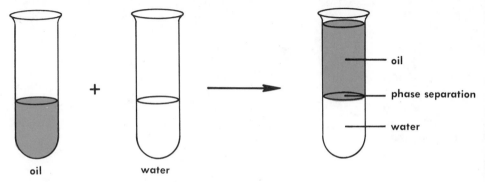

Figure 9–8   The immiscibility of oil and water.

Useful generalizations regarding the solubility of salts are found in Table 9–2. These are commonly called **solubility rules.**

When the components of a solution are liquids, we speak of the *miscibility* or degree of "mixability" of the different liquid compounds. Often two liquids, such as oil and water, are *immiscible* and a **phase** separation results (Fig. 9–8). A phase is a homogeneous system separated from other systems by a physically distinct boundary.

"We should be thankful. What if oil and water *did* mix!" (American Scientist, November-December, 1972. p. 745.)

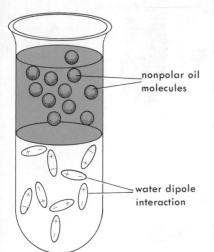

nonpolar oil
molecules

water dipole
interaction

Figure 9–9   The phase separation of oil and water, illustrating the effect of polarity.

The phase separation between liquids can be observed as two distinct layers. This is caused by the different light-refractive properties of the two liquids. Immiscibility can be explained on the basis of polarity. In the case of oil and water, the water dipoles interact among themselves much more effectively than they do with the nonpolar oil molecules. Figure 9–9 illustrates the dipole interactions among water molecules leading to the phase separation with oil.

Nonpolar liquids are quite miscible in each other. The liquid phase separation naturally produces an upper layer and a lower layer, the order depending on the comparative densities. In the oil and water system, the water is the lower layer, since it is more dense than oil. If an even denser nonpolar liquid, such as carbon tetrachloride, were poured into the tube, a three-phase system would be observed (Fig. 9–10).

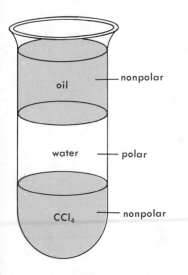

oil — nonpolar

water — polar

CCl_4 — nonpolar

Figure 9–10   A three-phase separation resulting from differences in polarity and density.

Some general rules regarding the solubility of a solute in a solvent are as follows:

1. Most ionic compounds of Groups IA and IIA elements are water-soluble.
2. Polar covalent compounds, such as $HCl(g)$ and $NH_3(g)$, tend to be water-soluble.
3. Nonpolar compounds are more soluble in nonpolar solvents than in polar solvents (e.g., I_2 is soluble in CCl_4 and C_6H_6).

Rules 2 and 3 can be combined in the simplest axiom "like dissolves like," meaning that polar solutes dissolve in polar solvents and nonpolar solutes dissolve in nonpolar solvents.

9.5 IONIC EQUATIONS

Aqueous solutions may contain ions (from strong electrolytes), molecules (from nonelectrolytes), or mixtures of both ions and molecules (from weak electrolytes). It is often desired to write equations showing only the net changes in a reaction. The purpose of any equation is to represent in a symbolic way the changes that occur in a chemical reaction. Ionic equations show only those species (ions or molecules) that undergo a change in the reaction. For example, in the reaction between $NaCl(aq)$ and $AgNO_3(aq)$, the precipitate $AgCl(s)$ is formed. Only the silver ions and chloride ions have participated. While the $Ag^+(aq)$ and $Cl^-(aq)$ started out as submicroscopic invisible ions, they end up as the clearly insoluble compound, $AgCl(s)$. The sodium and nitrate ions that were present at the start remain unchanged. Ions that are present in a system without undergoing a change are descriptively labeled **spectator ions.** As in the case of any "spectator," they are present at the scene of the action but are not involved in it. When writing ionic equations we must consider the relative solubilities of salts (Table 9–2) and the distinctions we made among strong and weak electrolytes and nonelectrolytes (Table 9–1).

Writing an ionic equation for the reaction of solutions of $AgNO_3$ and $NaCl$ can be shown in three steps. As experience is gained, we will write the **net ionic equation** for reactions occurring in aqueous solution. A net ionic equation is one in which all the spectator ions are omitted. When directions call for the addition of chloride ion to silver ion, it will be understood that any soluble ionic metal chloride solution could be added to a soluble ionic silver compound and produce the same change—the precipitation of $AgCl(s)$. To illustrate:

Step 1. Write the *molecular* equation.

$$AgNO_3\,(aq) + NaCl(aq) \rightarrow AgCl(s) + NaNO_3\,(aq)$$

Step 2. Write ions for each compound that exists predominately as ions in solution. This is known as the *total ionic equation*. Balancing is not necessary at this step.

$$Ag^+(aq) + NO_3^-(aq) + Na^+(aq) + Cl^-(aq) \rightarrow AgCl(s) + Na^+(aq) + NO_3^-(aq)$$

Step 3. Eliminate the unchanged species (i.e., the spectator ions) and write the *net ionic equation*. At this stage, balancing is essential.

$$Ag^+(aq) + Cl^-(aq) \rightarrow AgCl(s)$$

Rules for writing net ionic equations:

1. **Strong electrolytes** and **soluble salts** are written in **ionic form:** $Na^+(aq)$, $Cl^-(aq)$, $H^+(aq)$, $Cl^-(aq)$.
2. **Nonelectrolytes** and **weak electrolytes** are written in **molecular form:** $C_{12}H_{22}O_{11}(s)$; $C_2H_5OH(\ell)$; $HOAc(aq)$.
3. **Insoluble species, precipitates, solids,** and **gases** are written in **molecular form:** $AgCl(s)$; $NaCl(s)$; $NH_3(g)$; $HCl(g)$.
4. Spectator ions are not shown.
5. Equations must demonstrate conservation of **mass** and **charge.**

Example 9.1

Write a net ionic equation for the following reaction:

$$BaCl_2(aq) + Na_2SO_4(aq) \rightarrow BaSO_4(s) + NaCl(aq)$$

1. Write the total ionic equation.

$$Na^+(aq) + SO_4^{2-}(aq) + Ba^{2+}(aq) + Cl^-(aq) \rightarrow BaSO_4(s) + Na^+(aq) + Cl^-(aq)$$

2. Eliminate Na^+ and Cl^- as spectator ions, and write the net ionic equation. Pay careful attention to the balancing of both the number of moles of each species involved and the net ionic charge.

$$Ba^{2+}(aq) + SO_4^{2-}(aq) \rightarrow BaSO_4(s)$$

Notice the conservation of charge:

$$(2+) + (2-) = 0$$

Example 9.2

Write the net ionic equation for the following reaction:

$$KOH(aq) + HClO_4(aq) \rightarrow KClO_4(aq) + H_2O(l)$$

1. Note that $H_2O(l)$ is a nonelectrolyte (or at best a very weak electrolyte) and that it is predominately molecular. Write the total ionic equation.

$$K^+(aq) + OH^-(aq) + H^+(aq) + ClO_4^-(aq) \rightarrow K^+(aq) + ClO_4^-(aq) + H_2O(l)$$

2. Since only the H^+ and OH^- undergo a change, the K^+ and ClO_4^- spectator ions are omitted in the writing of the net ionic equation.

$$H^+(aq) + OH^-(aq) \rightarrow H_2O \ (l)$$

Example 9.3

Write the net ionic equation for $NaBr \ (aq) + Cl_2 \ (g) \rightarrow NaCl(aq) + Br_2(l)$.
1. Write the total ionic equation.

$$Na^+(aq) + Br^- \ (aq) + Cl_2 \ (g) \rightarrow Na^+(aq) + Cl^- \ (aq) + Br_2 \ (l)$$

2. The sodium ion is obviously the spectator ion. Both the $Cl_2(g)$ and the $Br^-(aq)$ are changed in the reaction. Write the net ionic equation.

$$2Br^-(aq) + Cl_2 \ (g) \rightarrow 2Cl^-(aq) + Br_2 \ (l)$$

Example 9.4

Write the net ionic equation for the reaction:

$$Zn(s) + 2HCl \ (aq) \rightarrow ZnCl_2 \ (aq) + H_2(g)$$

1. Write the total ionic equation.

$$Zn(s) + H^+(aq) + Cl^-(aq) \rightarrow Zn^{2+}(aq) + Cl^-(aq) + H_2(g)$$

2. Eliminate the spectator ion, $Cl^-(aq)$, and write the net ionic equation.

$$Zn(s) + 2H^+(aq) \rightarrow Zn^{2+}(aq) + H_2(g)$$

Example 9.5

Write the net ionic equation for: aqueous carbonate ion + aqueous hydrogen ion → carbon dioxide + water.
1. Write a skeleton ionic equation:

$$CO_3^{2-}(aq) + H^+(aq) \rightarrow CO_2(g) + H_2O(l)$$

2. Balance the hydrogens (conservation of mass):

$$CO_3^{2-}(aq) + 2H^+(aq) \rightarrow CO_2(g) + H_2O(l)$$

3. Check for conservation of charge.

$$CO_3^{2-}(aq) + 2H^+(aq) \rightarrow CO_2(g) + H_2O(l)$$

$$(2-) + (2+) = 0$$

The net ionic equation is complete.

Example 9.6

Will a reaction occur if aqueous solutions of lead(II) nitrate and potassium chromate are mixed? If so, write the net ionic equation.
1. Consider the fact that there will be Pb^{2+}, NO_3^-, K^+, and CrO_4^{2-} ions in solution. Table 9–2 reveals that CrO_4^{2-} and Pb^{2+} from insoluble $PbCrO_4(s)$. Therefore, the reaction should occur.
2. Write the balanced equation.

$$Pb(NO_3)_2(aq) + K_2CrO_4(aq) \rightarrow PbCrO_4(s) + 2KNO_3(aq)$$

3. Since only the Pb^{2+} and CrO_4^{2-} are involved in the reaction, the net ionic equation is:

$$Pb^{2+}(aq) + CrO_4^{2-}(aq) \rightarrow PbCrO_4(s)$$

Exercise 9.1

Write net ionic equations for the following reactions:
1. $NaOH(aq) + H_2SO_4(aq) \rightarrow Na_2SO_4(aq) + H_2O(l)$
2. $FeCl_3(aq) + Na_2S(aq) \rightarrow Fe_2S_3(aq) + NaCl(aq)$
3. $Na_3PO_4(aq) + Pb(NO_3)_2(aq) \rightarrow NaNO_3(aq) + Pb_3(PO_4)_2(s)$
4. $Cu(NO_3)_2(aq) + H_2S(g) \rightarrow CuS(s) + NHO_3(aq)$
5. $BaCl_2(aq) + Na_2SO_4(aq) \rightarrow BaSO_4(s) + NaCl(aq)$
6. Hydrochloric acid + sodium carbonate solution → sodium chloride solution + carbon dioxide gas + water
7. Barium chloride solution + sulfuric acid → solid barium sulfate + hydrochloric acid
8. Solid calcium + water → calcium hydroxide precipitate + hydrogen gas
9. Solutions of nickel(II) nitrate + potassium hydroxide → solid nickel (II) hydroxide + potassium nitrate solution
10. Solutions of zinc sulfate + ammonium sulfide → solid zinc sulfide + ammonium sulfate solution

9.6 CONCENTRATIONS OF SOLUTIONS

Any scientific discipline that is involved in obtaining and reporting experimental data in a form suitable for mathematical processing must be concerned with accuracy. It is often unsatisfactory to use relative terms such as "dilute," "concentrated," "medium," "small," "hot," and "large." These loosely qualitative terms might be adequate on some occasions, but for quantitative work such as in the case of stoichiometric calculations, the chemist must know precisely the concentration or dilution of a solution. One chemist's dilute acid (relative to metal being cleaned) might be deadly concentrated to a colleague working with living organisms. At the same time, the terms "dilute" and "concentrated" must not be confused with "weak" and "strong." In Chapter 10, the distinction between weak and strong acids and bases will be discussed in detail. A strong acid such as hydrochloric acid can be prepared as a very *dilute* solution by the addition of a great quantity of water. Conversely, the relatively weak acetic acid can be obtained as a highly *concentrated* solution if many moles of it are dissolved in a minimal volume of water. We will consider several common methods for expressing concentrations of solutions.

9.7 MOLAR CONCENTRATIONS

The most useful way of expressing exact concentrations of solutions is in terms of *moles of solute per liter of solution.* Such concentrations are called **molar** and are symbolized by the capital letter **M.** The utility of this concentration is that it expresses with considerable accuracy the actual number of particles (ions or molecules) per unit of solution volume. Expressing the definition of molar concentration in the form of an equation is a nice "shorthand" statement that we can remember.

$$\text{Molarity} = \text{moles of solute per liter of solution} = \frac{n}{V}$$

It must be emphasized that V represents the volume in *liters of solution, not liters of solvent.* One liter of 1 molar solution contains 1 mole of solute plus enough solvent to make a final volume of solution (solute +

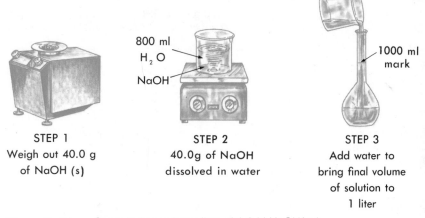

STEP 1	STEP 2	STEP 3
Weigh out 40.0 g of NaOH (s)	40.0g of NaOH dissolved in water	Add water to bring final volume of solution to 1 liter

Figure 9–11   Steps in preparing a liter of 1.0 M NaOH(aq).

solvent) of one liter. The exact final volume of solution is measured in a volumetric flask (Fig. 9–11).

For the actual preparation of solutions, calculations can be most simply done by using dimensional analysis.

Example 9.7

What mass of NaOH(s) is needed to prepare a liter of 1.00 M NaOH(aq)?
1. Calculate the molar mass of NaOH.

$$NaOH = 40.0 \text{ g/mole}$$

2. Using dimensional analysis, make the calculation:

$$mass = \frac{40.0 \text{ g}}{1 \text{ mole}} \times \frac{1 \text{ mole}}{1 \text{ } l} \times 1 \text{ } l$$

$$\boxed{mass = 40.0 \text{ g}}$$

3. To make the solution, place 40.0 g of NaOH(s) in a volumetric flask, add enough water to dissolve the NaOH(s), and then add enough water to bring the final volume of solution up to the 1000 mL mark on the flask, as shown in Figure 9–11.

Example 9.8

What mass of NaOH(s) is needed to prepare 250 mL of a 0.20 M solution of NaOH?
1. Calculate the molar mass of NaOH = 40.0 g/mole and then the mass of NaOH actually used.

$$mass = \frac{40.0 \text{ g}}{1 \text{ mole}} \times \frac{0.20 \text{ mole}}{1 \text{ } l} \times 0.250 \text{ } l$$

$$\boxed{mass = 2.0 \text{ g}}$$

2. Dissolve 2.0 g of NaOH(s) in about 200 mL of water in a volumetric flask and then add enough water to reach the final volume of 250 mL of solution.

Example 9.9

Calculate the number of moles of KCN in 400.0 mL of 0.50 M solution.

$$n = 400.0 \text{ mL} \times \frac{1 \text{ } l}{10^3 \text{ mL}} \times \frac{0.50 \text{ mole}}{1 \text{ } l}$$

number of moles

$$n = 200 \times 10^{-3} \text{ mole}$$

or. $\boxed{n = 0.20 \text{ mole}}$

SOLUTIONS

Example 9.10

How many molecules of glucose ($C_6H_{12}O_6$) are there per milliliter of a 2.5 M solution?

$$\text{Number of molecules} = 1.00 \text{ mL} \times \frac{2.5 \text{ mole}}{1000 \text{ mL}} \times \frac{6.02 \times 10^{23} \text{ molecules}}{1 \text{ mole}}$$

$$\text{Number of molecules} = \boxed{1.5 \times 10^{21}}$$

Example 9.11

A bottle of H_2SO_4 (aq) is labeled 92.0 per cent by mass (92.0 g H_2SO_4 per 100.0 g solution) with a specific gravity of 1.83. What is the molarity of the H_2SO_4 (aq)?

1. Convert sp. gr. to density and calculate the molar mass of H_2SO_4.

$$d = 1.83 \text{ g/mL} \quad H_2SO_4 = 98.1 \text{ g/mole}$$

2. Find mass of H_2SO_4 per mL solution.

$$92.0\% \text{ by mass means } \frac{92.0 \text{ g } H_2SO_4}{100 \text{ g solution}}$$

3. Find the molarity:

$$\frac{92.0 \text{ g } H_2SO_4}{100 \text{ g solution}} \times \frac{1.83 \text{ g solution}}{1 \text{ mL solution}} = \frac{1.68 \text{ g } H_2SO_4}{1 \text{ mL solution}}$$

$$\text{molarity} = \frac{1.68 \text{ g } H_2SO_4}{1 \text{ mL solution}} \times \frac{10^3 \text{ mL}}{L} \times \frac{1 \text{ mole } H_2SO_4}{98.1 \text{ g } H_2SO_4}$$

$$\text{molarity} = 17.1 \text{ mole/L} = \boxed{17.1 \text{ M}}$$

Exercise 9.2

1. How many grams of solute are needed to prepare each of the following solutions?
   (a) 35.0 mL of 2.0 M $AgNO_3$
   (b) 160.0 mL of 0.50 M $Na_2SO_4 \cdot 10 \, H_2O$

   Note: The mass of the hydrated water is included in calculating the molar mass because it makes up part of the actual mass of salt used.

2. What molar concentration results when 10.0 g of KOH(s) is dissolved in water to a final volume of 200.0 mL?
3. Find the molarity of HNO_3(aq), which is 50.0 per cent by mass and has a sp. gr. of 1.3.
4. If 30.0 g of acetone (C_3H_6O) is dissolved in water to a final volume of 0.50 liter, approximately how many molecules of acetone are there in 10.0 mL?

Dilution of Solutions

The molarity problems presented thus far have dealt mainly with solutes weighed in grams. However, it is often necessary or convenient to prepare large volumes of dilute solutions from small volumes of concentrated solutions (called "stock" solutions). For example, a large volume of 0.10 M HCl(aq) could be prepared easily by measuring out a few milliliters of 12.0 M (concentrated) HCl and adding it to enough water so that the desired molarity and volume are obtained.

We must be careful to distinguish between the *concentration* (molarity) and the *amount* (moles) of a solute in a solution. The concentration of salt in a drop of sodium chloride solution is the same as its concentration in a liter of the same solution. Obviously, the *amount* of NaCl in the liter of solution is much greater than it is in the drop. If the drop were diluted to a liter of solution, the *moles* of solute would remain unchanged even though the *concentration* is greatly reduced (Fig. 9–12).

If we begin with the understanding that in a dilution process,

number of moles before dilution
= number of moles after dilution,

and if we bear in mind that

number of moles = molarity $\times$ volume

$$n = MV$$

we can substitute "equals for equals" and write:

$$M_1 V_1 = M_2 V_2$$

where

$$\frac{mole}{\cancel{L}} \times \cancel{L} = \frac{mole}{\cancel{L}} \times \cancel{L}$$

$M_1 V_1 =$ the number of moles before dilution
$M_2 V_2 =$ the number of moles after dilution

large volume
of solvent

8 particles

small volume
of solvent

8 particles

Concentrated

Dilute

Figure 9–12   Dilution does not alter the number of moles of solute.

SOLUTIONS

Example 9.12

What volume of 1.00 M stock solution of KCl is needed to prepare 500.0 mL of 0.200 M solution?
1. Moles of KCl before dilution = moles KCl after dilution:

$$M_1V_1 = M_2V_2$$

2. Organize the data:
   $M_1 = 1.00$ M
   $V_1 = ?$
   $M_2 = 0.200$ M
   $V_2 = 500.0$ mL
3. Notice that the 500.0 mL does not have to be changed to liters, because the answer for V_1 will also be in milliliters. This is possible because multiplication of both sides of the equation by 10^3 converts liters to milliliters. Substitute and solve for V_1.

$$(1.00 \text{ M})(V_1) = (0.200 \text{ M})(500.0 \text{ mL})$$

$$V_1 = \frac{(0.200 \text{ M})(500.0 \text{ mL})}{1.00 \text{ M}}$$

$$V_1 = 100 \text{ mL}$$

4. The final step is to pour 100 mL of the 1.00 M stock into a volumetric flask and add sufficient water to yield 500.0 mL of 0.200 M KCl(aq).

Example 9.13

What volume of 96.0 per cent commercial H_2SO_4(aq) is needed to prepare 500 mL of 0.20 M solution? The sp. gr. of 96.0 per cent acid is 1.84.
1. The first step is to calculate the molar concentration of the commercial acid. The answer is 18.0 M.
2. Organize the data and solve.
   $M_1 = 18.0$ M
   $V_1 = ?$
   $M_2 = 0.20$M
   $V_2 = 500$ mL
   $M_1V_1 = M_2V_2$
   $(18.0 \text{ M})(V_1) = (0.20 \text{ M}) (500 \text{ mL})$

   $$\boxed{V_1 = 5.6 \text{ mL}}$$

3. **Caution:** The 5.6 mL of concentrated H_2SO_4 should be added slowly and with stirring to about 100 mL of water and then diluted to 500 mL. The great heat of hydration of H_2SO_4 could cause dangerous splattering if the less dense water is poured into the acid. It's a good rule always to add acid to water, never the reverse.

Exercise 9.3

How should the following dilutions be carried out?
1. 250.0 mL of 0.340 M KNO_3 from 3.20 M stock solution.
2. 0.650 L of 0.0700 M NaOH from 5.20 M stock solution.
3. 3.00 L of 0.150 M HCl from 6.00 M stock solution.
4. 500.0 mL of 0.350 M KCl from 6.00 M stock solution.
5. 350.0 mL of 0.020 M KCN from 1.2 M stock solution.

The **normal** concentration of a solution, symbolized by the capital letter N, is another method for expressing the concentration of solutions. This method is based on the *equivalent* mass of the solute. The equivalent mass of a substance is that mass which can combine with or produce 8.00 g of oxygen, 1.01 g of hydrogen, or the equivalent.

Since we are usually applying the concept of normality to acids and bases, let us define an equivalent in terms of an acid or a base. An equivalent mass (simply called equivalent) of an acid is the mass of an acid that provides one mole of hydrogen ions. Thus, H_3PO_4 contains three equivalents per mole, H_2SO_4 contains two equivalents per mole, and HCl contains one equivalent per mole.

Examples:

$$HCl(aq) + NaOH(aq) \rightarrow NaCl(aq) + HOH(l)$$

$$H_2SO_4(aq) + 2\,NaOH(aq) \rightarrow Na_2SO_4(aq) + 2HOH(l)$$

$$H_3PO_4(aq) + 3\,NaOH(aq) \rightarrow Na_3PO_4(aq) + 3HOH(l)$$

An equivalent of a base is defined as the mass required to supply one mole of hydroxide ions in a reaction.

The equivalent mass of a salt is obtained by dividing the molar mass by the total positive (or negative) valence. Examples of equivalents of some acids, bases, and salts are tabulated below

Compound	Molar Mass (g/mole)	Equivalent Mass (g/equiv)	
H_2SO_4	98.1	49	$(98.1 \div 2)$
HNO_3	63	63	
H_3PO_4	98.0	32.7	$(98.0 \div 3)$
NaOH	40	40	
$Ca(OH)_2$	74	37	$(74 \div 2)$
KCl	74.5	74.5	
Na_2SO_4	142	71	$(142 \div 2)$
$AlCl_3$	133.5	44.5	$(133.5 \div 3)$
$Al_2(SO_4)_3$	342	57	$(342 \div 6)$

For the time being, it is sufficient to define the equivalent mass of a metal as the molar mass divided by its total positive valence in a compound.

$$\text{eq mass} = \frac{\text{molar mass}}{\text{total positive valence}}$$

Example 9.14

Calculate the equivalent masses of the following: (a) NaOH, (b) H_2SO_4, (c) $Zn(NO_3)_2$, (d) Al_2S_3.

(a) 1. Molar mass of NaOH = 40.0 g/mole
   2. total positive valence of sodium in NaOH is 1.

$$3. \text{ eq mass} = \frac{40.0 \text{ g}}{1 \text{ mole}} \times \frac{1 \text{ mole}}{1 \text{ eq}} = 40.0 \text{ g/eq}$$

(b) 1. Molar mass of H_2SO_4 = 98.1 g/mole
   2. total positive valence of hydrogen in H_2SO_4 is 2.

$$3. \text{ eq mass} = \frac{98.1 \text{ g}}{1 \text{ mole}} \times \frac{1 \text{ mole}}{2 \text{ eq}} = 49.0 \text{ g/eq}$$

(c) 1. Molar mass of $Zn(NO_3)_2$ = 189.4 g/mole
   2. total positive valence of zinc in $Zn(NO_3)_2$ is 2.

$$3. \text{ eq mass} = \frac{189.4 \text{ g}}{1 \text{ mole}} \times \frac{1 \text{ mole}}{2 \text{ eq}} = 94.5 \text{ g/eq}$$

(d) 1. Molar mass of Al_2S_3 = 150.2 g/mole
   2. total positive valence of aluminum is Al_2S_3 is 6.

$$3. \text{ eq mass} = \frac{150 \text{ g}}{1 \text{ mole}} \times \frac{1 \text{ mole}}{6 \text{ eq}} = 25.0 \text{ g/eq}$$

A normal solution is defined as that which contains one equivalent mass of solute per liter of solution. In other words, the normality is equal to the number of equivalent masses per liter of solution. In equation form:

$$N \text{ (normality)} = \frac{\text{\# equivalent of solute}}{\text{liter of solution}}$$

Example 9.15

How many grams of $Ca(OH)_2(s)$ are needed to prepare 150 mL of 0.25 N solution?
1. Calculate the equivalent mass of $Ca(OH)_2$ from the molar mass

$$\text{Molar mass of } Ca(OH)_2 = 74 \text{ g/mole}$$

Since the total positive valence of calcium in the compound is 2, we know that $Ca(OH)_2$ has 2 equivalents per mole.
2. Using dimensional analysis:

$$\text{mass} = \frac{74 \text{ g}}{1 \text{ mole}} \times \frac{1 \text{ mole}}{2 \text{ eq}} \times \frac{0.25 \text{ eq}}{1 \text{ L}} \times \frac{1 \text{ L}}{10^3 \text{ mL}} \times 150 \text{ mL}$$

$$\boxed{\text{mass} = 1.4 \text{ g}}$$

3. Place 1.4 g of $Ca(OH)_2(s)$ in 150 mL of solution.

While Example 9.15 illustrates a typical method for preparing a solution from a solid solute in which the concentration is expressed in normality units, most normal solutions are prepared by dilution of aqueous stock solutions. Dilution problems involving normal concentrations are the same as with molar concentrations.

equivalents before dilution = # equivalents after dilution

or

$$N_1 V_1 = N_2 V_2$$

$$\frac{eq}{\cancel{L}} \times \cancel{L} = \frac{eq}{\cancel{L}} \times \cancel{L}$$

Remember that the number of equivalents = NV, so it is simply a matter of substituting "equals for equals." The advantage of the normal system is that one mL of a 1 N solution of anything will react exactly with 1 mL of a 1 N solution of anything else and that solutions of equal normality will react volume for volume.

9.9 THE COLLIGATIVE PROPERTIES OF SOLUTIONS

The term **colligative** refers to properties depending only on the number of particles present without regard to their size or mass. As applied to solutions, colligative properties are affected by changes in the number of solute particles per given quantity of solvent but not on the specific nature of the solute. The introduction of solute particles into a solvent involves a variety of interactions among the solute particles themselves (ions, molecules, and conglomerates) and between the solute particles and the solvent molecules. Practically, the colligative effect is the sum of the ways in which solute particles affect the vapor pressure of the solvent. Particles of solute bind water molecules to some extent, so that fewer water molecules will shift from the liquid to the vapor phase at a given temperature. Furthermore, the even distribution of solute particles means that they are at the surface of the solvent too, so that the fraction of volatile solvent molecules at the liquid-air interface (where evaporation takes place) is reduced. Consequently, the vapor pressure of the solvent at a given temperature is reduced also. As more solute is added, the more the vapor pressure is lowered. This effective lowering of vapor pressure is illustrated in Figure 9–13, where we can see that fewer molecules of water vapor exert less pressure. Remember, from our study of the gas laws, that pressure is directly proportional to the number of gas molecules. Pressure is a colligate property of gases.

water
molecule

particle –
water molecule
attraction

Pure Water
at 20° C

Solution
at 20° C

Figure 9–13 The colligative effect of a solute-solvent system reduces the number of solvent molecules going into the vapor phase at a given temperature.

The lowering of the vapor pressure of a solvent has the effect of shifting its normal boiling and freezing points. However, if the shift of boiling point (ΔT_b) and the shift of freezing point (ΔT_f) are to be mathematically predictable, the solute must be a **nonvolatile** *nonelectrolyte*. A nonvolatile substance is one that has a relatively high boiling point and therefore will not move easily into the vapor phase.

Our reason for making a distinction between electrolytes and nonelectrolytes is that the dissolution of nonelectrolytes into nonionizing molecules results in a predictable number of particles that exhibit very weak intermolecular attractions, while ions tend to clump together because of a much stronger interionic attraction. Dissolved nonelectrolytes produce what may be described as an **ideal solution** because of their negligible molecular interactions. This is somewhat analogous to the concept of the ideal gas. When one mole of nonelectrolyte is dissolved, the dissociated molecules approach Avogadro's number of individual particles. However, this model is limited to relatively dilute solutions, since the molecular interactions do become significant at high concentrations.

The abnormal colligative properties of electrolytes were explained by the theory of Peter Debye and Erich Hückel in 1923. The **Debye-Hückel theory** says, in effect, that the interionic attractions among dissociated electrolytes create clumps of ions, so that one mole of an ionic compound will not produce as many moles of individual, free-moving particles in solution as expected. This does not mean that NaCl, for example, will not produce 2 moles of ions:

$$NaCl \text{ (s)} \xrightarrow{\text{dissociates}} Na^+ \ + \ Cl^-$$

1 mole   1 mole   1 mole

2 moles of ions

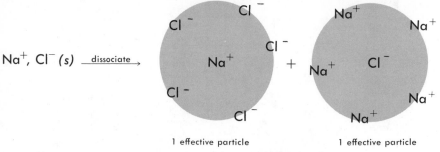

$$Na^+, Cl^- \ (s) \xrightarrow{\text{dissociate}}$$

1 effective particle   1 effective particle

Figure 9–14   Interionic attractions reduce the number of effective particles.

But, the Na^+ and Cl^- ions will interact so that the *effective* number of moles exhibits a colligative effect that is characteristic of fewer particles. In Figure 9–14, the clumping of ions is illustrated and shows the interionic effect on the number of solute particles.

It is re-emphasized that the colligative effect in dilute solutions is related to the number of individual particles, and not to the size, mass, or shape of the particles. One large particle composed of many clumped ions will have the same effect on the vapor pressure as one separate molecule.

In this text, the discussion of vapor pressure lowering and the effect of vapor pressure lowering on the ΔT_b and ΔT_f of solvents will be restricted to dilute solutions of nonelectrolytes.

9.10   THE MOLE FRACTION AND RAOULT'S LAW

In 1886, François Raoult observed that the vapor pressure lowering of a solvent was directly proportional to the **mole fraction** of solute. In an ideal solution, an increase in the number of grams of nonvolatile, nonelectrolyte solutes per gram of solution will lower the pressure according to the equation:

$$P = P^0 X$$

where

P = vapor pressure lowering of solution

P^0 = vapor pressure of pure solvent

X = mole fraction of solute

In case there should be more than one solute, the total effect on the vapor pressure of the solution is the sum of the separate effects due to the mole fraction of each component. The equation $P = P^0X$, which expresses Raoult's law, may be modified as follows:

$$P_{total} = (P^0X)_A + (P^0X)_B \cdots$$

where

P_{total} = total solution vapor pressure lowering

P^0 = vapor pressure of pure solvent

X_A = mole fraction of solute A

X_B = mole fraction of solute B

Example 9.16

Calculate the vapor pressure of a solution of glucose ($C_6H_{12}O_6$) in which 30.00 g of glucose is dissolved in 200.0 g of water at 23.0°C. (Vapor pressure of water at 23.0°C = 21.0 torr.)

1. Calculate the moles of glucose and water:

$$n_{C_6H_{12}O_6} = 30.00 \text{ g } C_6H_{12}O_6 \times \frac{1 \text{ mole } C_6H_{12}O_6}{180 \text{ g } C_6H_{12}O_6} = 0.167 \text{ mole glucose}$$

$$n_{H_2O} = 200.0 \text{ g } H_2O \times \frac{1 \text{ mole } H_2O}{18.0 \text{ g } H_2O} = 11.1 \text{ moles } H_2O$$

2. The mole fraction, X, of glucose is:

$$X = \frac{\text{moles of solute}}{\text{moles of solute} + \text{moles of solvent}}$$

$$X = \frac{0.167 \text{ mole}}{0.167 \text{ mole} + 11.1 \text{ moles}} = \frac{0.167 \text{ mole solute}}{11.3 \text{ moles sol'n}} = 0.0148$$

3. The vapor pressure lowering is calculated from the equation:

$$P = P^0X$$
$$P = (21.0 \text{ torr})(0.0148)$$
$$P = 0.311 \text{ torr}$$

4. The vapor pressure of the solution at 23.0°C is:

$$P = 21.0 \text{ torr - 0.31 torr}$$
$$P = 20.7 \text{ torr}$$

The ΔT_b and ΔT_f of solvent are a direct result of the lowering of vapor pressure. Figure 9–15 graphically represents the relationship between the vapor pressure of water and the shifting of the normal boiling and freezing points of water when it functions as a solvent.

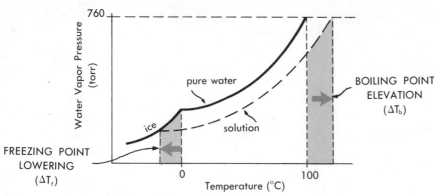

Figure 9–15 The effect of vapor pressure lowering on the freezing and boiling points of water.

When vapor pressure lowering data are collected using other solvents, the colligative effect is observed to be similar. However, the extent to which the freezing point is lowered and the boiling point is raised may vary considerably. The quantitative investigation of the colligative effect requires a standardization of the way in which concentrations are expressed. Since the relationship between the lowering of vapor pressure and the mole fraction of solute has been demonstrated, the best way to express concentrations of solution for uniform comparisons is to adopt a system in which all solutes of the same concentration can have the same mole fraction. The molar solution concentration cannot serve this purpose. While molar solutions of sugar and urea have equal numbers of moles, they will not necessarily be dissolved in equal numbers of moles of water when their volumes are equal. The objective of comparing equal fractions is achieved by the use of molar solutions.

9.11 MOLALITY

A **molal** *solution, symbolized by* **m,** *is one in which there is a concentration of one mole of solute per kilogram of solvent.*

$$\text{molality} = \frac{\text{moles solute}}{\text{kg solvent}}$$

Example 9.17

What is the molality of a solution in which 6.20 g of NaCl (s) is dissolved in 86.0 g of water?
1. Calculate the moles of NaCl:

$$NaCl = 58.5 \text{g/mole}$$

$$n = 6.20 \text{ g} \times \frac{1 \text{ mole}}{58.5 \text{ g}} \times 0.106 \text{ mole}$$

SOLUTIONS

2. Converting 86.00 g to 0.086 kg, solve for molality:

$$m = \frac{0.106 \text{ mole solute}}{0.0860 \text{ kg solvent}} = \boxed{1.23 \text{ molal}}$$

Example 9.18

What mass of urea, $CO(NH_2)_2$, is needed to prepare a 0.100 m solution in 500.0 g of benzene, C_6H_6?

1. Organize the data, bearing in mind that we can obtain the molar mass of the solute from the formula

$$\text{molar mass of solute} = CO(NH_2)_2 = 60.0 \text{ g/mole}$$

$$\text{kg mass of solvent} = 500.0 \text{ g} = 0.5000 \text{ kg}$$

$$\text{molality} = 0.100 \text{ m} = \frac{0.100 \text{ mole solute}}{\text{kg solvent}}$$

2. Using dimensional analysis to obtain the necessary mass of urea:

$$\text{mass} = \frac{60.0 \text{ g solute}}{\text{mole solute}} \times \frac{0.100 \text{ mole solute}}{\text{kg solvent}} \times 0.500 \text{ kg solvent}$$

$$\boxed{\text{mass} = 3.00 \text{ g urea}}$$

Exercise 9.4

1. To what extent will the water vapor pressure be lowered at 29.0°C if 10.00 g of urea, $CO(NH_2)_2$, is dissolved in 250.0 g of water?
2. What is the molality of a solution in which 20.00 g of CCl_4 is dissolved in 400.0 mL of benzene (density = 0.880 g/mL)?
3. How many grams of CS_2 (l) are required to prepare a 0.25 molar solution in 0.30 kg of chloroform, $CHCl_3$?

9.12 RELATING ΔT_f, ΔT_b, AND MOLAR MASS

Figure 9–15 graphically represents the colligative effect of solute particles on the vapor pressure of the solvent. The vapor pressure curve of the solution shows that the freezing point of the solvent is lowered and the boiling point is elevated. On the basis of these observations, chemists find it useful to standardize the mole fraction to the extent that the ΔT_f and the ΔT_b of a variety of solvents are measured for 1.0 molal concentrations. For any solvent, the ΔT_f in a 1.0 m solution is called the **molal freezing point depression constant**, symbolized K_f, and the 1.0 m ΔT_b is called the **molal boiling point elevation constant, K_b**. If the molality of the nonelectrolyte solution is more than 1.0 m, the ΔT_f will be greater than the K_f for a 1.0 m solution and the ΔT_b will be higher than the K_b for the pure solvent. A concentration less than 1.0 m will have the same effect but to a lesser degree (i.e., the

$1.0 m$
$= \dfrac{1 \text{ mol solute}}{Kg \text{ solvent}}$

ΔT_f and the ΔT_b will be less than the K_f and K_b values). The equations relating these terms are:

$$T_f = mK_f$$

and

$$T_b = mK_b$$

where

ΔT_f and ΔT_b = number of degrees deviation from boiling and freezing points of pure solvent

m = molality

K_f and K_b = molal T_f and T_b constants

Analysis will show the dimensions of both K_f and K_b to be $\dfrac{\text{kg }^\circ\text{C}}{\text{mole}}$ (i.e., kilogram degrees per mole).

These equations may be used to calculate molar masses of unknown solutes after measuring the ΔT_f or ΔT_b values. An example of a very practical application is the use of water-soluble nonelectrolytes for antifreeze solutions, in which low molecular mass, nonvolatile substances (like Prestone) are used to lower the freezing point of automobile radiator water to a safe level. Table 9–3 lists freezing and boiling point data that may be helpful in solving related problems.

TABLE 9–3 **Boiling and Freezing Point Data for Selected Solvents**

Solvent	Formula	$T_f(^\circ C)$	K_f (kg $^\circ C$/mole)	T_b ($^\circ C$)	K_b (kg $^\circ C$/mole)
benzene	C_6H_6	5.48	5.12	80.15	2.53
carbon disulfide	CS_2	—	—	46.25	2.34
carbon tetrachloride	CCl_4	−22.8	29.8	76.8	5.02
acetic acid	CH_3COOH	16.6	3.90	118.1	3.07
chloroform	$CHCl_3$	−63.5	4.68	61.2	3.63
water	H_2O	0	1.86	100	0.52
naphthalene	$C_{10}H_8$	80.2	6.80	—	—
camphor	$C_{10}H_{16}O$	178.4	37.7	208.3	5.95
cyclohexane	C_6H_{12}	6.5	20.0	80.9	2.79

Example 9.19

Find the boiling point of a solution of 2.40 g urea in 50.0 g water at standard pressure. (Refer to Table 9–3.)

1. Organize the data

   mass solute = 2.40 g

   kg mass solvent = 0.0500 kg

   K_b for water = 0.52 $\dfrac{\text{kg }^\circ\text{C}}{\text{mole}}$

   molar mass solute (urea) = 60.0 g/mole

   ΔT_b = ?

2. Using dimensional analysis to find ΔT_b (i.e., boiling point elevation in degrees Celsius):

$$\Delta T_b = \frac{0.52 \text{ kg }^\circ\text{C}}{\text{mole solute}} \times \frac{2.40 \text{ g solute}}{0.0500 \text{ g solvent}} \times \frac{\text{mole solute}}{60.0 \text{ g solute}} = 0.42^\circ\text{C}$$

3. The boiling point of the solution is 0.42 degree higher than the solvent alone.

$$100.00^\circ\text{C} + 0.42^\circ\text{C} = \boxed{100.42^\circ\text{C}}$$

Example 9.20

What is the molar mass of a nonvolatile nonelectrolyte if 0.4900 g dissolved in 10.00 g of benzene, C_6H_6, results in a solution that has a freezing point 2.93 degrees lower than pure benzene? (Refer to Table 9–3.)

1. Organize the data:
   mass of solute = 0.4900 g
   mass of benzene = 0.0100 kg
   ΔT_f benzene = 5.48°C
   ΔT_f = 5.48°C – 2.93°C = 2.55°C
   molar mass of nonelectrolyte = ?
2. Use dimensional analysis to find the molar mass:

$$\text{molar mass} = \frac{0.4900 \text{ g solute}}{0.0100 \text{ kg solvent}} \times \frac{5.12 \text{ kg }^\circ\text{C}}{\text{mole}} \times \frac{1}{2.55^\circ\text{C}}$$

$$\text{molar mass} = \quad 98.4 \text{ g/mole}$$

QUESTIONS & PROBLEMS

9.1   What is the difference between a true solution and a colloidal suspension?

9.2   Define the following terms:
(a) solute
(b) solvent
(c) miscible
(d) electrolyte
(e) saturated
(f) ionization
(g) solubility
(h) dissociation

9.3   Write equations that predict the species in solution when the following compounds are dissolved in water. Estimate the number of moles of particles in solution. Assume that there is no inter-ionic attraction.

Example:

$$\text{Ba(NO}_3)_2 \text{ (s)} \xrightarrow{\text{H}_2\text{O}} \underbrace{\text{Ba}^{2+}\text{(aq)} + 2 \text{ NO}_3{}^-\text{(aq)}}_{\text{3 moles of particles}}$$

(a) $SrCl_2(s)$
(b) $HCN(g)$
(c) $C_{12}H_{22}O_{11}$
(d) $CO(NH_2)_2(s)$
(e) $Fe_2(SO_4)_3(s)$

9.4   Write net ionic equations for the following reactions:
(a) $(NH_4)_2S(aq) + Pb(NO_3)_2(aq) \rightarrow$
    $NH_4NO_3(aq) + PbS(s)$
(b) $AlCl_3(aq) + LiOH(aq) \rightarrow$
    $Al(OH)_3(s) + LiCl(aq)$
(c) $H_2CrO_4(aq) + Al(s) \rightarrow$
    $Al_2(CrO_4)_3(s) + H_2(g)$
(d) $Hg_2(NO_3)_2(aq) + CaCl_2(aq) \rightarrow$
    $Hg_2Cl_2(s) + Ca(NO_3)_2(aq)$
(e) $NiSO_4(aq) + Ba(OAc)_2(aq) \rightarrow$
    $Ni(OAc)_2(aq) + BaSO_4(s)$

9.5   Balance the following net ionic equations:
(a) $C_2O_4{}^{2-}(aq) + Fe^{3+}(aq) \rightarrow Fe_2(C_2O_4)_3(s)$
(b) $Zn(s) + H^+(aq) \rightarrow Zn^{2+}(aq) + H_2(g)$
(c) $PO_4{}^{3-}(aq) + Ag^+(aq) \rightarrow Ag_3PO_4(s)$

(d) $Al(s) + Cu^{2+}(aq) \rightarrow Al^{3+}(aq) + Cu(s)$

(e) $S^{2-}(aq) + Co^{3+}(aq) \rightarrow Co_2S_3(s)$

9.6 What mass of $KBr(s)$ is needed to prepare 250.0 mL of a 0.050 M solution?

9.7 What mass of $KOH(s)$ is required to prepare 0.20 liter of 0.040 M solution?

9.8 What is the molarity of a solution when 3.0 g of $CaSO_4 \cdot 5 \, H_2O$ is dissolved to make 75.0 mL of solution?

9.9 What is the molarity of an $HCl(aq)$ solution that is 60 per cent by mass HCl and has a sp. gr. of 1.10?

9.10 How many milliliters of 6.0 M $HCl(aq)$ are needed to prepare 250.0 mL of 0.02 M solution?

9.11 What volume of 0.20 M $KCN(aq)$ is needed to prepare 300 mL of a 0.0025 M solution?

9.12 What volume (mL) of 80.0 per cent $HNO_3(aq)$, sp. gr. 1.24, is needed to prepare 0.5 liter of 0.1 M solution?

9.13 How many grams of $Ca(NO_3)_2(s)$ are needed to prepare 50.0 mL of 3.0 N solution?

9.14 What is the normal concentration of 210.0 mL of $Ba(NO_3)_2(aq)$ which contains 16.0 g of solute?

9.15 What volume of 3.0 N $H_2SO_4(aq)$ is needed to prepare 0.75 liter of 4.0×10^{-3} N solution?

9.16 Concentrated nitric acid has a density of 1.42 g/mL and is 72.0 per cent by mass. What is its molar concentration?

9.17 What is the molarity of a solution made by dissolving 20.0 g of NaOH in 250 mL of solution?

9.18 What is the molarity of a solution having 5.5 mg of $CaCl_2$ per mL of solution?

9.19 Calculate the grams of barium nitrate needed to make 500 mL of 0.100 M solution.

9.20 How many moles of $CuSO_4$ are there in 200 mL of 0.20 M solution?

9.21 Calculate the number of mL of 0.20 M $CuSO_4$ solution that will contain 0.56 g of solute.

9.22 Calculate the volume of 0.500 M cesium iodide stock solution that is needed to make 250 mL of 0.200 M solution.

9.23 A sodium hydroxide solution has a density of 1.43 g/mL and is 40.0 per cent NaOH by mass. How many mL of this stock solution are needed to make a liter of 1.00 M solution?

9.24 How many grams of solute are in one liter of 40.0 per cent H_2SO_4 solution?

9.25 How many liters of 12.0 M hydrobromic acid should be added to 3.00 L of 2.00 M hydrobromic acid to make 8.00 L of 8.00 M hydrobromic acid on dilution with water? Assume that volumes are additive.

9.26 What is the final concentration of KCl in a solution made by adding 250 mL of 0.50 M KCl to 750 mL of 1.50 M KCl?

9.27 Concentrated sulfuric acid has a density of 1.86 g/mL and is 98.0 per cent H_2SO_4. Calculate:

(a) The molarity of concentrated $H_2SO_4(aq)$

(b) The volume of concentrated $H_2SO_4(aq)$ required to prepare 250 mL of 3.00 M solution

(c) The molarity of a solution made by diluting 30.0 mL of concentrated acid to a volume of 500 mL

9.28 A solution of acid is labeled 3.00 M. How many grams of nitric acid are contained in 250 mL of solution?

9.29 Calculate the molarity and normality of the following solutions:

(a) A zinc nitrate solution containing 50.0 mg Zn^{2+} per mL of solution

(b) A magnesium sulfate solution containing 0.67 mg Mg^{2+} per mL of solution

(c) An aluminium chloride solution containing 0.313 g of $AlCl_3$ per L of solution

9.30   What is the molarity of Na^+ ions and SO_4^{2-} ions in a 0.44 M solution of Na_2SO_4? Assume that the ions are completely dissociated in solution.

9.31   What is the mole fraction of ethanol, CH_3CH_2OH, when 20.00 g is dissolved in 100.00 g of water?

9.32   Calculate the vapor pressure of the solution in Problem 9.31 at 23°C.

9.33   How many grams of glucose, $C_6H_{12}O_6$, are required to prepare a 0.200 molal solution in 80.00 g of water?

9.34   What is the boiling point elevation of an acetic acid solution if 6.20 g of urea, $CO(NH_2)_2$, is dissolved in 31.0 g of acetic acid?

9.35   When 5.30 g of a nonelectrolyte is dissolved in 200.00 g of benzene, the freezing point of the solution is 4.63°C. If the percentage composition of the nonelectrolyte is 9.4 per cent hydrogen and 90.6 per cent carbon, find the true molecular formula.

9.36   Approximately what mass of ethanol, CH_3CH_2OH, is needed to lower the freezing point of 200 g of CCl_4 by 25°C? From Table 9–3, the K_f of CCl_4 is 29.8 $\dfrac{kg \; °C}{mole}$.

9.37   What is the molar mass of an unknown electrolyte if 68.40 g dissolved in 600.00 g of benzene lowers the freezing point from +5.53°C to −2.15°C?

9.38   How many grams of naphthalene ($C_{10}H_8$) must be dissolved in 5.00 g of CS_2 (l) to raise the boiling point of the solution to 48.00°C?

9.39   If 1 mole of $CaCl_2$ were assumed to have no interionic attractions upon dissociation in 1.0 kg of water, what ΔT_f would be expected?

9.40   A carbohydrate was assayed to be 40.0 per cent carbon, 6.7 per cent hydrogen, and 53.3 per cent oxygen. If dissolving 2.50 g of the carbohydrate in 50.00 g of chloroform raises the boiling point of the solution to 62.40°C, find the true molecular formula of the carbohydrate.

9.41   When 0.65 g of phenylenediamine is dissolved in 20.00 g of water, the ΔT_f of the solution is −0.56°C. If the percentage composition of phenylenediamine is 66.64 per cent carbon, 7.46 per cent hydrogen, and 25.90 per cent nitrogen, what is the true molecular formula?

9.42   A solution contains 15.0 g of silver nitrate per 250 g of water. What is its molality?

9.43   When 487 g of solution was evaporated to dryness, it was found that 65 g of KCl(s) residue remained. Calculate the molality of the original solution.

9.44   It is desired to make 350.00 g of a 0.500 m solution of alcohol (C_2H_5OH). How many grams of water and alcohol are required to make the solution?

9.45   What mass of C_2H_5OH must be added to a kilogram of water to yield a solution with a freezing point of −10° F?

9.46   Which solution contains a greater mass of solute, 150 g of 0.450 m NaCl or 300 g of 0.200 m KCl?

9.47   How many grams of $CaCl_2$ should be added to 350 g of water to make a 1.20 m solution?

9.48   Calculate the molar mass of a nonvolatile nonelectrolyte if a solution of 1.110 g dissolved in 9.858 g of water freezes at −2.64°C.

9.49   What is the boiling point of a solution containing 208 g of $C_{12}H_{22}O_{11}$ in 525 g of water?

9.50   How many grams of glycerol ($C_3H_8O_3$) must be added to 1.510 kg of water to produce a solution that will not freeze above −10.0°C?

9.51   Calculate the molar mass of a nonvolatile nonelectrolyte if 6.00 g in 75.0 g of water gives a solution that boils at 100.41°C.

CHAPTER TEN

At the completion of this chapter, you should be able to:

1. List four prominent characteristics of acids and bases.
2. Summarize the Arrhenius concept of acids and bases.
3. Define acids and bases in terms of the Brønsted-Lowry theory.
4. Distinguish between strong and weak acids.
5. Write conjugate acid-base pairs of common acids and bases.
6. Use a table of relative acid-base strengths to predict products of acid-base reactions.
7. Write and balance ionic equations for acid-base reactions.
8. Define amphiprotism (amphoterism) and identity amphiprotic substances.
9. Explain and give examples of neutralization and hydrolysis reactions.
10. Explain and illustrate the meaning of polyprotic acids.
11. Predict the acidity or alkalinity of the solution resulting from a hydrolysis reaction.
12. Write complete and balanced ionic equations illustrating neutralization and hydrolysis reactions.
13. State the Lewis concept of acids and bases.
14. Identify Lewis acids and bases.
15. Write complete and balanced equations for Lewis acid-base reactions.
16. Calculate hydrogen ion and hydroxide ion concentrations in solution.
17. Calculate pH values of solutions.
18. Explain the significance of pH in terms of hydrogen ion concentration and of relative acidity or alkalinity.
19. Determine hydrogen ion or hydroxide ion concentrations from pH values.
20. Solve stoichiometric problems involving solutions of acids or bases.
21. State the theory and application of acid-base titrations.
22. Calculate normal concentrations of acids or bases from titration data.

ACIDS AND BASES

In previous chapters we have referred to acids and bases by describing an acid as a compound that produces hydrogen ions in water and a base as one producing hydroxide ions in water. The relative strength of a particular acid is related to the vigor and extent to which the ionization of molecules produce hydrogen ions. Acids were further characterized as compounds in which the symbol for hydrogen usually appears first in the formula.

Simple common bases were described as compounds that furnish hydroxide ions in aqueous solution. The strong bases were said to be those ionic compounds that dissociate extensively into metal ions and hydroxide ions in water solution. Weak bases dissociate only slightly.

Acids and bases of the types discussed above are often characterized by a set of working definitions.

Acids

1. Acids have a sour taste (careful! indiscriminate tasting of chemicals is not recommended). The word *acid* means "sour" in Latin.
2. Acids turn blue litmus paper red.
3. Many acids will react with metals above hydrogen in the activity series to produce hydrogen gas. For example,

$$Zn(s) + 2\,H^+(aq) \rightarrow Zn^{2+}(aq) + H_2(g)$$

4. Basic solutions will be *neutralized* by acids. Neutralization is defined in the Arrhenius theory as the reaction between $H^+(aq)$ and OH^- (aq) to form water. The net ionic equation for the reaction between a strong acid and a strong base is

$$H^+(aq) + OH^-\,(aq) \rightarrow H_2O\,(\ell)$$

Bases

1. Bases have a bitter taste (again; tasting is not advised). Unsweetened cooking chocolate is an example of a bitter taste.
2. Bases turn red litmus paper blue. The dye, phenolphthalein, turns from colorless to red in the presence of sufficient $OH^-(aq)$ concentration.

3. Bases react with heavy-metal ions to form insoluble hydroxides or oxides. For example,

$$Fe^{3+}(aq) + 3\,OH^-(aq) \rightarrow Fe(OH)_3(s)$$

4. Basic solutions feel slippery to the touch.
5. Acid solutions are neutralized by bases.

An understanding of operational definitions and a familiarity with the classical properties of simple common acids and bases are very useful. These two major classes of solutions, acids and bases, have been known and used for centuries. But the inadequacy of the classical definitions became apparent when it was discovered that some metal ions in aqueous solution test as acids and that a number of anions, such as $CN^-(aq)$, $CO_3^{2-}(aq)$, $PO_4^{3-}(aq)$, and $S^{2-}(aq)$, behave like moderately strong bases. Obviously, the classical theory of acids and bases proposed by Svante Arrhenius in the latter part of the nineteenth century had to be revised.

10.2 THE ARRHENIUS CONCEPT OF ACIDS AND BASES

The Arrhenius concept considers an acid as a substance that produces hydrogen ions in water solution and a base as one that produces hydroxide ions in water. In Arrhenius terms, an acid reacts with a base to form a salt and water:

$$\text{acid} \quad + \quad \text{base} \quad \rightarrow \quad \text{salt} \quad + \quad \text{water}$$
$$HCl(aq) + KOH(aq) \rightarrow KCl(aq) + H_2O(\ell)$$

The reaction of the acid with the base is said to be one of neutralization; the reverse reaction is called hydrolysis. The terms neutralization, hydrolysis, and salt are applicable to the Arrhenius theory only.

The Arrhenius theory is still used because of its simplicity. Its weakness is that it is limited to hydroxide-containing compounds for bases and that it applies only to water (aqueous) solutions. It fails to account for the observed acidity of dissolved $CO_2(g)$ in water or the basic reaction of $NH_3(g)$ in water.

A major advance occurred in 1923 when Johannes Brønsted in Denmark and Thomas Lowry in England independently proposed a broader theory of acids and bases that was not restricted to water solutions.

10.3 THE BRØNSTED-LOWRY THEORY

The Brønsted-Lowry theory is the most generally used theory of acids and bases. Unlike the Arrhenius theory, almost any solvent

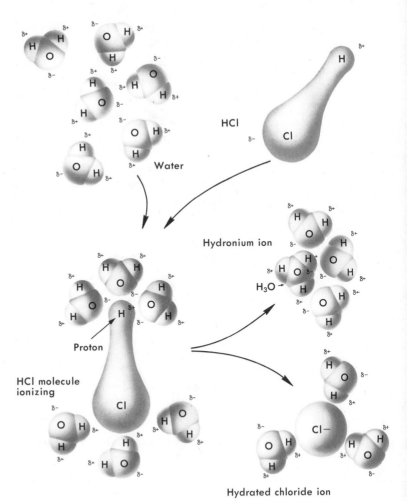

Figure 10–1 Proton transfer between water and HCl(g).

medium is permissible. The major liability of the theory is that it restricts acids to proton-containing species (i.e., species able to provide hydrogen ion, H^+, since the terms proton and hydrogen ion are synonymous). According to the Brønsted-Lowry theory, an acid is any species that can donate a proton, and a base is defined as a proton acceptor. An acid reacts with a base to form another acid and another base. For example, when HCl gas is placed in contact with water, hydrochloric acid is formed. The proton transfer is illustrated by the models in Figure 10–1).

The equation that summarizes the proton transfer illustrated in Figure 10–1 is:

$$\overset{\text{proton transfer}}{\underset{\text{acid} \qquad \text{base} \qquad \text{hydronium ion}}{HCl(g) + H_2O(\ell) \rightarrow H_3O^+(aq) + Cl^-(aq)}}$$

The hydronium ion is usually written as the empirical formula, H_3O^+, but it should be pointed out that experimental evidence indicates that the transferred proton bonds to more than one water molecule in the process of hydration. The actual formula is likely to be $(H_2O)_4H^+$ (or $H_9O_4^+$) or an even larger aggregate. Because of the variability in the hydronium ion structure, the empirical H_3O^+ is usually used.*

Notice in the proton transfer between HCl(g) and $H_2O(l)$ that the water is a proton acceptor. This means that in this particular type of acid-base interaction, water acts as a base. This is not to suggest that water is a strong base. In fact, the lack of vigor demonstrated by water in donating and accepting protons classifies it as both a weak acid and weak base.

Look at the same equation again.

$$HCl(g) + H_2O(l) \rightarrow H_3O^+(aq) + Cl^-(aq)$$

There is a question we may ponder. Can the $H_3O^+(aq)$ formed act as a proton donor to the $Cl^-(aq)$ to allow the reverse reaction to occur?

---

*Many chemists prefer to indicate the hydrated proton as $H^+(aq)$ rather than to use the empirical formula for the hydronium ion. Good arguments can be made for either formula, both of which are used in this text.

TABLE 10–1 **Relative Acid-Base Strength in 0.10 M Aqueous Solution**

	Acid	*Base*	
Strongest acid	$HClO_4$	ClO_4^-	Weakest base
	HI	I^-	
	HBr	Br^-	
	H_2SO_4	HSO_4^-	
	HNO_3	NO_3^-	
	HCl	Cl^-	
	H_3O^+	H_2O	
	H_2SO_3	HSO_3^-	
	HSO_4^-	SO_4^{2-}	
	H_3PO_4	$H_2PO_4^-$	
	HF	F^-	
	HNO_2	NO_2^-	
	HOAc	OAc^-	
	$Al(H_2O)_6^{3+}$	$Al(H_2O)_5(OH)^{2+}$	
	H_2CO_3	HCO_3^-	
	H_2S	HS^-	
	$H_2PO_4^-$	HPO_4^{2-}	
	HSO_3^-	SO_3^{2-}	
	NH_4^+	NH_3	
	HCN	CN^-	
	HCO_3^-	CO_3^{2-}	
	HPO_4^{2-}	PO_4^{3-}	
	HS^-	S^{2-}	
	H_2O	OH^-	
	CH_3OH	CH_3O^-	
	NH_3	NH_2^-	
	OH^-	O^{2-}	
Weakest acid	H_2	H^-	Strongest base

Left margin: Increasing strength of acid ↑

Right margin: Increasing strength of base ↓

ACIDS AND BASES

$$\text{HCl(g)} + \text{H}_2\text{O}(l) \leftarrow \overset{\frown}{\text{H}_3\text{O}^+(aq)} + \text{Cl}^-(aq)$$

Experimental evidence shows that the reverse reaction does occur to a slight extent. Chemists have collected and organized data of this type so that a listing of acids and bases according to strength is available for reference (Table 10–1). That is to say, there is a listing of proton donors and acceptors that indicates, by position on the list, the tendency to lose or gain protons. On this list of relative strengths of acids, HCl is observed to be a stronger acid than H_3O^+, and $\text{Cl}^-(aq)$ is a weaker base than water. This means that the proton transfer from HCl to H_2O occurs to a greater extent than does the reverse reaction (shown as a longer arrow toward the right, $\rightleftharpoons$). At equilibrium, the species in solution is almost 100 per cent H_3O^+ and Cl^-:

$$\text{HCl(g)} + \text{H}_2\text{O}\ (l) \rightleftharpoons \text{H}_3\text{O}^+(aq) + \text{Cl}^-(aq)$$

In summary, we can say that strong acids and bases ionize greatly, whereas weak acids and bases ionize slightly:

$$\text{HX} \xrightarrow{\ \sim\ 100\%\ } \text{H}^+ + \text{X}^-$$

general strong acid formula   in water

$$\text{HA} \underset{\rightleftharpoons}{\overset{\text{slight degree}}{=\!=\!=\!=}} \text{H}^+ + \text{A}^-$$

general weak acid formula   in water

$$\text{M OH} \underset{=\!=}{\overset{\sim\ 100\%}{\longrightarrow}} \text{M}^+ + \text{OH}^-$$

general strong base formula   in water

$$\text{M OH} \overset{\text{slight degree}}{\rightleftharpoons} \text{M}^+ + \text{OH}^-$$

general weak base formula   in water

Now let us investigate the meaning of the term *conjugate acid-base pair*. Notice that the transfer of a proton by the HCl changes the structure from a strong acid molecule to a weak base ion (a proton acceptor).

$$\overset{\frown}{\text{HCl(g)}} + \text{H}_2\text{O}(l) \rightleftharpoons \overset{\frown}{\text{H}_3\text{O}^+(aq)} + \text{Cl}^-(aq)$$

strong   weak
acid   base   acid   base

At the same time, the weak molecular base H_2O becomes a strong acid, H_3O^+. Generalization of this observation expresses the essence of the Brønsted-Lowry concept:

1. When an acid transfers a proton, it becomes a base.
2. When a base accepts a proton, it becomes an acid.

The two related species in each proton transfer operation are known as a conjugate acid-base pair. Furthermore:

3. A strong acid will have a weak conjugate base.
4. A weak acid will have a strong conjugate base. For example:

$$HCl \rightarrow Cl^-$$

strong acid   weak conjugate base

$$H_2O \rightarrow H_3O^+$$

weak base   strong conjugate acid

In Brønsted acid-base reactions the conjugate acid-base pairs are easy to recognize. *The only difference between an acid and its conjugate base is the transferred proton.* For example, in the reaction between HCl(aq) and $CO_3{}^{2-}$(aq), the reaction is:

$$HCl(aq) + CO_3{}^{2-}(aq) \rightleftharpoons Cl^-(aq) + HCO_3{}^-(aq)$$

acid$_1$ base$_2$ base$_1$ acid$_2$

The only difference between acid$_1$ and base$_1$ (a conjugate pair) is the proton in HCl, which is absent in Cl^-. The difference between the carbonate ion and the hydrogen carbonate ion is also a proton. Once again, we must pay special attention to the conservation of charge in ionic equations. In the equation:

$$HCl(aq) + CO_3{}^{2-}(aq) \rightleftharpoons Cl^-(aq) + HCO_3{}^-(aq)$$

the net charge on both sides of the equation is observed to be $2-$.

Table 10–1 is a list of Brønsted-Lowry acid-base conjugate pairs. The direction of the increasing tendency for an ion to accept or donate a proton is indicated. This list applies to dilute (0.10 M) aqueous solutions. In concentrated aqueous solution or in nonaqueous solvents, the order of acid strength is different.

Now let us examine the Brønsted acid-base reaction for a weak acid, acetic acid (HOAc), with water.

$$HOAc(\ell) + H_2O(\ell) \rightleftharpoons H_3O^+ + OAc^-(aq)$$

In the forward reaction, the acetic acid molecule donates a proton to the water molecule. In so doing, the HOAc becomes the conjugate base (OAc⁻), and the water molecule accepts a proton and becomes the conjugate acid. Thus the forward reaction shows the ionization of molecular HOAc(l) in molecular water. In the reverse reaction the acid H_3O^+ reacts with the base OAc⁻ to produce HOAc and water. Experiments show that the forward reaction proceeds only slightly and that at equilibrium the relative amount of H_3O^+ and OAc⁻ will be much less than that of HOAc and H_2O.

Example 10.1

Write the Brønsted-Lowry equation for the reaction between $HNO_3(aq)$ and $OH^-(aq)$. Identify the conjugate acid-base pairs.

1. Write the reactants:

$$HNO_3(aq) + OH^-(aq) \rightleftharpoons$$

2. Obviously, the strong acid HNO_3 will be the proton donor.

$$HNO_3(aq) + OH^-(aq) \rightleftharpoons NO_3^-(aq) + H_2O(l)$$
$$\text{acid}_1 \qquad \text{base}_2 \qquad \text{base}_1 \qquad \text{acid}_2$$

The distinction made above between pair 1 and pair 2 is completely arbitrary. The equation could just as well be written:

$$HNO_3(aq) + OH^-(aq) \rightleftharpoons NO_3^-(aq) + H_2O(l)$$
$$\text{acid}_2 \qquad \text{base}_1 \qquad \text{base}_2 \qquad \text{acid}_1$$

3. Notice that the only difference between HNO_3 and NO_3^- is a proton, and the same is true when comparing OH^- and H_2O.
4. The net charge of $1-$ on both sides of the equation satisfies the conservation law.
   (From this point on we will use double arrows of equal length ($\rightleftharpoons$) to represent reversibility.)

Example 10.2

Write the equation for the first proton transfer in the acid-base reaction between $H_3PO_4(aq)$ and NH_2^- (amide ion).

1. Find the reacting species in Table 10–1 in order to determine the proton donor. H_3PO_4 definitely appears to be the acid.
2. Write the equation.

$$H_3PO_4(aq) + NH_2^-(aq) \rightleftharpoons H_2PO_4^-(aq) + NH_3(aq)$$

3. Again, notice the conservation of charge.

Example 10.3

1. Complete and balance the following acid-base equations:
   (a) $HClO_4(aq) + NO_2^-(aq) \rightleftharpoons$
   (b) $HOAc(aq) + CN^-(aq) \rightleftharpoons$
   (c) $OH^-(aq) + HCO_3^-(aq) \rightleftharpoons$
   (d) $Al(H_2O)_6{}^{3+}(aq) + NH_2^-(aq) \rightleftharpoons$
   (e) $H_2SO_4(l) + HF(g) \rightleftharpoons$
2. List the stronger proton donor in each reaction:
   $HClO_4$, $HOAc$, HCO_3^-, $Al(H_2O)_6{}^{3+}$, H_2SO_4
3. Complete the equations, paying careful attention to the conservation of charge.

   (a) $\overset{\frown}{HClO_4}(aq) + \overset{\frown}{NO_2^-}(aq) \rightleftharpoons ClO_4^-(aq) + HNO_2(aq)$

   (b) $\overset{\frown}{HOAc}(aq) + \overset{\frown}{CN^-}(aq) \rightleftharpoons HCN(g) + OAc^-(aq)$

   (c) $\overset{\frown}{OH^-}(aq) + \overset{\frown}{HCO_3^-}(aq) \rightleftharpoons H_2O(l) + CO_3{}^{2-}(aq)$

   (d) $\overset{\frown}{Al(H_2O)_6{}^{3+}}(aq) + \overset{\frown}{NH_2^-}(aq) \rightleftharpoons Al(H_2O)_5(OH)^{2+}(aq) + NH_3(g)$

Note: In the hydrated aluminum ion, a proton is transferred from one of the hydrated water molecules to the amide ion (base). This leaves the aluminum ion with 5 water molecules and the remaining hydroxide ion from the water molecule that lost the proton. In addition, the proton transfer reduces the net charge from $3+$ to $2+$, since the proton is a unit of positive charge.

Exercise 10.1

Complete and balance the following Brønsted-Lowry equations. All solutions are 0.10 M. Label the conjugate acid base pairs.
1. $HI(aq) + NH_3(aq) \rightleftharpoons$
2. $HCN(aq) + NH_2^-(aq) \rightleftharpoons$
3. $O^{2-}(aq) + HF(aq) \rightleftharpoons$
4. $HS^-(aq) + H_3O^+(aq) \rightleftharpoons$
5. $S^{2-}(aq) + HSO_4^-(aq) \rightleftharpoons$
6. $HCO_3^-(aq) + H^-(aq) \rightleftharpoons$
7. $Fe(H_2O)_6{}^{3+}(aq) + OH^-(aq) \rightleftharpoons$
8. $CH_3O^-(aq) + Zn(H_2O)_4{}^{2+}(aq) \rightleftharpoons$
9. $HClO_4(aq) + HSO_4^-(aq) \rightleftharpoons$
10. $HNO_2(aq) + CO_3{}^{2-}(aq) \rightleftharpoons$

10.4 AMPHOTERISM

Amphoterism, or **amphiprotism** as a more descriptive label, refers to the ability of some substances, most notably water, to behave either as acids or as bases; that is, to be either proton donors or acceptors.

$H_2SO_4(aq)$ is a diprotic (two-proton) acid, and $H_3PO_4(aq)$ is a triprotic acid. Acids that can supply more than one proton per molecule are called **polyprotic acids.** In each case the first proton is released more readily than are subsequent protons. Diprotic acids undergo ionization in two separate steps, and triprotic acids have three steps in their ionization.

ACIDS AND BASES

When the diprotic acid H_2SO_4 ionizes in water it does so in two steps.

1. $H_2SO_4(aq) + H_2O(l) \rightleftharpoons HSO_4^-(aq) + H_3O^+(aq)$

(first ionization)

2. $HSO_4^-(aq) + H_2O(l) \rightleftharpoons SO_4^{2-}(aq) + H_3O^+(aq)$

(second ionization)

Analysis of the species in solution shows that the first step is dominant and that the HSO_4^- ion concentration is much greater than that of the SO_4^{2-} ion. However, in dilute solutions the proton-accepting properties of bases push the ionization process to completion so that the product is $2H_3O^+(aq) + SO_4^{2-}(aq)$. If salts such as $NaHSO_4$ or $KHSO_4$ are dissolved, there is an abundance of $HSO_4^-(aq)$ ions. Similarly structured ions such as $H_2PO_4^-$, HPO_4^{2-}, and HCO_3^- behave much alike in their capacity to regain the lost proton (acting as conjugate bases) or to donate still another proton. Clearly, these ions are amphiprotic, and their behavior as acids or bases depends on the nature of the species with which they react. Some examples of amphiprotic ions are:

$HSO_4^-(aq) + H_3O^+(aq) \rightleftharpoons H_2SO_4(aq) + H_2O(l)$
base

$HSO_4^-(aq) + CN^-(aq) \rightleftharpoons HCN(g) + SO_4^{2-}(aq)$
acid

$H_2PO_4^-(aq) + CO_3^{2-}(aq) \rightleftharpoons HPO_4^{2-}(aq) + HCO_3^-(aq)$
acid

$H_2PO_4^-(aq) + HNO_3(aq) \rightleftharpoons H_3PO_4(aq) + NO_3^-(aq)$
base

$HCO_3^-(aq) + HOAc(aq) \rightleftharpoons OAc^-(aq) + H_2CO_3(aq)*$
base

$HCO_3^-(aq) + OH^-(aq) \rightleftharpoons CO_3^{2-}(aq) + H_2O(l)$
acid

---

*There is no experimental evidence supporting the existence of an H_2CO_3 molecule or an H_2SO_3 molecule or NH_4OH. These species are often, and perhaps preferably, written as $[CO_2 (aq)]$, $[SO_2 (aq)]$ and $[NH_3 (aq)]$.

10.5 HYDROLYSIS

When a few drops of the indicator phenolphthalein are added to
an aqueous solution of sodium acetate, a pink color shows us that the
solution is basic. Further tests on other aqueous solutions show that
$NaCl(aq)$ is neutral, $NH_4Cl(aq)$ is acidic, and $NH_4CN(aq)$ is basic.
Since salts are made of ions, we suspect that reactions between water
and the ions must produce H_3O^+ or OH^- in sufficient amounts to
make the solutions acidic or basic. We say that when $NaOAc(s)$ is dissolved in water, the salt (actually the OAc^- ion; the Na^+ ion is a spectator) undergoes *hydrolysis. Hydrolysis is any proton transfer reaction in which
water is one of the reactants.* According to the Arrhenius concept, hydrolysis is the reverse of neutralization; that is, the combination of a salt
with water to form an acid and a base. From the Brønsted view it is
merely an acid-base reaction in which water acts as either an acid or a
base.

Now we can account for the acidic or basic or neutral solutions
formed when the above salts are dissolved in water.

1. NaCN + water: the CN^- ion is a strong base (a strong proton
   acceptor) so that sufficient OH^- remains in solution to make it
   basic (see Fig. 10–2 for mechanism):

$$
\begin{array}{cccc}
& & \text{conj.} & \text{conj.} \\
\text{base} & \text{acid} & \text{acid} & \text{base}
\end{array}
$$

$$CN^-(aq) + HOH(l) \rightleftharpoons HCN(aq) + OH^-(aq)$$

2. NH_4Cl + water: the Cl^- is a spectator ion in this case, but the
   NH_4^+ ion can donate a proton to the base HOH:

$$
\begin{array}{cccc}
& & \text{conj.} & \text{conj.} \\
\text{acid} & \text{base} & \text{base} & \text{acid}
\end{array}
$$

$$NH_4^+(aq) + HOH(l) \rightleftharpoons NH_3(aq) + H_3O^+(aq)$$

3. NH_4CN + water: both the anion and cation hydrolyze as follows:

| | | conj. | conj. |
| base | acid | acid | base |

$$CN^-(aq) + HOH(l) \rightleftharpoons HCN(aq) + OH^-(aq)$$

| | | conj. | conj. |
| acid | base | base | acid |

$$NH_4^+(aq) + HOH(l) \rightleftharpoons NH_3(aq) + H_3O^+$$

Both an acid (H_3O^+) and a base (OH^-) are produced, but the hy-drolysis of the CN^- ion proceeds to a greater extent than that of the NH_4^+ ion, so the overall reaction results in a basic solution.

4. NaCl + water: here neither the anion nor cation hydrolyzes, no reaction occurs, and the solution retains the neutrality of water.

An example of a hydrolysis reaction is:

$$H_2O(l) + CN^- (aq) \rightleftharpoons HCN (aq) + OH^- (aq)$$
$$\text{weak acid}$$

Note: The solution tests decidedly basic. Although there is a considerable buildup of $OH^-(aq)$, the $H^+ (aq)$ is bound to CN^- in the formation of the weak electrolyte, HCN. The mechanism is shown in Figure 10–2.

Other examples of hydrolysis are:

$$H_2O(l) + CO_3^{2-}(aq) \rightleftharpoons HCO_3^-(aq) + OH^-(aq)$$

$$H_2O(l) + CH_3O^- (aq) \rightleftharpoons CH_3OH (aq) + OH^- (aq)$$

$$H_2O (l) + S^{2-} (aq) \rightleftharpoons HS^-(aq) + OH^- (aq)$$

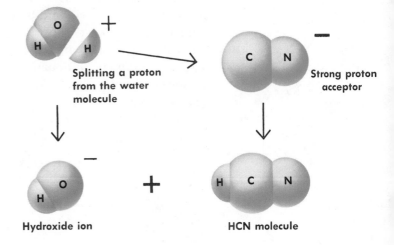

Splitting a proton from the water molecule

Strong proton acceptor

Figure 10–2 The hydrolysis of the cyanide ion.

Hydroxide ion

HCN molecule

10.6 THE LEWIS CONCEPT

In 1923, Gilbert N. Lewis proposed a still broader concept of acids and bases involving electron pairs. The Lewis concept is more inclusive than the Brønsted-Lowry theory because the definitions of acid and base are not restricted to the proton-transfer factor. There is no contradiction, however, because the essential mechanism in a proton transfer is the formation of a covalent bond. A Lewis acid reacts with a Lewis base to form an addition product (called an adduct) that is covalently bonded. When CN^- accepts a proton, it is in effect donating a pair of electrons that form a covalent bond between the hydrogen nucleus and the cyanide ion (Fig. 10–3). In terms of the Lewis concept, the proton, or hydrogen ion, is the acid, and the cyanide ion is the base. The *Lewis concept defines an acid as an electron-pair acceptor and a base as an electron-pair donor.* The covalent bond formation is the difference between the Lewis and Brønsted classifications. The Lewis concept is broader, because there are many ions, atoms, and molecules that can function as electron-pair acceptors in forming covalent bonds.

An example of a Lewis acid-base reaction that does not fit the Brønsted theory is the reaction between BF_3 and NH_3. While NH_3 is a Brønsted base consistent with Brønsted's proton concept, the BF_3 molecule has no protons to donate and hence cannot be a Brønsted acid. However, the BF_3 molecule can be an electron-pair acceptor, which classifies it as a Lewis acid:

$$
\begin{array}{c}
\text{F} \\
\cdot{}^{\times} \\
\text{F}{}^{\times}\!\text{B} \\
{}^{\times}\!\cdot \\
\text{F}
\end{array}
\quad + \quad
\begin{array}{c}
\text{H} \\
\cdot{}^{\times} \\
:\text{N}{}^{\times}\!\text{H} \\
{}^{\times}\!\cdot \\
\text{H}
\end{array}
\quad \rightarrow \quad
\begin{array}{cc}
\text{F} & \text{H} \\
\cdot{}^{\times} & \cdot{}^{\times} \\
\text{F}{}^{\times}\!\text{B}: & \text{N}{}^{\times}_{\times}\text{H} \\
{}^{\times}\!\cdot & {}^{\times}\!\cdot \\
\text{F} & \text{H}
\end{array}
$$

electron pair   electron pair   covalent bond
acceptor (acid)   donor (base)   (adduct)

Figure 10–3 The donation of an electron pair by the cyanide ion.

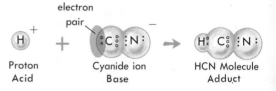

Proton Acid   Cyanide ion Base   HCN Molecule Adduct

Lewis acids include atoms, ions, and molecules that do not have complete octets of valence electrons. Other examples of Lewis acid-base reactions are:

$$\text{Ca}^{2+} + \overset{..}{\underset{..}{\text{O}}}{:}^{2-} \quad + \quad \overset{:\overset{..}{\text{O}}:}{\underset{:\overset{..}{\text{O}}:}{\text{S}\text{O}:}} \quad \longrightarrow \quad \text{Ca}^{2+},\ \text{SO}_4{}^{2-}$$

calcium oxide   + sulfur trioxide   → calcium sulfate
electron pair donor   electron pair acceptor
(base)   (acid)

$$\text{Cu}^{2+} \quad + \quad 4{:}\overset{\text{H}}{\underset{\text{H}}{\text{N}}}\text{H} \quad \longrightarrow \quad \text{Cu(NH}_3)_4{}^{2+}$$

copper(II) ion   + ammonia   → copper(II) tetrammine ion
electron pair acceptor   electron pair donor
(acid)   (base)

10.7 HYDROGEN ION CONCENTRATION

Pure water is the standard for acid-base neutrality. Water molecules do ionize, but only to a slight degree. In chemically pure water, the hydrogen ion and hydroxide ion concentrations are each found to be 1.0×10^{-7} moles per liter at 25°C. (The ionization of water varies slightly with the temperature.) The Brønsted equation,

$$\text{H}_2\text{O}(l) + \text{H}_2\text{O}(l) \rightleftharpoons \text{H}_3\text{O}^+ (\text{aq}) + \text{OH}^- (\text{aq})$$

may be simplified to:

$$\text{H}_2\text{O}(l) \rightleftharpoons \text{H}^+ (\text{aq}) + \text{OH}^- (\text{aq})$$

The product of the ion concentrations is:

$$\text{K}_\text{W} = [\text{H}^+][\text{OH}^-],$$

where the brackets, [], around the ion formulas indicate molar concentration.

Since the concentrations of H^+ and OH^- are 1.0×10^{-7} molar at 25°C,

$$K_w = [1.0 \times 10^{-7}][1.0 \times 10^{-7}]$$

$$K_w = 1.0 \times 10^{-14} \quad \text{for any aqueous solution at 25°C.}$$

When an acid is added to the water, the $[H^+]$ becomes larger as the $[OH^-]$ becomes smaller (owing to the reaction of OH^- with the excess H^+ to form water), but the ion product, K_w, remains constant.

Dilute solutions of strong acids are essentially 100 per cent ionized, so their H^+ concentration can be easily determined from their molar concentration. Thus, a 0.01 M HCl solution contains 0.01 mole per liter of H^+, and a 0.25 M HNO_3 solution is 0.25 M with respect to H^+ concentration. In an acid solution the $[H^+]$ exceeds the $[OH^-]$, and in basic solutions the reverse is true. From this relationship, we may calculate $[H^+]$ if we know the $[OH^-]$.

Example 10.4

Calculate the hydroxide-ion concentration in a 0.10 M solution of HCl at 25°C.
1. Since HCl is a strong electrolyte, 100 per cent ionization is assumed. Therefore, the $[H^+]$ is 0.10 M. (The $[H^+]$ of water is negligible.) In a liter of solution:

$$\underset{\text{0.10 mole}}{HCl(aq)} \xrightarrow[\text{ionization}]{\sim 100\%} \underset{\text{0.10 mole}}{H^+(aq)} + \underset{\text{0.10 mole}}{Cl^-(aq)}$$

2. $K_w = [H^+][OH^-] = 1.0 \times 10^{-14}$
$[0.10][OH^-] = 1.0 \times 10^{-14}$
$[OH^-] = 1.0 \times 10^{-13}$
The $[OH^-]$ (aq) is very small, as expected.

Example 10.5

What is the $[H^+]$ in a 0.0020 M solution of NaOH?
1. The $[OH^-]$ will be 0.0020 M, since the ionic NaOH dissociates completely into $Na^+(aq)$ and $OH^-(aq)$.
2. $K_w = [H^+][OH^-]$
$[H^+] (2.0 \times 10^{-3}) = 1.0 \times 10^{-14}$

$$[H^+] = \frac{1.0 \times 10^{-14}}{2.0 \times 10^{-3}}$$

$[H^+] = 5.0 \times 10^{-12}$
The $[H^+]$ (aq) is small, as expected.

ACIDS AND BASES

Because of the great importance attached to the control of acidity and alkalinity in chemistry and biology, a system proposed by Peter Sørensen in 1909 is used to express the degree of acidity. This system converts the values for $[H^+]$ from the cumbersome values just observed to small numbers between 0 and 14. This is accomplished by expressing the molar concentrations of hydrogen ion as *exponents of 10* and eliminating the negative sign. For example, $[H^+] = 10^{-5}$ M is written as pH = 5.

Most chemists refer to the acidity of a solution by its pH. The symbol pH means the power (exponent) of the hydrogen ion concentration and hence the "strength" of the solution. The pH is mathematically defined as the negative logarithm (base 10) of the $[H^+]$. So a 0.10 M HCl solution has a $[H^+]$ of 1.0×10^{-1} and a pH of 1; and a 0.0010 HNO_3 solution has a pH of 3 (the $-\log$ of 1.0×10^{-3}).

Even a highly basic solution has a $[H^+]$ and a pH. For example, if a solution is 0.010 M KOH (a strong base), it has a $[OH^-]$ of 1.0×10^{-2} and a $[H^+]$ of 1.0×10^{-12}. The pH of a 0.010 M KOH solution is 12. Table 10-2 lists several corresponding values for $[H^+]$, $[OH^-]$, and pH.

Example 10.6

Calculate the pH of chemically pure water in which the $[H^+]$ and $[OH^-]$ are each 1.0×10^{-7} M.
1. Write the equation:

$$pH = -\log [H^+]$$

2. Substitute the 1.0×10^{-7} M for $[H^+]$, and note the logarithms of 1.0 and 10^{-7}. (Calculations with log values is explained in Appendix II.) The log of $1.0 = 0$, and the log of $10^{-7} =$ the exponent, -7. Since the log of a product of two numbers equals the sum of the two separate logs (i.e., $\log ab = \log a + \log b$)

$$pH = -[\log 1.0 + \log 10^{-7}]$$
$$pH = -[0 + (-7)]$$
$$\boxed{pH = 7.0}$$

TABLE 10–2 **Some Comparative Values of $[H^+]$, $[OH^-]$, and pH**

	$[H^+]$	$[OH^-]$	pH
Acidic	1×10^{-2}	1×10^{-12}	2
Neutral	1×10^{-7}	1×10^{-7}	7
Basic	1×10^{-11}	1×10^{-3}	11
Basic	1×10^{-13}	1×10^{-1}	13

Example 10.7

   Calculate the pH of a solution in which $[H^+] = 4.6 \times 10^{-3}$ M.
1. Write the equation:

$$pH = -\log [H^+]$$

2. Substitute the numerical value for $[H^+]$.
3. Solve:

$$pH = -\log [4.6 \times 10^{-3}]$$
$$= -[\log 4.6 + \log 10^{-3}]$$
$$= -[0.66 - 3]$$
$$= 2.34$$

Example 10.8

   Find the pH of a solution of sodium hydroxide in which the $[OH^-] = 5.1 \times 10^{-2}$.
1. Calculate the $[H^+]$ in order to obtain the pH.
   $[H^+][OH^-] = 1.0 \times 10^{-14}$
   $[H^+] (5.1 \times 10^{-2}) = 1.0 \times 10^{-14}$
   $$[H^+] = \frac{1.0 \times 10^{-14}}{5.1 \times 10^{-2}} = \frac{10 \times 10^{-15}}{5.1 \times 10^{-2}}$$
   $[H^+] = 1.96 \times 10^{-13} M$
2. Now find the pH.
   $pH = -\log [H^+]$
   $pH = -(0.29 - 13)$
   $pH = -(-12.71)$

   $$\boxed{pH = 12.71}$$

Note that the more basic the solution, the *higher* the pH.

   The concept of pH has greatest usefulness when the application is made to the measurement of hydrogen ion concentrations that are relatively low. The useful range of the scale of pH values is between 0 and 14, with the pH of chemically pure water, 7, taken as both the midpoint of the scale and the measure of perfect neutrality at 25°C. Figure 10–4 illustrates the relative pH values of a number of common substances in water solution at 25°C. Acid solutions have pH values below 7, whereas basic solutions have pH values greater than 7; the more basic the solution, the higher the pH value.
   There are many situations that require that we convert measured pH values to hydrogen ion or hydroxide ion concentrations. The following example is an illustration of how this is done.

ACIDS AND BASES

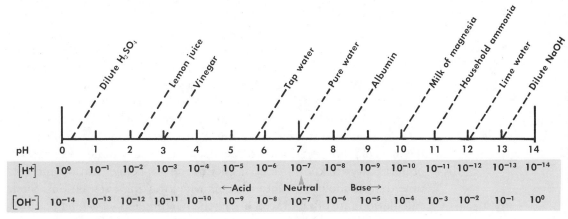

Figure 10–4   pH, ([H+] and [OH−]) values of some common solutions.

Example 10.9

Calculate the hydrogen ion concentration of a solution having a pH of 5.85.
1. Write the equation:

$$pH = -\log [H^+]$$

2. Substitute and solve:

$$[H^+] = \text{antilog of } -pH$$
$$= \text{antilog } -(5.85)$$
$$= \text{antilog } (0.15 - 6)$$

$$\boxed{[H^+] = 1.41 \times 10^{-6} \text{ M}}$$

Exercise 10.4

1. If a water solution has a hydrogen ion concentration of 0.0032 M, what is the hydroxide ion concentration?
2. Calculate the pH of the following solutions:
   (a) 0.2 M HCl (assume 100% ionization)
   (b) 0.04M H_2SO_4 (assume complete ionization, $H_2SO_4 \rightarrow 2\,H^+ + SO_4^{2-}$)
   (c) 4.4×10^{-3} M[H+]
   (d) 1.7×10^{-9} M[H+]
   (e) 8.2×10^{-4} M[OH−]
3. Calculate the pH of the following solutions:
   (a) 0.0002 M NaOH
   (b) 0.05 M $Sr(OH)_2$ (assume 100% dissociation, $Sr(OH)_2 \rightarrow Sr^{2+} + 2\,OH^-$)
   (c) 3.7×10^{-8} M[OH−]
   (d) 5.3×10^{-2} M[OH−]
   (e) 7.2×10^{-11} M[H+]

10.9 SOLUTION STOICHIOMETRY

We have found that a correctly balanced equation for a stoichiometric reaction indicates the quantitative relationships among reactants and products. A number of problems, classified as mole-mole, mole-mass, mass-mass, mass-volume, and volume-volume, have been presented. However, many reactants, especially those involving volumes of solutions, are not conveniently expressed in grams. Of course, the mass of solute in grams can be calculated if the volume and molarity or normality are known. It is simpler to convert concentration and volume data to moles or to equivalents when making calculations. Such calculations are often placed under the heading of *solution stoichiometry*.

Example 10.10

Calculate the mass of zinc needed to react completely with 50.0 mL of 0.50 M HCl.
1. Write a balanced equation for the reaction in order to determine the stoichiometry.

$$Zn(s) + 2\ HCl(aq) \rightarrow ZnCl_2(aq) + H_2(g)$$

2. The equation indicates that 1 mole of zinc will react with 2 moles of HCl.
3. Since the molarity and volume of the HCl is given, the number of moles of dissolved HCl can be found:

$$n = \frac{0.50\ mole}{1L} \times 0.050\ L$$

$$n = 0.025\ mole\ HCl$$

4. Zinc is 65.4 g/mole.
5. Solve:

moles HCl → moles Zn → mass Zn

$$x = 0.025\ \cancel{mole\ HCl} \times \frac{1\ \cancel{mole\ Zn}}{2\ \cancel{moles\ HCl}} \times \frac{65.4\ g\ Zn}{1\ \cancel{mole\ Zn}}$$

$$\boxed{x = 0.82\ g\ of\ zinc}$$

Example 10.11

What volume of 0.020 M HCl is necessary to react with an excess of zinc in order to produce 600.0 mL of hydrogen gas at STP?
1. Write the balanced equation for the reaction stoichiometry.
$$Zn(s) + 2\ HCl(aq) \rightarrow ZnCl_2(aq) + H_2(g)$$
2. The presence of an excess of zinc makes HCl the reaction-limiting factor; that is, the volume of gas depends directly on the number of moles of HCl reacting.
3. The stoichiometry indicates that 2 moles of HCl will produce 1 mole of hydrogen gas. Solve for moles of HCl

---

ACIDS AND BASES

$$\text{volume } H_2 \rightarrow \text{moles } H_2 \rightarrow \text{moles HCl}$$

$$x = 600.0 \text{ mL} \times \frac{1 \text{ L}}{10^3 \text{ mL}} \times \frac{1 \text{ mole } H_2}{22.4 \text{ L}} \times \frac{2 \text{ moles HCl}}{1 \text{ mole } H_2}$$

$$x = 0.0536 \text{ mole HCl}$$

4. The volume of HCl(aq) can then be calculated:

$$V = 0.0536 \text{ mole} \times \frac{1 \text{ L}}{0.020 \text{ mole}}$$

$$\boxed{V = 2.7 \text{ L of HCl(aq)}}$$

Example 10.12

Suppose that 1.68 g of zinc were added to 2.68 liters of 0.020 M HCl. Would 600 mL of H_2 gas at STP be obtained as it was in Example 10.11?

1. The data in this problem do not indicate whether the zinc is in excess as it was before. Therefore, we must determine whether the HCl(aq) or the Zn(s) is the reaction-limiting substance in this case. This means that the reaction stops when either reactant, the Zn or HCl, is used up.

2. The molar ratio, indicated by the balanced equation, shows that one mole of zinc is needed to react completely with two moles of HCl. Calculate the number of moles of each reactant in order to see whether the molar ratio of Zn to HCl is 1 to 2.
   For zinc:

$$n = 1.68 \text{ g} \times \frac{1 \text{ mole}}{65.4 \text{ g}} = 0.0257 \text{ mole}$$

For HCl:

$$n = \frac{0.020 \text{ mole}}{1 \text{ L}} \times 2.68 \text{ L} = 0.054 \text{ mole}$$

3. Compare the whole-number molar ratios by dividing both values by the smaller number:

$$Zn = \frac{0.0257}{0.0257} = 1.00 \text{ mole}$$

$$HCl = \frac{0.054}{0.0257} = 2.1 \text{ mole}$$

4. The ratio is not 1 mole of zinc to 2 moles of HCl. The calculations indicate an excess of HCl. Since there is not enough zinc to react completely with the acid, the final volume of gas would be less than 600 mL at STP.

In reality, chemical reactions are rarely performed at STP. A laboratory maintained at 0°C is not a comfortable environment. In order to adjust calculated results to actual conditions, the gas laws are commonly applied.

10.9   SOLUTION STOICHIOMETRY

Example 10.13

What volume of 0.20 M H_2SO_4 would have to react with an excess of calcium in order to produce 0.300 liter of hydrogen gas measured at 23°C and 750 torr?
1. Write the balanced equation for the reaction.
   $Ca(s) + H_2SO_4 (aq) \rightarrow CaSO_4 (s) + H_2(g)$
2. The stoichiometry indicates that 1 mole of H_2SO_4 will produce 1 mole of hydrogen.
3. A simple method of expressing the number of moles of $H_2(g)$ at 23°C and 750 torr is to make use of the equation of state.
   $PV = nRt$
   where
   $n = \dfrac{PV}{RT}$
4. Organize the data:
   $p = 750$ torr
   $V = 0.300$ L
   $R = \dfrac{62.4 \text{ L-torr}}{\text{mole °K}}$
   $T = (23 + 273)°K = 296°K$
5. Solve:
   volume $H_2 \rightarrow$ moles $H_2 \rightarrow$ moles H_2SO_4

$$\text{moles of } H_2SO_4 = \frac{(\cancel{\text{mole}} \ \cancel{°K})(0.300 \ \cancel{L})(750 \ \cancel{\text{torr}}) \ \cancel{H_2}}{(62.4 \ \cancel{L\text{-torr}})(296 \ \cancel{°K})} \times \frac{1 \text{ mole } H_2SO_4}{1 \ \cancel{\text{mole } H_2}}$$

$$= 1.22 \times 10^{-2} \text{ mole } H_2SO_4$$

6. The volume of H_2SO_4 can now be calculated.

$$V = 1.22 \times 10^{-2} \ \cancel{\text{mole}} \times \frac{1 \text{ L}}{0.20 \ \cancel{\text{mole}}} = \boxed{6.1 \times 10^{-2} \text{ L of } H_2SO_4}$$

7. In other words, 24 milliliters of 0.20 molar H_2SO_4 would be sufficient to produce 0.300 L of hydrogen gas when measured at 23°C and 750 torr.

A variation of Example 10.13 would be the calculation of the volume of gas produced under nonstandard conditions. If, for example, an excess of zinc metal were added to 50.0 mL of 0.050 M H_2SO_4, the number of moles of hydrogen produced could be determined from the balanced equation. The equation of state once again would provide the answer for the volume of gas under the stated conditions of temperature and pressure.

Example 10.14

What volume of hydrogen gas, measured at 36°C and 1.02 atm pressure, would be obtained by the reaction of an excess of zinc with 50.0 mL of 0.0500 M H_2SO_4?
1. Determine the molar ratios from the balanced equation.

$$Zn(s) + H_2SO_4(aq) \rightarrow ZnSO_4(aq) + H_2(g)$$

The stoichiometry indicates that the acid and gas concentrations are equimolar.

2. Calculate the number of moles of H_2 produced:

$$\text{moles of } H_2SO_4 \rightarrow \text{moles } H_2$$

$$\text{moles of } H_2 = 0.0500 \cancel{L\ H_2SO_4} \times \frac{0.0500 \cancel{\text{mole } H_2SO_4}}{1 \cancel{L\ H_2SO_4}} \times \frac{1 \text{ mole } H_2}{1 \cancel{\text{mole } H_2SO_4}}$$

$$= 2.50 \times 10^{-3} \text{ mole } H_2$$

3. Now find the volume occupied by 2.5×10^{-3} mole of H_2 at $36.0°C$ and 1.02 atm.

$$PV = nRT \text{ becomes } V = \frac{nRT}{P}$$

4. Organize the data and solve:
$P = 1.02$ atm
$V = ?$
$n = 2.50 \times 10^{-3}$ mole

$$R = \frac{0.0821 \text{ L-atm}}{\text{mole } °K}$$

$T = 309°K$

$$V = \frac{(2.50 \times 10^{-3} \cancel{\text{mole}})(0.0821 \text{ L-}\cancel{atm})(309 \cancel{°K})}{(\cancel{\text{mole } °K})(1.02 \cancel{atm})}$$

$$V = 62.2 \times 10^{-3} \text{ L} = \boxed{62.2 \text{ mL}}$$

Exercise 10.5

1. How many grams of magnesium are required to react completely with 75.0 mL of 1.4 M HCl?
2. What volume of 2.3 M hydrochloric acid is needed to react completely with 50.0 mg of iron in the formation of iron(II) chloride?
3. If 2.0 g of CaO is added to 25.0 mL of 0.10 M H_2SO_4, what mass of $CaSO_4(s)$ could be obtained? (Hint: check the molar ratios of the reactants in order to find the reaction-limiting component.)
4. What volume of gas (at STP) can be obtained if an excess of magnesium is added to 50.0 mL of 2.5 M HCl?
5. How many milliliters of 0.075 M H_2SO_4 react with an excess of aluminum to produce 450.0 mL of hydrogen gas at $27.0°C$ and a pressure of 735 torr?

10.10 ACID-BASE TITRATION

A fast and accurate technique for the determination of the concentration of an unknown solution is to mix it with a measured volume of a solution of known concentrations. The process of adding a solution of one reactant to a solution of another reactant until the number of equivalents of each reactant is the same is called **titration.** The reacting solutions are called **titrants.** They are added through accurately calibrated burets (Fig. 10–5) into a constantly stirred Erlenmeyer flask until an indicator in the solution changes color at the **end point.**

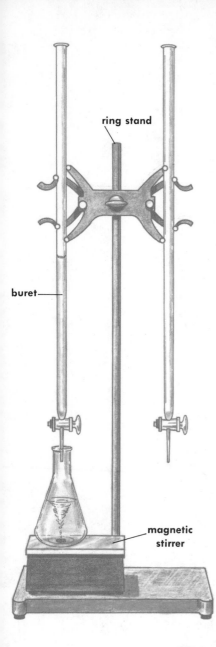

ring stand

buret

magnetic
stirrer

Figure 10–5   A common arrangement of burets, flask, and magnetic stirrer used to perform titration.

The concentration of an unknown acid may be determined by measuring the volume of a basic solution of known concentration needed to neutralize it. The most commonly used indicator in the titration of acids and bases is phenolphthalein, an organic substance that is red in base and colorless in acid. A pH meter may also be used to determine the end point. The data obtained from a titration is used to calculate the unknown concentration.

Since neutralization is the reaction between equivalents of H^+ (aq) and OH^- (aq) to form water, many chemists find it desirable to use normal concentrations rather than molar concentrations. While a liter of 1.0 M NaOH (aq) will *not* neutralize a liter of 1.0 M H_2SO_4 (aq), equal volumes of 1.0 N solutions of any acid or base *will* neutralize each other. Since one mole of H_2SO_4 can produce two moles of H^+ (aq), a half-mole of H_2SO_4 (the equivalent mass) produces 1 mole of H^+ (aq):

$$H_2SO_4 \rightarrow 2\,H^+ \quad + SO_4{}^{2-}$$

$$\text{1 mole} \qquad \text{2 moles} \qquad \text{1 mole}$$

$$\text{1/2 mole} \rightarrow \text{1 mole} \qquad \text{1/2 mole}$$

In effect, neutralization occurs when the number of equivalents of acid equals the number of equivalents of base.

$$\# \text{ eq acid} = \# \text{ eq base}$$

The practical application of this equivalence is demonstrated in the examples that follow.

Example 10.15

What is the normality of an unknown acid if 30.2 mL is required to neutralize 61.5 mL of 0.500 N NaOH(aq)?
1. Since both titrants are in solution, the relationship is

$$\# \text{ equivalents of acid} = \# \text{ equivalents of base}$$
$$(\text{normality acid} \times \text{vol. acid}) = (\text{normality base} \times \text{vol. base})$$
$$N_A V_A = N_B V_B$$

2. Organize the data and solve:
 $N_A = ?$
 $V_A = 30.2$ mL
 $N_B = 0.500$ N
 $V_B = 61.5$ mL

$$N_A(30.2 \text{ mL}) = (0.500 \text{ N})(61.5 \text{ mL})$$

$$N_A = \frac{(0.500 \text{ N})(61.5 \text{ mL})}{(30.2 \text{ mL})}$$

$$\boxed{N_A = 1.02 \text{ N}}$$

Exercise 10.6

1. Find the normal concentration of an unknown acid if 126.2 mL is required to neutralize 43.8 mL of 1.4 N KOH (aq.)
2. If 84.0 mg of oxalic acid crystals is dissolved in water, what is the normal concentration of NaOH(aq) when 43.7 mL is required to adjust the pH of the resulting solution to 7.00? Assume that at this pH, the NaOH was added in an amount sufficient to neutralize the oxalic acid to oxalate.
3. If on three successive titrations 20.0 mL samples of 0.22 N KOH(aq) are apparently neutralized by 31.3 mL, 30.9 mL, and 31.4 mL of H_2SO_4 (aq), calculate the average normal concentration of the acid. Assume that H_2SO_4 was neutralized to K_2SO_4.

10.1   Define acids and bases in terms of the Arrhenius concept.

10.2   Define acids and bases in terms of the Brønsted-Lowry concept

10.3   What is meant by an amphoteric or amphiprotic substance? Give illustrations.

10.4   Write formulas and names of the conjugate bases of these acids:
(a) HF
(b) H_2S
(c) HSO_4^-
(d) NH_3
(e) $HClO_3$

10.5   Write formulas and names of the conjugate acids of these bases:
(a) HCO_3^-
(b) NH_3
(c) OH^-
(d) S^{2-}
(e) CN^-

10.6   Complete and balance the following acid-base equations. Label the conjugate acid-base pairs.
(a) $HSO_4^-(aq) + OH^-(aq) \rightleftharpoons$
(b) $H_2S(aq) + CN^-(aq) \rightleftharpoons$
(c) $HCO_3^-(aq) + CH_3O^-(aq) \rightleftharpoons$
(d) $O^{2-}(aq) + Zn(H_2O)_4^{2+}(aq) \rightleftharpoons$
(e) $OAc^-(aq) + HI(aq) \rightleftharpoons$

10.7   Complete and balance the following equations:
(a) $HF(aq) + H_2PO_4^-(aq) \rightleftharpoons$
(b) $CH_3O^-(aq) + NaHSO_4(s) \rightleftharpoons$
(c) $NH_2^-(aq) + HSO_4^-(aq) \rightleftharpoons$
(d) $HClO_4(aq) + HCO_3^-(aq) \rightleftharpoons$
(e) $HNO_3(aq) + HPO_4^{2-}(aq) \rightleftharpoons$

10.8   Define hydrolysis. Illustrate by means of net ionic equations.

10.9   Write equations for possible reactions of the following with water. Identify the conjugate acid-base pairs:
(a) CO_3^{2-}
(b) HPO_4^{2-}
(c) NH_3
(d) NH_4^+
(e) F^-

10.10   Complete and balance the following equation for hydrolysis reactions:
(a) $H_2O(l) + CO_3^{2-}(aq) \rightleftharpoons$
(b) $CH_3O^-(aq) + H_2O(l) \rightleftharpoons$
(c) $Al(H_2O)_6^{3+}(aq) + H_2O(l) \rightleftharpoons$
(d) $H_2O(l) + OAc^-(aq) \rightleftharpoons$
(e) $S^{2-}(aq) + H_2O(l) \rightleftharpoons$

10.11   HCN(aq) is a weaker acid than HOAc (aq). Which aqueous solution would you expect to have the higher pH (i.e., be more basic): 0.10M NaOAc or 0.10 M NaCN? Explain why.

10.12   Define acids and bases in terms of the Lewis concept.

10.13   Select Lewis acids from the following list: S^{2-}, H^+ (aq), H_2O, NH_3, BF_3, O^{2-}, Al^{3+} (aq).

10.14   Define pH. Express the relationship between pH and hydrogen ion concentration.

10.15   If a water solution has a hydrogen ion concentration of 0.068 M, what is the pH?

10.16   Calculate the pH of the following solutions:
(a) 0.04 M HBr
(b) 1.3×10^{-4} M H^+
(c) 0.0032 M HCl
(d) 6.2×10^{-8} M H^+
(e) 4.5×10^{-5} M OH^-

10.17   Given the following pH values, calculate the hydrogen ion and hydroxide concentrations of the following solutions:
(a) pH = 4.2
(b) pH = 1.7

(c) pH = 11.6
(d) pH = 8.3
(e) pH = 0.8

10.18   How many grams of barium are needed to react completely with 60.0 mL of 0.75 M HBr?

10.19   What volume of 0.650 M HCl is required to react completely with 0.0802 g of calcium?

10.20   If 35.0 mL of 0.0120 M HNO_3 reacts with an excess of zinc, what volume of hydrogen gas will be produced when measured at $40.0°C$ and 1.20 atm?

10.21   Calculate the normal concentration of an acid if 23.2 mL neutralizes 81.5 mL of 0.830 N NaOH(aq) in a titration.

10.22   Find the molar and normal concentrations of a solution of H_2SO_4 (aq) if 216.0 mL is required to neutralize 2.7 g of CaO that has been dissolved in water, one product being $CaSO_4$.

10.23   What is the molar concentration of $Ca(OH)_2$(aq) if 43.4 mL is titrated with 0.360 g of oxalic acid (in water solution) to provide CaC_2O_4?

CHAPTER ELEVEN

LEARNING OBJECTIVES

At the completion of this chapter, you should be able to:

1. Define equilibrium in terms of a reversible reaction.

2. List five factors that may affect the rate of chemical reactions.

3. Predict changes in equilibrium systems related to alterations of concentration, temperature, and pressure.

4. State LeChatelier's principle.

5. Apply LeChatelier's principle to the control of reversible reactions.

6. State the law of chemical equilibrium.

7. Write an equilibrium expression for a reversible reaction.

8. Explain the origin and meaning of an equilibrium constant.

9. Identify the variety of symbols used as equilibrium constants.

10. Describe the relative strengths of acids in terms of their ionization constants.

11. Calculate hydrogen ion concentrations and pH values of weak acids, given their ionization constants.

12. Explain the meaning of the term *solubility product constant* (K_{sp}).

13. Describe the relative solubilities of slightly soluble compounds, given a list of their K_{sp} values.

14. Determine the solubility of slightly soluble compounds in terms of grams of solute per 100 g of water, or grams per liter of solution.

CHEMICAL EQUILIBRIUM AND REACTION RATE

In reversible reactions, the products can react to form the starting materials. Such reactions are usually written with a double arrow between the starting materials and the products, to show that the reaction goes in both directions at once.

Since the rate of reaction at a given temperature depends on the concentration of the reactants, at first the reverse reaction will be very slow. Then, as the amount of product increases, the rate will increase, until at some point the rate of the reverse reaction exactly equals the rate of the forward reaction. At this point, there is no net change in the amount of each substance present; as fast as a compount is formed, other molecules of the same compound are used up.

If, for example, A and B react to form C and D, and then C and D react to re-form A and B, the equilibrium reaction would be,

$$A + B \rightleftharpoons C + D$$

Conventionally, the initial reaction between A and B is called the **forward reaction** and may be so indicated by the lowercase letter f over the arrow pointing toward the right:

$$A + B \xrightarrow{f} C + D$$

The secondary reaction between C and D is called the **reverse reaction** and may be symbolized by the lowercase letter r placed below the arrow pointing to the left:

$$A + B \xleftarrow[r]{} C + D$$

The reversibility may be clearly indicated by combining the forward and reverse notations.

$$A + B \xrightleftharpoons[r]{f} C + D$$

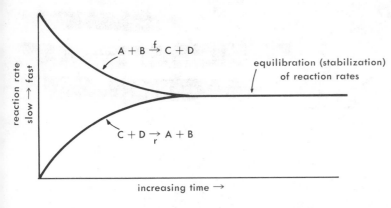

equilibration (stabilization)
of reaction rates

Figure 11–1 The stabilization of the reaction rates in an equilibrium system.

When the *rates* of the forward and reverse reactions become stabilized and equal, the number of moles of each species present in the system also becomes constant, because A and B continue to react to form C and D just as fast as C and D reconstitute A and B. This balance in a state of constant change is known as **dynamic equilibrium.** Although the rates of the two reactions are equal, the *concentrations* of reactants and products will seldom be equal. They vary widely.

differ, but constant at equil.

The rate of a chemical reaction refers to the number of moles of reactants that undergo chemical change per unit of time. The progress of the reversible reaction $A + B \rightleftharpoons C + D$ toward equilibrium is illustrated in Figure 11–1, where it can be seen that the rate of the forward reaction decreases until equilibrium is reached. Notice that the rates of the forward and reverse reactions become equal and constant at equilibrium.

Since the quantitative considerations of equilibrium reactions deal with comparisons between the stabilized rate of the forward and reverse reactions of particular systems, we will review the primary factors that affect reaction rates.

11.2 FACTORS AFFECTING REACTION RATES (KINETICS)

In order for a chemical reaction to occur, three requirements must be satisfied:

(1) The reacting particles must collide.
(2) Their collisions must take place with sufficient energy.
(3) The colliding particles must have favorable geometry.

The first requirement is obvious, but mere collision of particles is insufficient to cause a chemical reaction. Not all reacting materials possess the same energy at a given temperature. Collisions between the slower particles are ineffective, i.e., they do not result in the breaking of old (reactant) bonds and the formation of new (product) bonds.

Only the more energetic particles may undergo effective collisions, i.e., those that result in a chemical change.

The third requisite for a chemical reaction to occur pertains to the spatial orientation of the colliding particles relative to one another. If orientation is favorable, a large number of effective collisions is possible. For unfavorable orientation only the most energetic particles will produce effective collisions.

The study of the factors that affect the rates of chemical reactions is often called **kinetics.** The *collision theory,* which is central to the kinetic molecular theory of gases (Chapter 6), is a properly related concept. However, the elastic collisions between noninteracting gas molecules are not the same as the collisions between chemically reacting molecules, atoms, and ions. The factors that affected only the volume, temperature, and pressure characteristics of gases now affect the rates of chemical change. Five principal factors will be considered: **concentration, temperature, pressure,** the **nature of the reactants,** and **catalysis.**

Concentration

The rate of the reaction between A and B depends on the number of moles of A and B present per unit volume. The initial concentrations of A and B are always expressed in terms of *moles per liter,* or as *molarity* if they are in solution. The symbolism that indicates moles per liter is a set of brackets, [], as was described in the previous chapter. The more particles of A and B that are present per unit volume, the more collisions that will result. More collisions per unit of time naturally enhance the probability that more particles will collide with sufficient energy and favorable geometry to result in a chemical change. An example of the necessity for sufficient energy and favorable geometry when reactive species collide may be observed in the forward reaction between $A_2(g)$ and $B_2(g)$ in the formation of $AB(g)$.

$$A_2(g) + B_2(g) \rightleftharpoons 2 AB(g)$$

Figure 11–2 illustrates some possible interactions between the molecules of A and the molecules of B.

The rates of many common chemical reactions are sometimes empirically found to be directly proportional to the *product* of the molar concentrations. The rate expression for the reaction between A and B in such cases may be represented as

$$\text{rate} \propto [A][B]$$

This relationship will be expanded mathematically a little later in this chapter when we discuss the law of chemical equilibrium.

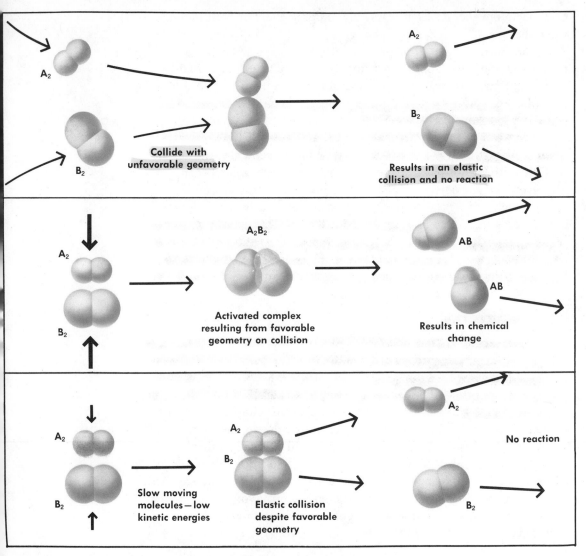

Figure 11–2   Critical factors affecting the collisions between reactants.

Temperature

Another factor affecting reaction rates is temperature. Many reactions that apparently do not occur or that happen slowly at room temperature will proceed vigorously when the reactants are heated. For example, a mixture of powdered iron and sulfur will glow with an intense heat of reaction once activated by a burner flame.

$$\text{Fe (s)} + \text{S (s)} \rightarrow \text{FeS (s)} \qquad \Delta H = -22.7 \text{ kcal}$$

The flame supplies enough energy (called the **energy of activation**) to jostle sluggish reactants into an active condition, which allows the breaking of the original bonds of attraction and the formation of new

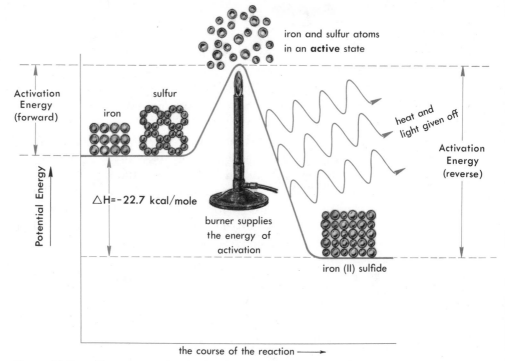

Iron and sulfur atoms in an **active** state

Activation Energy (forward)

sulfur

iron

Potential Energy

$\triangle H = -22.7$ kcal/mole

burner supplies the energy of activation

heat and light given off

Activation Energy (reverse)

iron (II) sulfide

the course of the reaction ⟶

Figure 11-3 The energy of activation breaks the bonds between iron atoms, and also the bonds between sulfur atoms, so that the new iron-sulfur bonds can form.

ones. For example, Figure 11–3 diagrammatically illustrates the reaction between silvery gray iron and yellow sulfur in the formation of dull gray iron(II) sulfide. Notice that the energy barrier for this exothermic reaction (activation energy forward) is less than for the reverse reaction, and that the ΔH (see Chapter 5) is *negative*. $\Delta H = -22.7$ kcal per mole of FeS produced.

However, an important consideration to be retained, especially as it will be demonstrated to relate to equilibrium systems, is that *the application of heat favors endothermic reactions, while the removal of heat (cooling) favors exothermic reactions, once they are started.*

Temperature is therefore a factor to be considered in equilibrium reactions. The symbolic reaction

$$A + B \underset{r}{\overset{f}{\rightleftharpoons}} C + D + \text{heat}$$

indicates that the forward reaction (f) is exothermic, whereas the reverse reaction (r) is endothermic. The rates of these reactions and the equilibrium concentrations of all the components decided by the rates will depend on the control of the temperature of the system.

In nonvapor systems, the effect of pressure is small and often negligible. However, if some of the interacting species in a reaction are in the gaseous state, pressure may be a significant factor. An increase in external pressure on the system allows the gas molecules to interact at a faster rate because of an increase in the number of collisions per unit of time.

In an equilibrium system, the changes in the rates of the forward or reverse reactions depend on the number of moles of gas per unit volume among the initial reactants and products. If there are more moles of gas per liter present as initial reactants than as products, the rate of the forward reaction will be increased when pressure is applied. When the initial products consists of more moles of gas per liter of gas than do the reactants, the reverse reaction is favored. Some examples are:

1.   A (g)   +   B (g) $\rightleftharpoons$ C (s) + D (s)

    1 mole of gas   1 mole of gas   no gas

The rate of the *forward* reaction increases with pressure.

2.   A (s)   +   B (g) $\rightleftharpoons$ C (g) + D (g)

        1 mole of gas   2 moles of gas

The *reverse* reaction is favored by increase of pressure.

3.   A (s)   +   B (g) $\rightleftharpoons$ C (g) + D (*l*)

        1 mole of gas   1 mole of gas

The increase of pressure has no significant effect on either the forward or reverse reaction rate; thus neither reaction is favored, because reaction rates of both sides are affected equally.

Catalysis

The rates of many reactions may be affected markedly by the use of a catalyst. Although the mechanisms of many catalyzed reactions are not fully understood, a catalyst seems to increase the rate of a reaction by lowering the activation energy required by providing a faster alternate pathway for the reacting species as they undergo chemical change. Another factor that partially defines a catalyst is that the catalytic agent is recoverable at the end of the reaction, often in its original condition.

Some substances are described as *contact catalysts,* whereas others are classified as *carrier catalysts.* A contact catalyst provides a focal point for interacting gas molecules. For example, when ammonia reacts with oxygen in the presence of platinum, a contact catalyst, the change is rapid:

$$2\ NH_3(g) + \tfrac{5}{2}\ O_2(g) \xrightarrow[\text{heat}]{\text{Pt}} 2\ NO(g) + 3\ H_2O\ (l)$$

This catalyzed oxidation of ammonia is actually part of a commercial method for producing nitric acid.

Carrier catalysts apparently enter into the reaction as they form temporary unstable structures that rapidly decompose while the final products are produced. Although the carrier catalyst does take part in the reaction in that it provides an alternate and more rapid pathway, it is reconverted to its original form during the next stage of reaction and so apparently is not changed in the final analysis. An example of a carrier catalyst is found in the use of copper when ethanol (ethyl alcohol, or grain alcohol) is oxidized to acetic acid. Normally, ethanol will change slowly to acetic acid when it is exposed to the oxygen in the air. However, the rapid catalyzed reaction occurs when copper temporarily enters the reaction and is oxidized to copper(II) oxide, which proceeds to decompose and reconstitute the original free copper.

$$CH_3CH_2OH(l) + O_2(g) \xrightarrow{(Cu \rightleftharpoons CuO)} CH_3COOH(l) + H_2O(l)$$
ethanol   acetic acid

It must be emphasized that catalysts cannot change the molar concentrations at equilibrium. A catalyst provides a different reaction mechanism, which requires less activation energy. Thus a catalyst alters equally the rates of the forward and reverse reactions, so that less time is required for the reversible reaction to attain equilibrium.

FUNKY WINKERBEAN   Tom Batiuk

The Nature of the Reactants

The nature of the reactants is probably the most obvious condition affecting rates. The fact that some compounds and elements are extremely reactive and that they undergo chemical change very rapidly is readily observed and is a matter affecting laboratory safety. The addition of copper to concentrated sulfuric acid results in a very slow reaction at room temperature whereas a mixture of $KClO_3$ and sulfuric acid produces a dangerously explosive reaction because of the rapidity of the change.

11.3 PREDICTING CHANGES IN EQUILIBRIUM REACTIONS— LECHATELIER'S PRINCIPLE

In a reversible reaction, the molar concentrations of the initial reactants and products reach certain fixed values when a state of dynamic equilibrium is attained. However, since concentration, temperature, and possibly pressure (when gases are involved) may affect forward and reverse reaction rates unequally, these factors must be taken into account as agents of change in equilibrium systems.

The French chemist Henri LeChatelier, in 1884, proposed an explanation of the altering effects of changes in concentration, temperature, and pressure on equilibrium systems. **LeChatelier's principle** says in effect that *when a stress is applied to a system in equilibrium, the concentrations of the interacting species will change in such a way as to relieve or cancel the effect of the stress.* The special significance of LeChatelier's principle to the chemist is that by thoughtfully selecting a particular type of stress, such as changing the temperature, selectively increasing or decreasing concentrations of some of the reacting species, or altering the pressure on gases, the forward or reverse reaction can be favored decisively as the reaction moves toward a restoration of equilibrium.

Example 11.1

In the reaction $N_2(g) + 3 H_2(g) \overset{Pt}{\rightleftharpoons} 2 NH_3(g)$ what kinds of changes in concentration, temperature, and pressure could be used to favor the *forward* reaction? What would be the effect of the catalyst platinum? The ΔH is negative for the forward reaction.
1. List all pertinent observations regarding this equilibrium system.
    (a) All of the reactants and products are in the gaseous state.
    (b) The forward reaction is exothermic, and the reverse reaction is endothermic.
    (c) The addition of more N_2 and/or H_2 to the same reaction volume would increase the collision rate in the forward reaction.
2. Pressure change:
    Since $3 H_2(g)$ and $1 N_2(g)$ represent 4 moles of gas that react to form 2 moles of ammonia, the forward reaction could be favored by increasing the pressure so that more collisions between H_2 and N_2 molecules would occur. In terms of

---

LeChatelier's principle, the response to the stress of pressure increase would be a favoring of the reaction that produces the products that occupy less volume. Two moles of ammonia occupy half the volume that would be occupied by a combined total of 4 moles of H_2 and N_2.

3. Temperature change:
Since the forward reaction is exothermic, any heat that exceeds the minimal activation energy would aid the reverse, or endothermic, reaction. Therefore, the forward reaction would be favored by keeping the temperature at the threshold activation level and by removing *heat* as it is formed.

4. Concentration changes:
The constant addition of both $H_2(g)$ and $N_2(g)$ would favor the forward reaction by increasing the number of collisions of reactants while simultaneously increasing the pressure on the system. If ammonia could be removed (perhaps by cooling and condensation to the liquid state), the forward reaction would be favored even more, as LeChatelier's principle would relate to increased production of ammonia in response to the stresses in equilibrium restoration.

5. The only effect resulting from the use of the platinum catalyst would be a proportional speeding up of both reactions—favoring neither the forward nor reverse direction. There is no such thing as a one-way catalyst, because that would allow us to get something for nothing.

Example 11.2

In the reaction,

$$CO(g) + Cl_2(g) \rightleftharpoons COCl_2(g) \qquad \Delta H = negative$$

what would be the effects on the equilibrium system of an increase in temperature, decrease in pressure, and a removal of $COCl_2(g)$?

1. Temperature increase:
Since the forward reaction is exothermic, an increase in temperature would favor the reverse, or endothermic, reaction. In other words, the equilibrium would be shifted to the left so that the concentrations of CO and Cl_2 would be higher at the new equilibrium.

2. Pressure decrease:
Since the CO and Cl_2 represent 2 moles of gas and the $COCl_2$ is only 1 mole of gas, a decrease in pressure would favor the formation of those products that occupy more volume. Therefore, the equilibrium would be shifted to the left (reverse action) to relieve the decrease in pressure.

3. Removal of $COCl_2(g)$:
The reduction of the concentration of the $COCl_2$ places a stress on the system such that the forward reaction would be favored as a move to restore the original equilibrium concentrations.

Exercise 11.1

1. In the following reactions, indicate whether an increase or decrease of pressure will favor the forward reaction:
   (a) $H_2(g) + I_2(g) \rightleftharpoons 2 HI(g)$
   (b) $H_2(g) + \frac{1}{2} O_2(g) \rightleftharpoons H_2O(l)$
   (c) $2 NO_2(g) \rightleftharpoons N_2O_4(g)$
   (d) $CuO(s) + H_2(g) \rightleftharpoons Cu(s) + H_2O(l)$
   (e) $COCl_2(g) \rightleftharpoons Cl_2(g) + CO(g)$
2. What effect would heating have on the following reactions?
   (a) $CO_2(g) + C(s) \rightleftharpoons 2 CO(g) \qquad \Delta H = +$
   (b) $\frac{1}{2} N_2(g) + \frac{1}{2} O_2(g) \rightleftharpoons NO(g) \qquad \Delta H = +$

(c) $(NH_4)_2 CO_3(s) \rightleftharpoons 2 NH_3(g) + CO_2(g) + H_2O(l)$ $\Delta H = +$

(d) $PCl_3(g) + Cl_2(g) \rightleftharpoons PCl_5(g)$ $\Delta H = -$

3. Indicate an equilibrium shift to the right (forward) or left (reverse) in the following reactions for each of the specified conditions.

(a) $SO_2(g) + \frac{1}{2} O_2(g) \rightleftharpoons SO_3(g)$ $\Delta H = -$
   (1) temperature is raised
   (2) pressure is reduced
   (3) a catalyst is added

(b) $CO(g) + H_2(g) \rightleftharpoons CH_4(g) + H_2O(g)$ $\Delta H = +$
   (1) temperature is lowered
   (2) pressure is increased
   (3) CH_4 is removed from the system

(c) $NOCl(g) \rightleftharpoons NO(g) + \frac{1}{2} Cl_2(g)$ $\Delta H = +$
   (1) temperature is raised
   (2) pressure is increased
   (3) extra chlorine gas is added

11.4 THE LAW OF CHEMICAL EQUILIBRIUM

A dynamic equilibrium clearly implies that in a simple reversible reaction, the rates of the forward and reverse reactions are equal, since the initially faster reaction slows down, and the slower reaction speeds up. When the rates of the reactions are equal, the concentrations of all the interacting species are constant. It must be remembered, however, that a change in temperature, or in pressure if gases are involved, will change the equilibrium concentrations. When equilibrium data are treated mathematically, the temperature and pressure factors must be taken into account. The quantitative treatment of equilibrium concentrations is governed by the **law of chemical equilibrium,** sometimes called the law of mass action and mathematically described as the **equilibrium expression.**

In the symbolic reaction, $A + B \rightleftharpoons C + D$, it was pointed out previously that the rate of the forward reaction is proportional to the **product** of the molar concentrations.

$$\text{rate}_f \propto [A][B]$$

Conversely, the rate of the reverse reaction is proportional to the product of the molar concentrations of the reverse reactants:

$$\text{rate}_r \propto [C][D]$$

If the generalized reaction had a different form, such as:

$$A_2 + 2 B_2 \rightleftharpoons 2 AB_2$$

$$1 \text{ mole} + 2 \text{ moles} \qquad 2 \text{ moles}$$

CHEMICAL EQUILIBRIUM AND REACTION RATE

the expression of the rate proportionalities would be:

$$\text{rate}_f \propto [A_2][B_2]^2$$

and

$$\text{rate}_r \propto [AB_2]^2$$

   Notice that the *coefficients in the equation become exponential values* in the expressions of rate proportionality. The reasoning used to explain these empirical observations is that presence of more than one mole of a reacting species creates a much greater probability for favorable collisions to occur per unit of time. This reasoning experience is a good example of the historical interplay between experiment and theory. It was long ago, in 1864, that the Norwegian chemists Guldberg and Waage discovered experimentally the law of mass action, which we now explain in terms of the law of chemical equilibrium. The proportionality of rates and the products of molar concentrations can easily be expressed mathematically for a simple one-step process. For the equation,

$$nA + oB \rightleftharpoons pC + qD$$
$$\text{rate}_f = k_f[A]^n[B]^o$$

where k_f is the proportionality constant for the forward reaction.

$$\text{rate}_r = k_r[C]^p[D]^q$$

where k_r is the proportionality constant for the reverse reaction. At equilibrium, the equalization of the rates,

$$\text{rate}_f = \text{rate}_r$$

permits an equating of the expressions for the rates:

$$k_f[A]^n[B]^o = k_r[C]^p[D]^q$$

Conventionally, the two rate constants are combined into a single constant called the **equilibrium constant,** and symbolized K_c. The right-hand side of the equation is written as a *numerator,* the left-hand side becomes the *denominator,* and the coefficients become *exponents.*

$$\frac{k_f}{k_r} = K_c = \frac{[C]^p[D]^q \text{ (products)}}{[A]^n[B]^o \text{ (reactants)}}$$

In the second generalized equation, $A_2 + 2 B_2 \underset{r}{\overset{f}{\rightleftharpoons}} 2 AB_2$, the equilibrium expression is:

$$K_c = \frac{[AB_2]^2}{[A_2][B_2]^2}$$

From the preceding examples, the law of chemical equilibrium can be extracted. *In an equilibrium system at constant temperature, the ratio of the product of the molar concentrations of the products compared to the product of the molar concentrations of the reactants, each raised to a suitable power, is equal to a constant value called the equilibrium constant.*

What is the *significance* of an equilibrium constant? The ability to calculate or use an equilibrium constant is meaningless if it is interpreted as being nothing more than a number. For example, in the reaction $H_2(g) + I_2(g) \rightleftharpoons 2 HI(g)$, the equilibrium expression is

$$K_c = \frac{[HI]^2}{[H_2][I_2]}$$

The value of K_c is found to be different at various temperatures:

$$K_c = 64 \text{ at } 445°C$$

$$K_c = 46 \text{ at } 490°C$$

$$K_c = 808 \text{ at } 25°C$$

The significance of these K_c values is that they indicate the temperature range most appropriate to produce a large amount of HI. Remember, the K_c is related to the molar concentrations of products to reactants at equilibrium. The value 808 for the K_c at 25°C means that a low temperature provides the greatest yield of product. In effect, the relationship

$$\frac{[HI]^2}{[H_2][I_2]} = 808$$

clearly demonstrates a greater yield than the K_c at 445°C, in which case,

$$K_c = \frac{[HI]^2}{[H_2][I_2]} = 64$$

Now let us examine the significance of the symbol K_W and its equivalence to 1.0×10^{-14} as it was introduced in the previous chapter

CHEMICAL EQUILIBRIUM AND REACTION RATE

in the introduction to pH. The K_W is derived from the equilibrium constant for the reaction

$$H_2O(l) \rightleftharpoons H^+(aq) + OH^-(aq)$$

The equilibrium expression is

$$K_c = \frac{[H^+][OH^-]}{[H_2O]}$$

Since so little $H_2O(l)$ actually ionizes, the molar concentration is essentially constant (55.6 moles/liter). Combining the constants K_c and $[H_2O]$, the equilibrium expression may be simplified:

$$K_c[H_2O] = K_c[55.6] = K_W$$

and

$$K_W = [H^+][OH^-] = 1.0 \times 10^{-14} \text{ at } 25°C$$

The value 1.0×10^{-14} illustrates how small the ion concentration is in pure water and the limited degree of ionization of water.

The K_c for a weak acid such as acetic acid is found to have an equilibrium dissociation constant at $25°C$ of 1.8×10^{-5}. Probably the most widely used symbol for the equilibrium constant of a weak acid is K_a, although K_i and K_{diss} are also used. The value for the K_a calculated from experimentally determined data indicates the ratio of the product of the molar concentration of the ions compared to the molar concentration of the unionized molecules. For example, the equation for the dissociation (or ionization) of a mole of HOAc per liter of solution is

$$HOAc(aq) \rightleftharpoons H^+(aq) + OAc^-(aq)*$$

The equilibrium expression is:

$$K_a = \frac{[H^+][OAc^-]}{[HOAc]} \quad \text{ions}$$

un-ionized molecules

The value at $25°C$ for the K_a is 1.8×10^{-5}, which means that the degree of ionization of molecules is very small. We are disregarding the $[H^+]$ from the ionization of water.

$$K_a = \frac{[H^+][OAc^-]}{[HOAc]} = 1.8 \times 10^{-5}$$

---

*An alternative to the abbreviated formula for acetic acid is CH_3COOH. The equation is: $CH_3COOH(aq) \rightleftharpoons H^+(aq) + CH_3COO^-(aq)$.

Therefore, by any definition of an acid, HOAc(aq) must be described as weak. It should be mentioned that water is not included in the equilibrium expression, despite the fact that the ionization of acetic acid takes place in water solution. This is because the concentration of water molecules remains relatively constant, and its value is automatically included in the K_c.

Table 11–1 shows a comparison of some acids with regard to a quantitative description of their strengths in terms of K_a values.

The equilibrium constants (also referred to as dissociation or ionization constants) of weak bases, symbolized K_b, communicate the same type of information as do the K_a values for acids. For example, the value for the K_b of NH_3(aq) is 1.8×10^{-5}. The equation,

$$NH_3(g) + H_2O(\ell) \rightleftharpoons NH_4{}^+(aq) + OH^-(aq)$$

appears in the equilibrium expression as,

$$K_b = \frac{[NH_4{}^+][OH^-]}{[NH_3]}$$

The K_b value means that the ratio of the iron product concentration to the ammonia concentration is very small. In other words, the aqueous solution of ammonia yields relatively few hydroxide ions in solution. Again, water contributes few OH^- ions and is disregarded.

$$K_b = \frac{[NH_4{}^+][OH^-]}{[NH_3]} = 1.8 \times 10^{-5}$$

ions

molecules

It is only a coincidence that K_a for HOAc(ℓ) + $H_2O(\ell)$ is numerically the same as K_b for NH_3(g) + $H_2O(\ell)$. The great majority of metal hydroxides, such as $Mg(OH)_2$, $Fe(OH)_3$, $Fe(OH)_2$, $Ca(OH)_2$, $Cd(OH)_2$, and $Zn(OH)_2$, tend to be only slightly soluble in water. The equilibrium constants that reflect the tendency of these weak bases to

TABLE 11–1 **Equilibrium Constants of Weak Acids at 25°C**

	Name	First Ionization Reaction	K_a
increasingly weak acids	oxalic	$H_2C_2O_4 \rightleftharpoons H^+ + HC_2O_4{}^-$	5.6×10^{-2}
	sulfurous	$H_2SO_3 \rightleftharpoons H^+ + HSO_3{}^-$	1.3×10^{-2}
	phosphoric	$H_3PO_4 \rightleftharpoons H^+ + H_2PO_4{}^-$	5.9×10^{-3}
	hydrofluoric	$HF \rightleftharpoons H^+ + F^-$	6.7×10^{-4}
	nitrous	$HNO_2 \rightleftharpoons H^+ + NO_2{}^-$	5.1×10^{-4}
	acetic	$HOAc \rightleftharpoons H^+ + OAc^-$	1.8×10^{-5}
	hydrocyanic	$HCN \rightleftharpoons H^+ + CN^-$	4.8×10^{-10}

dissociate to a slight degree will be introduced in our discussion of **solubility product** (Section 11.7).

When the ionization of diprotic or triprotic acids is considered, the meaning of the K_a values becomes more involved. For example, the ionization of diprotic H_2SO_3 can be shown to occur in two steps,

(1) $H_2SO_3(aq) \rightleftharpoons H^+(aq) + HSO_3^-(aq)$
(2) $HSO_3^-(aq) \rightleftharpoons H^+(aq) + SO_3^{2-}(aq)$

Each step in the ionization will naturally have its own K_a value. A reference table lists the folowing K_a values at 25°C:

$$H_2SO_3 \qquad K_{a_1} = 1.3 \times 10^{-2}$$

$$HSO_3^- \qquad K_{a_2} = 5.6 \times 10^{-8}$$

If these K_a values are viewed within the framework of the equilibrium expression, the species of ions and molecules in greatest abundance at equilibrium can be determined.

$$K_{a_1} = \frac{[H^+][HSO_3^-]}{[H_2SO_3]} = 1.3 \times 10^{-2}$$

$$K_{a_2} = \frac{[H^+][SO_3^{2-}]}{[HSO_3^-]} = 5.6 \times 10^{-8}$$

The K_{a_1} value for any diprotic acid is much larger than the K_{a_2} indicating that the first ionization step is far more extensive than the second. We may also conclude the following:

1. Most of the H_2SO_3 does not ionize. This is consistent with the labeling of H_2SO_3 as a weak acid.
2. At equilibrium $H^+(aq)$ and $HSO_3^-(aq)$ will be vastly more abundant than $SO_3^{2-}(aq)$.
3. The $HSO_3^-(aq)$ could be characterized as an even weaker acid than $H_2SO_3(aq)$.

Example 11.2

Given the reaction: $2 PCl_3(g) + 2 Cl_2(g) \rightleftharpoons 2 PCl_5(g)$.
Calculate K_c if the equilibrium concentrations of reactants and products were found to be:

$$[PCl_3] = 1.2 \text{ M}; [Cl_2] = 1.4 \text{ M}; [PCl_5] = 1.8 \text{ M}$$

$$K_c = \frac{[PCl_5]^2}{[PCl_3]^2 [Cl_2]^2} = \frac{[1.8]^2}{[1.2]^2 [1.4]^2}$$

$$\boxed{K_c = 1.1}$$

Example 11.3

Given the reaction: $2 NO(g) + Cl_2(g) \rightleftharpoons 2 NOCl(g)$, the equilibrium concentrations of reactants and products were found to be [NOCl] = 0.66M; [NO] = 0.33 M; [Cl$_2$] = 0.66 M.

Write the equilibrium expression and calculate the numerical value of K_c.

$$K = \frac{[NOCl]^2}{[NO]^2[Cl_2]} = \frac{[0.66]^2}{[0.33]^2[0.66]}$$

$$\boxed{K_c = 6.1}$$

Exercise 11.2

1. Write the equilibrium expressions for the following equilibrium reactions:
   (a) $3 H_2(g) + N_2(g) \rightleftharpoons 2 NH_3(g)$
   (b) $4 NH_3(g) + 5 O_2(g) \rightleftharpoons 6 H_2O(g) + 4 NO(g)$
   (c) $H_2C_2O_4(aq) \rightleftharpoons HC_2O_4^-(aq) + H^+(aq)$
   (d) $CH_3NH_2(l) + H_2O(l) \rightleftharpoons CH_3NH_3^+(aq) + OH^-(aq)$
       (Hint: Do not include H_2O in the equilibrium expression.)
   (e) $H_2PO_4^-(aq) \rightleftharpoons H^+(aq) + HPO_4^{2-}(aq)$
2. Organize the following acids in the order of *decreasing* strength.
   (a) H_3PO_4 $K_{a_1} = 5.9 \times 10^{-3}$
   (b) HCNO $K_a' = 2.0 \times 10^{-4}$
   (c) HS$^-$ $K_{a_2} = 1.3 \times 10^{-13}$
   (d) $H_2C_2O_4$ $K_{a_1} = 5.6 \times 10^{-2}$
   (e) HOAc $K_a = 1.8 \times 10^{-5}$
3. K_c for the reaction $N_2(g) + 3 H_2(g) \rightleftharpoons 2 NH_3(g)$ is 6.6 Calculate the equilibrium concentration of NH_3 when the equilibrium concentration of N_2 is 2.0 M and that of H_2 is 3.0 M.
4. In the equilibrium: $2 SO_2(g) + O_2(g) \rightleftharpoons 2 SO_3(g)$, equilibrium concentrations are: [SO$_2$] = 0.22 M; [O$_2$] = 0.40 M; [SO$_3$] = 0.80 M. Calculate K_c. What is the concentration of SO_3 when the concentrations of SO_2 and O_2 are 0.12 M and 0.20 M, respectively?

11.5 CALCULATIONS INVOLVING EQUILIBRIUM CONSTANTS OF WEAK ELECTROLYTES IN SOLUTION

When a 1.0 M solution of HCl is prepared, it may be assumed that the HCl(g) *ionizes* almost 100 per cent. This observation characterizes HCl(aq) as a strong acid. The resulting [H$^+$] may be assumed to be approximately 1.0 M and the pH approximately 0. However, the pH of a weak acid such as acetic acid is another story. A 1.0 M solution of HOAc will not form anywhere near a 1 molar concentration of hydrogen ion. The law of chemical equilibrium is required to calculate the pH of solutions of weak electrolytes.

CHEMICAL EQUILIBRIUM AND REACTION RATE

Example 11.4

Calculate the pH of a 1.0 M solution of HOAc.
1. Write the chemical reaction for the ionization of HOAc.

$$HOAc(aq) \rightleftharpoons H^+(aq) + OAc^-(aq)$$

2. Write the equilibrium expression:

$$K_a = \frac{[H^+][OAc^-]}{[HOAc]}$$

3. Organize the data. Let x = moles per liter of HOAc that ionize.

$$K_a \text{ (from Table 11-1)} = 1.8 \times 10^{-5}$$
$$[HOAc] = [1.0-x]$$

At equilibrium, the 1.0 molar concentration of HOAc is then diminished by x moles per liter. The number of moles of HOAc that ionize will form x moles of H^+ and, equally, x moles of OAc^- per liter.

$$HOAc \xrightarrow{\text{ionization}} H^+ + OAc^-$$

at equilibrium:

$$[H^+] = [OAc^-] = [x]$$

and

$$[HOAc] = [1.0 - x]$$

4. Substitute in the equilibrium expression to solve for $[H^+]$.

$$1.8 \times 10^{-5} = \frac{[x][x]}{[1.0 - x]}$$

Since the K_a for acetic acid indicates a very low value for the ion concentrations compared to the 1.0 M concentration of the HOAc molecules, the equation is usually simplified by ignoring the difference between 1.0 and $[1.0 - x]$. To maintain the denominator as $[1.0 - x]$ would involve the solution of a quadratic equation by the formula, $x = \frac{-b \pm \sqrt{b^2 - 4ac}}{2a}$. This represents a great deal of trouble for a correction that would be analogous to subtracting two cents from ten thousand dollars. However, when the calculated value for x exceeds 5 per cent of the original molar concentration, it must be subtracted in order to avoid an excessive error.

5. Simplify the equilibrium expression, and solve for x.

$$1.8 \times 10^{-5} = \frac{x^2}{1.0}$$

$x^2 = 1.8 \times 10^{-5}$ (calculate the square root of the equation)
$x = \sqrt{1.8 \times 10^{-5}} = \sqrt{18 \times 10^{-6}}$
$x = 4.2 \times 10^{-3}$
$x = [H^+] = 4.2 \times 10^{-3}$ M
Note: 4.2×10^{-3} M is considerably less than 5 per cent of 1.0 M and is therefore acceptable. If the hydrogen ion concentration were more than 5 per

cent of original concentration of the HOAc, it would still be possible to avoid the quadratic equation by the method of "successive approximations" that you will learn in a more advanced chemistry course.

6. Calculate the pH of the solution.

$$pH = -\log (H^+)$$
$$pH = -(\log 4.2 + \log 10^{-3})$$
$$pH = -(0.63 - 3)$$

$$pH = 2.4$$

The ionization constant of a particular weak electrolyte can be calculated from its measured pH when the original molar concentration is known.

Example 11.5

Calculate the K_a of a 0.005 M hydrocyanic acid solution where the pH is measured at 5.81 at 25°C.

1. Write the chemical equation for the ionization of HCN.

$$HCN(aq) \rightleftharpoons H^+(aq) + CN^-(aq)$$

2. Write the equilibrium expression:

$$K_a = \frac{[H^+][CN^-]}{[HCN]}$$

3. Organize the data that will produce the $[H^+]$ and $[CN^-]$ concentrations, both of which are equal, since each ionizing molecule of HCN produces 1 molecule of H^+ and 1 molecule of CN^-.

$$pH = 5.81$$
$$pH = -\log[H^+]$$
$$[H^+] = \text{antilog of } -pH$$
$$= \text{antilog } (-5.81)$$
$$= \text{antilog } (0.15 - 6)$$
$$[H^+] = 1.55 \times 10^{-6}$$
$$[CN^-] = 1.55 \times 10^{-6} \text{ M}$$
$$[HCN] = 0.005 \text{ M} = 5 \times 10^{-3} \text{ M}$$
$$K_a = ?$$

4. Substitute in the equilibrium expression and solve for the K_a of HCN.

$$K_a = \frac{(1.55 \times 10^{-6})^2}{(5 \times 10^{-3})} = \frac{2.40 \times 10^{-12}}{5 \times 10^{-3}}$$

$$K_a = \frac{2.40 \times 10^{-13}}{5 \times 10^{-3}} = 4.8 \times 10^{-10}$$

$$K_a \text{ of } HCN(aq) = 4.8 \times 10^{-10}$$

5. Note how closely the calculated value for K_a agrees with the generally accepted value listed in Table 11–1.

CHEMICAL EQUILIBRIUM AND REACTION RATE

11.6 CALCULATIONS INVOLVING EQUILIBRIUM CONSTANTS OF SLIGHTLY SOLUBLE COMPOUNDS

The Solubility Product Principle

When an excess of a slightly soluble compound is added to water, an equilibrium develops between the undissolved solid compound and the relatively small concentration of ions in solution. This is true of many salts and hydroxide compounds that normally appear as precipitates in reactions.

For example, the addition of sufficient chloride ion to a solution containing silver ion produces the precipitate AgCl(s). Silver chloride obviously is not very soluble in water. In fact, a concentration in the millimolar range represents a saturated solution. It would be quite impossible to prepare a 1 molar, 1/10 molar, or even 1/1000 molar solution of AgCl in water.

The equilibrium equation for the dissociation of AgCl(s) is:

$$AgCl(s) \rightleftharpoons Ag^+(ag) + Cl^-(aq)$$

The equilibrium expression would appear to be:

$$K_{diss} = \frac{[Ag^+][Cl^-]}{[AgCl]}$$

However, since so little AgCl(s) actually dissociates, the original number of moles of AgCl may be considered to remain constant. This is analogous to the conditions that lead to the development of a K_W value for water ionization. Combining the two constants, K_{diss} and [AgCl], the equilibrium expression can be simplified:

$$K_{diss} \times [AgCl] = \underbrace{[Ag^+][Cl^-]}_{\text{ion product}} = K_{sp}$$

The combined constants are provided with a new symbol, K_{sp}, and are called the **solubility product constant**. The term **solubility product** is quite appropriate, since the value for K_{sp} is indeed the product of the ion concentrations in a saturated solution at equilibrium.

11.7 THE SIGNIFICANCE OF K_{sp} values

A glance at a table of solubility product constants (Table 11–2) quickly points out markedly varying degrees of ion dissociation in water. A compound in which the K_{sp} is 10^{-5} for example, is much more soluble than another compound that has a K_{sp} of 10^{-20}.

With the availability of K_{sp} values for slightly soluble compounds, it is possible to calculate the molar concentration of a selected ion that would be necessary to cause precipitation. That is to say, if the ion product were adjusted to exceed the K_{sp}, a precipitate would result. In the same manner, ion additions may be calculated so that precipitation can be avoided. Such calculations are commonly used in analytical chemistry when it is advantageous to separate ions by precipitation techniques.

11.8 FINDING THE K_{sp}

Solubility product constants, like all equilibrium constants, vary with temperature. In the development of any table of K_{sp} values, the temperature at which the ion product was calculated must be specified. An illustration of this importance is the case of $PbCl_2$. At $25°C$, the K_{sp} is 1.7×10^{-5}, which indicates a great contrast in behavior to its extensive solubility in hot water.

The K_{sp} values of slightly soluble compounds can be calculated after experimentally measuring their solubility in terms of grams of

TABLE 11–2 Equilibrium Expressions of Some Slightly Soluble Compounds at 25°C

Name	Dissociation Reaction	Equilibrium Expression
silver acetate	$AgOAc \rightleftharpoons Ag^+ + OAc^-$	$K_{sp} = [Ag^+][OAc^-]$
lead(II) chloride	$PbCl_2 \rightleftharpoons Pb^{2+} + 2Cl^-$	$K_{sp} = [Pb^{2+}][Cl^-]^2$
barium fluoride	$BaF_2 \rightleftharpoons Ba^{2+} + 2F^-$	$K_{sp} = [Ba^{2+}][F^-]^2$
magnesium carbonate	$MgCO_3 \rightleftharpoons Mg^{2+} + CO_3^{2-}$	$K_{sp} = [Mg^{2+}][CO_3^{2-}]$
silver chloride	$AgCl \rightleftharpoons Ag^+ + Cl^-$	$K_{sp} = [Ag^+][Cl^-]$
silver iodide	$AgI \rightleftharpoons Ag^+ + I^-$	$K_{sp} = [Ag^+][I^-]$
cadmium sulfide	$CdS \rightleftharpoons Cd^{2+} + S^{2-}$	$K_{sp} = [Cd^{2+}][S^{2-}]$
iron(III) hydroxide	$Fe(OH)_3 \rightleftharpoons Fe^{3+} + 3 OH^-$	$K_{sp} = [Fe^{3+}][OH^-]^3$
mercury(II) sulfide	$HgS \rightleftharpoons Hg^{2+} + S^{2-}$	$K_{sp} = [Hg^{2+}][S^{2-}]$
bismuth(III) sulfide	$Bi_2S_3 \rightleftharpoons 2Bi^{3+} + 3S^{2-}$	$K_{sp} = [Bi^{3+}]^2[S^{2-}]^3$

decreasing solubility (arrow pointing down alongside the table)

CHEMICAL EQUILIBRIUM AND REACTION RATE

solute per liter of water in saturated solutions. For example, the solubility of AgCl at room temperature is found to be 0.000186 g per 100 g of water. This may be considered to be 0.00186 g per liter of solution, since the small amount of soluble AgCl hardly affects the volume of water. The molar concentration of the dissociated AgCl is the same as the molar concentrations of $Ag^+(aq)$ and $Cl^-(aq)$, since each dissociating unit of AgCl forms one ion each of Ag^+ and Cl^-, and since the concentration of undissociated AgCl, a constant at a given temperature, may be considered negligible. The molar concentration may be calculated:

$$\text{molarity} = \frac{\text{mass}}{\text{liters}} \times \frac{1}{\text{molar mass}} = \text{mole/L}$$

The molar mass of AgCl = 143 g/mole.

$$\text{molarity} = \frac{0.00186 \text{ g}}{1 \text{ L}} \times \frac{1 \text{ mole}}{143 \text{ g}}$$

$$\text{molarity} = 1.30 \times 10^{-5} \text{ mole/L}$$

Since the equilibrium expression is

$$K_{sp} = [Ag^+][Cl^-]$$

the K_{sp} value can be calculated from the molar concentrations of Ag^+ and Cl^-, both of which are equal to 1.30×10^{-5} M.

$$K_{sp} = (1.3 \times 10^{-5})^2$$

$$K_{sp} = 1.7 \times 10^{-10} \text{ for AgCl at } 25°C$$

Table 11–3 presents a selected list of K_{sp} values (approximate) at 25°C that can be used for handy reference.

11.9 SAMPLE CALCULATIONS INVOLVING THE SOLUBILITY PRODUCT PRINCIPLE

It is sometimes useful to be able to calculate the solubility of a compound. The question of whether a solution can be made as concentrated as desired can be answered in this way.

TABLE 11–3 Approximate K_{sp} Values at 25°C

		K_{sp}				K_{sp}
Acetate	$AgC_2H_3O_2$	2×10^{-3}	Iodides	AgI		1.5×10^{-16}
				PbI_2		1×10^{-8}
Bromides	AgBr	1×10^{-13}				
	$PbBr_2$	5×10^{-6}	Sulfates	$BaSO_4$		1.5×10^{-9}
				$CaSO_4$		3×10^{-5}
Carbonates	$BaCO_3$	1×10^{-9}		$PbSO_4$		1×10^{-8}
	$CaCO_3$	5×10^{-9}				
	$MgCO_3$	2×10^{-8}	Sulfides	Ag_2S		1×10^{-49}
	$PbCO_3$	1×10^{-13}		CdS		1×10^{-26}
				CoS		1×10^{-21}
Chlorides	AgCl	1.7×10^{-10}		CuS		1×10^{-25}
	Hg_2Cl_2	1×10^{-18}		FeS		2×10^{-17}
	$PbCl_2$	1.7×10^{-5}		HgS		1×10^{-52}
				MnS		1×10^{-13}
Chromates	Ag_2CrO_4	1×10^{-12}		NiS		1×10^{-22}
	$BaCrO_4$	2×10^{-10}		PbS		1×10^{-27}
	$PbCrO_4$	2×10^{-14}		ZnS		1×10^{-23}
Fluorides	BaF_2	2×10^{-6}				
	CaF_2	2×10^{-10}				
	PbF_2	4×10^{-8}				
Hydroxides	$Al(OH)_3$	5×10^{-33}				
	$Cr(OH)_3$	1×10^{-30}				
	$Fe(OH)_2$	1×10^{-15}				
	$Fe(OH)_3$	5×10^{-38}				
	$Mg(OH)_2$	1×10^{-11}				
	$Zn(OH)_2$	5×10^{-17}				

Example 11.6

Is it possible to prepare a 1.0×10^{-3}M solution of $MgCO_3$ at 25°C? What is the greatest molar concentration possible? What is the solubility of $MgCO_3$ in g per 100 mL of solution?

1. Write the chemical equation for the dissociation of $MgCO_3$.

$$MgCO_3(s) \rightleftharpoons Mg^{2+}(aq) + CO_3^{2-}(aq)$$

2. Write the equilibrium expression.

$$K_{sp} = [Mg^{2+}][CO_3^{2-}]$$

3. Find the K_{sp} value in Table 11–3 and organize the data:

$$K_{sp} \text{ of } MgCO_3 = 2 \times 10^{-8}$$

$$[Mg^{2+}] = [x]$$

$$[CO_3^{2-}] = [x]$$

4. Substitute in the equilibrium expression and solve for the concentration of $MgCO_3$, which is the same as Mg^{2+} (aq) and CO_3^{2-} (aq).

$$2 \times 10^{-8} = x^2$$

$$x = \sqrt{2 \times 10^{-8}} = 1.4 \times 10^{-4} \text{ M } MgCO_3$$

CHEMICAL EQUILIBRIUM AND REACTION RATE

5. Since the concentration of a saturated solution of $MgCO_3$ is $1.4 \times 10^{-4}M$, it is *not* possible to prepare a 1.0×10^{-3} M solution at 25°C. The largest concentration possible is 1.4×10^{-4} M.
6. 1.4×10^{-4} mole per liter is equivalent to 1.4×10^{-5} moles per 100 mL of solution. The number of grams in 1.4×10^{-5} mole of $MgCO_3$ can be calculated:

$$\text{molar mass of } MgCO_3 = 84.3 \text{ g/mole}$$

$$\text{mass} = \frac{1.4 \times 10^{-5} \text{ mole}}{100 \text{ mL}} \times \frac{84.3 \text{ g}}{1 \text{ mole}}$$

$$\boxed{\text{mass of } MgCO_3 \text{ per 100 mL} = 1.2 \times 10^{-3} \text{ g}}$$

Example 11.7 presents another common problem that can be solved by the application of the solubility product principle.

Example 11.7

What is the maximum concentration of chloride ion that can be added to a 0.060 M solution of $Pb(NO_3)_2$ at 25°C without causing the precipitation of $PbCl_2$?
1. Write the chemical equation for the dissociation of $PbCl_2$.

$$PbCl_2 \rightleftharpoons Pb^{2+}(aq) + 2 \ Cl^-(aq)$$

2. Write the equilibrium expression. (Remember that coefficients become exponents.)

$$K_{sp} = [Pb^{2+}][Cl^-]^2$$

3. Organize the data:

$$K_{sp} \text{ of } PbCl_2 = 1.7 \times 10^{-5} \text{ (from Table 11–3)}$$
$$[Pb^{2+}] = 0.060 \text{ M [from } Pb(NO_3)_2]$$
$$[Cl^-] = [x], \text{ so } [Cl^-]^2 = [x]^2$$

4. Substitute in the equation and solve for the chloride ion concentration.

$$1.7 \times 10^{-5} = [0.060][x]^2$$

$$x^2 = \frac{1.7 \times 10^{-5}}{0.060} = 2.8 \times 10^{-4}$$

$$x = \sqrt{2.8 \times 10^{-4}}$$

$$\boxed{= 1.7 \times 10^{-2} \text{ M chloride ion}}$$

5. The concentration of chloride ion must not exceed 1.7×10^{-2} M, at which point precipitation begins. Remember, precipitation will not occur until the ion product exceeds the K_{sp}.

1. Find the solubility of silver acetate in grams per 100 mL of solution, using the K_{sp} value listed in Table 11–3.
2. Calculate the K_{sp} of $Cd(OH)_2$ if the compound is found to have a solubility of 2.1×10^{-4} g per 100 mL of solution.
3. What minimum mass of $K_2CrO_4(s)$ must be added to 200 mL of 1.6×10^{-6} M $Pb(NO_3)_2$ solution to initiate the precipitation of $PbCrO_4$? $K_{sp} = 2 \times 10^{-14}$.

QUESTIONS & PROBLEMS

11.1 Define the term *dynamic equilibrium* as it applies to reversible chemical reactions.

11.2 List five factors that may affect the rate of a reaction.

11.3 In what types of equilibrium reaction does pressure become a necessary factor for calculation purposes?

11.4 If Pt acts as a catalyst in the formation of ammonia from nitrogen and hydrogen gas, what effect will the removal of Pt have on the equilibrium concentration of NH_3? Explain your answer.

11.5 In the following unbalanced reaction equations, explain the effects of pressure changes on the forward and reverse reactions.
(a) $H_2(g) + Cl_2(g) \rightleftharpoons HCl(g)$
(b) $CO(g) + CuO(s) \rightleftharpoons Cu(s) + CO_2(g)$
(c) $C_6H_6(l) + O_2(g) \rightleftharpoons CO_2(g) + H_2O(g)$

11.6 What effect would an increase in temperature be expected to have on the following reactions?
(a) $N_2O_4(g) \rightleftharpoons 2 NO_2(g) - heat$
(b) $N_2O(l) \rightleftharpoons H_2(g) + \frac{1}{2} O_2(g) - heat$
(c) $C_2H_4(g) + H_2(g) \rightleftharpoons C_2H_6(g) + heat$

11.7 Given the reaction,

$$CO_2(g) \rightleftharpoons CO(g) + \frac{1}{2} O_2(g) \qquad \Delta H = -$$

indicate the direction of shift (forward or reverse) when,
(a) temperature is raised
(b) pressure is increased
(c) oxygen leaks out of the system

11.8 Explain the meaning of the following symbols:
(a) K_c
(b) K_i
(c) K_a
(d) K_{diss}
(e) K_{sp}
(f) K_w
(g) K_b

11.9 Write a sample equation and equilibrium expression for each of the symbols listed in Question 11.8.

11.10 Write equilibrium expressions for the following reactions.
(a) $2 NOCl(g) \rightleftharpoons 2 NO(g) + Cl_2(g)$
(b) $HCO_3^-(aq) \rightleftharpoons H^+(aq) + CO_3^{2-}(aq)$
(c) $CO_3^{2-}(aq) + H_2O(l) \rightleftharpoons$
$\qquad\qquad HCO_3^-(aq) + OH^-(aq)$
(d) $HNO_2(aq) \rightleftharpoons H^+(aq) + NO_2^-(aq)$
(e) $Fe(OH)_2(s) \rightleftharpoons Fe^{2+}(aq) + 2 OH^-(aq)$

11.11 Given the following equation:

$$2 NO(g) + O_2(g) \rightleftharpoons 2 NO_2(g) + 62 kcal$$

(a) Calculate the K_c if the equilibrium concentrations are 0.6M NO_2, 0.2 M NO, and 0.3M O_2.
(b) List three ways in which the *reverse* reaction might be favored.

11.12 Given the equation:

$$2 C_2H_6(g) + 7 O_2(g) \rightleftharpoons$$
$$4 CO_2(g) + 6 H_2O (l) + heat$$

state whether the forward or reverse reactions would be favored by the following changes:
(a) an increase in temperature
(b) a lowering of the pressure
(c) an addition of $C_2H_6(g)$
(d) the use of a catalyst

11.13 The equilibrium constant for $2 NO_2(g) \rightleftharpoons N_2O_4(g)$ is 2.6. If the equilibrium concentration of NO = 2.2 M, what is the equilibrium concentration of N_2O_4?

11.14 For the reaction $2 HI(g) \rightleftharpoons H_2(g) + I_2(g)$, equilibrium concentrations of reactants and products are: [HI] = 0.21 M, $[H_2]$ = 0.60 M, and $[I_2]$ = 1.22 M. Calculate K_c.

11.15 Analysis shows that when $NO_2(g)$ and $SO_2(g)$ are allowed to react and reach equilibrium, $[SO_3]$ = 0.60 M, [NO] = 0.40 M, $[NO_2]$ = 1.4 M, and $[SO_2]$ = 1.2 M. Calculate the equilibrium constant for the reaction:

$$SO_2(g) + NO_2(g) \rightleftharpoons NO(g) + SO_3(g)$$

11.16 Equilibrium concentrations for the reaction $H_2(g) + I_2(g) \rightleftharpoons 2 HI(g)$ are: $[H_2]$ = 0.44 M, $[I_2]$ = 0.36 M, and [HI] = 0.32 M. Calculate K_c and the equilibrium concentration of HI when $[H_2]$ = 0.88 M and $[I_2]$ = 0.66 M.

11.17 Find the first ionization constant, K_{a_1}, for H_3BO_3 (boric acid) if the pH of a 0.0040 M solution is 5.82. The chemical expression is $H_3BO_3(aq) \rightleftharpoons H^+(aq) + H_2BO_3^-(aq)$.

11.18 What is the pH of a 0.02 M solution of HClO(aq)? The K_a of HClO(aq) is 3.2×10^{-8}.

11.19 Find the K_a of a 0.005 M solution of HCNO(aq) in which the pH is 2.70.

11.20 Calculate the solubility of PbI_2 in grams per liter, given the theoretical K_{sp} value for PbI_2 = 1.0×10^{-8}.

11.21 Write equilibrium expressions for the two steps in the ionization of H_2SO_4.

11.22 If the reaction $X + Y \rightleftharpoons XY$ has the values K_c at 20°C = 17 and K_c at 80°C = 42, what should be done about the temperature if the production of XY is undesirable?

11.23 If the solubility of As_2S_3 is 5.2×10^{-4}g per liter, calculate the theoretical K_{sp} value.

11.24 Calculate the solubility of $Zn(OH)_2$ in g per 100 mL of solution.

CHAPTER TWELVE

OXIDATION-REDUCTION AND AN INTRODUCTION TO ELECTROCHEMISTRY

The type of chemical reaction described as oxidation-reduction, often contracted to **redox,** is a topic of major importance in chemistry. The tendency of elements to lose and gain electrons in the drive toward stability and lower energy content was introduced in Chapter 5 under the heading *ionization energy.*

An understanding of the principles of oxidation and reduction is especially critical because of the physical hazards involved. The chronicles of scientific activity are filled with stories of explosions, damage to skin and eyes, fires, and billions of dollars' worth of loss in corrosion. On the other hand, science has permitted great strides forward in the development of electro-chemical cells, protection of metals against corrosion, modern dental and surgical methods, and a growing understanding of the use and production of energy from fuels and foods. All of these processes involve oxidation-reduction reactions.

Oxidation of a substance is defined as an *increase* in its **oxidation number** due to a *loss* of electrons. **Reduction** is a *decrease* in the oxidation number, which is achieved by a *gain* of electrons. Any substance that promotes the loss of electrons because of its own tendency to gain them is called an **oxidizing agent.** Any substance that readily gives up its electrons is called a **reducing agent.** In the course of their work, oxidizing agents become reduced and reducing agents become oxidized. Oxidation cannot occur without reduction. The total concept of oxidation-reduction implies a **transfer** of the electrons lost by the substance being oxidized to the substance being reduced. Figure 12–1 illustrates the hypothetical relationship between the change in oxidation number and the processes of oxidation and reduction.

As we try to understand what is happening in an oxidation-reduction reaction and proceed to write balanced equations that describe the event, it is very helpful to know what substances act as

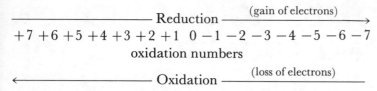

$$\xrightarrow{\text{Reduction} \quad \text{(gain of electrons)}}$$

$$+7 \quad +6 \quad +5 \quad +4 \quad +3 \quad +2 \quad +1 \quad 0 \quad -1 \quad -2 \quad -3 \quad -4 \quad -5 \quad -6 \quad -7$$

oxidation numbers

$$\xleftarrow{\text{Oxidation} \quad \text{(loss of electrons)}}$$

Figure 12–1 An increase in oxidation number is defined as oxidation (loss of electrons), and a decrease in oxidation (gain of electrons) is reduction.

oxidizing agents (acceptors of electrons) and those that act as reducing agents (donors of electrons).

A knowledge of common oxidizing and reducing agents can help us design reactions in which selected substances can be oxidized or reduced as we choose. It also helps us to know what substances ought to be avoided if we do not wish the oxidation or reduction of a particular species to take place. Furthermore, we can balance equations much more efficiently if the oxidizing and reducing agents can be identified quickly.

For your convenience, Table 12–1 lists some common oxidizing and reducing agents.

12.3 OXIDATION-REDUCTION REACTIONS

Consider the reaction between zinc metal and copper(II) sulfate in aqueous solution. When the zinc is placed in a tube of the blue solution, it darkens immediately as reddish clumps of solid form. The zinc

TABLE 12–1

Common Oxidizing Agents

Oxidizing Agent	Number of Electrons Gained	Common Products (in aqueous solution)
F_2 (g)	$+2$ e$^-$ $\longrightarrow$	2 F$^-$ (aq)
MnO_4^- (aq) (in acid)	$+5$ e$^-$ or $+3$ e$^-$ $\longrightarrow$	Mn^{2+} (aq), MnO$_2$ (s)
ClO_4^- (aq) (in acid)	$+8$ e$^-$ or $+7$ e$^-$ $\longrightarrow$	Cl$^-$ (aq), Cl$_2$ (g)
Cl_2 (g)	$+2$ e$^-$ $\longrightarrow$	2 Cl$^-$ (aq)
$Cr_2O_7^{2-}$ (aq) (in acid)	$+6$ e$^-$ $\longrightarrow$	2 Cr^{2+} (aq)
O_2 (g) (in acid)	$+4$ e$^-$ $\longrightarrow$	2 H$_2$O (l)
Br_2(l)	$+2$ e$^-$ $\longrightarrow$	2 Br$^-$ (aq)
NO_3^- (aq) (in acid)	$+3$ e$^-$ or $+1$ $\longrightarrow$	NO(g), NO$_2$ (g)

Common Reducing Agents

Reducing Agent	Number of Electrons Lost	Product
Li(s)	-1 e$^-$ $\longrightarrow$	Li$^+$(aq)
Ca(s)	-2 e$^-$ $\longrightarrow$	Ca^{2+}(aq)
Na(s)	-1 e$^-$ $\longrightarrow$	Na$^+$(aq)
Zn(s)	-2 e$^-$ $\longrightarrow$	Zn^{2+}(aq)
H$_2$(s)	-2 e$^-$ $\longrightarrow$	2 H$^+$(aq)
Cu(s)	-2 e$^-$ $\longrightarrow$	Cu^{2+}(aq)

OXIDATION-REDUCTION AND AN INTRODUCTION TO ELECTROCHEMISTRY

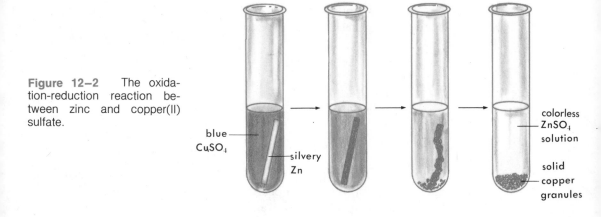

Figure 12–2 The oxidation-reduction reaction between zinc and copper(II) sulfate.

blue — CuSO₄

silvery Zn

colorless ZnSO₄ solution

solid copper granules

gradually disappears and the solution turns colorless (Fig. 12–2). This is an oxidation-reduction reaction. Zinc is a good reducing agent and becomes oxidized quite easily. The copper(II) ion gains electrons and is reduced to copper metal. The equation for the entire redox reaction is the sum of the oxidation part and the reduction part. The number of electrons lost must equal the number of electrons gained, since electrical charge is conserved.

In the oxidation part

$$
\begin{array}{ccc}
\text{oxidation number} & & \text{oxidation number} \\
0 & & +2 \\
\text{Zn(s)} & -2e^- & \to \text{Zn}^{2+}\text{(aq)} \\
\text{metallic zinc} & \text{two electrons lost} & \text{zinc ion}
\end{array}
$$

The rise in oxidation number from zero to "plus two" fits the definition of oxidation. The equation is said to represent a half-reaction, since only the oxidation half is given.

Next, in the reduction half-reaction

$$
\begin{array}{ccc}
\text{Cu}^{2+}\text{ (aq)} & +2e^- \to & \text{Cu(s)} \\
\text{copper(II) ion} & \text{two electrons} & \text{metallic copper} \\
\text{(from CuSO}_4\text{ solution)} & \text{gained} &
\end{array}
$$

we observed a *reduction half-reaction*. The copper gains two electrons, and so the lowering of the oxidation number from $+2$ to 0 is reduction.

$$\text{Cu}^{2+}\text{ (aq)} + 2e^- \to \text{Cu(s)}$$

When the two half-reactions (half oxidation and half reduction) are added, the net redox equation is obtained:

$$ox \qquad Zn - 2e^- \rightarrow \quad Zn^{2+}$$

$$\left.\begin{array}{l}\\ \end{array}\right\} \text{add}$$

$$\underline{red \qquad Cu^{2+} + 2e^- \rightarrow Cu}$$
$$Zn\,(s) + Cu^{2+}\,(aq) \rightarrow \quad Zn^{2+}\,(aq) + Cu\,(s)$$

zinc metal + copper(II) ion → zinc ion + copper metal

The resulting equation explains the observations. The formation of reddish clumps is the appearance of copper. The loss of blue color occurs as the copper(II) ion disappears and the zinc is "eaten away" as it changes to zinc ion.

In any redox equation, the species being oxidized and reduced can be determined by the increase and decrease in the oxidation numbers. The number of electrons lost and gained will always equal the changes in oxidation number of each species involved.

For example, an acid solution of permanganate ion is a good oxidizing agent. The manganese in the permanganate ion becomes reduced to Mn^{2+} (aq) in the process. Observe that the number of electrons gained (reduction) by the MnO_4^- (aq) is equal to the difference between the changes in oxidation number of the manganese:

$$\text{Skeleton equation:} \qquad \overset{+7}{MnO_4^-}(aq) \rightarrow \overset{+2}{Mn^{2+}}(aq)$$

The oxidation number of the manganese in MnO_4^- is $+7$, whereas that in the Mn^{2+} is obviously $+2$. The difference between $+7$ and $+2$ is $+5$. Therefore, 5 electrons must be gained when MnO_4^- is reduced to Mn^{2+},

$$\overset{+7}{MnO_4^-}\,(aq) + 5e^- \rightarrow \overset{+2}{Mn^{2+}}\,(aq)$$

Example 12.1

In an acid environment, $Cr_2O_7^{2-}$(aq) tends to react with many substances so that the chromium becomes changed to Cr^{3+} (aq). Label the change as oxidation or reduction, and calculate the total number of electrons gained or lost per unit of $Cr_2O_7^{2-}$.

1. Find the oxidation number of chromium before and after the change:

$$\text{oxidation numbers} \longrightarrow \underset{Cr_2O_7^{2-}}{\overset{+6 \quad -2}{}} \quad \begin{array}{l}-2 \text{ charge on the} \\ \text{dichromate ion}\end{array}$$

Total oxidation numbers: $+12 - 14$

OXIDATION-REDUCTION AND AN INTRODUCTION
TO ELECTROCHEMISTRY

2. **Note:** There are two atoms of chromium in $Cr_2O_7^{2-}$. Therefore, the conservation laws require that two ions of Cr^{3+} be produced:

$$Cr_2O_7^{2-} \rightarrow 2\ Cr^{3+}$$

3. The difference in oxidation number between *single* atoms of chromium is 3:

$$\overset{+6}{Cr_2O_7^{2-}} + 3e^- \rightarrow \overset{+3}{Cr^{3+}}$$

However, since *two* atoms of chromium change from $+6$ to $+3$, the difference must be doubled:

$$\overset{+6}{Cr_2O_7^{2-}} \rightarrow 2\ \overset{+6}{Cr^{3+}}$$

change in total oxidation numbers $+ 12 \rightarrow +6$

4. The decrease in the oxidation number of chromium indicates that it is a *reduction* change and that the number of electrons *gained* must be 6, the difference between $+12$ and $+6$.

5. The change may be summarized:

$$\overset{+6}{Cr_2O_7^{2-}}(aq) + 6e^- \rightarrow 2\ \overset{+3}{Cr^{3+}}(aq)$$

Exercise 12.1

Label each of the following changes as oxidation or reduction of the underlined species and indicate the number of electrons gained or lost:

(a) $\underline{Mn}^{2+}(aq) \rightarrow \underline{Mn}O_2\ (s)$
(b) $2\ \underline{I}O_3^-\ (aq) \rightarrow \underline{I}_2\ (s)$
(c) $2\ \underline{I}^-\ (aq) \rightarrow \underline{I}_2\ (s)$
(d) $\underline{S}O_4^{2-}\ (aq) \rightarrow \underline{S}^{2-}\ (aq)$
(e) $\underline{S}_2O_3^{2-}\ (aq) \rightarrow 2\ \underline{S}\ (s)$
(f) $\underline{S}O_4^{2-}\ (aq) \rightarrow \underline{S}O_3^{2-}\ (aq)$
(g) $2\ \underline{Cr}^{3+}\ (aq) \rightarrow \underline{Cr}_2O_7^{2-}\ (aq)$
(h) $\underline{Mn}^{2+}\ (aq) \rightarrow \underline{Mn}O_4^-\ (aq)$
(i) $\underline{Cl}O_4^-\ (aq) \rightarrow \underline{Cl}^-\ (aq)$
(j) $\underline{S}O_4^{2-}\ (aq) \rightarrow \underline{S}\ (s)$

12.4 BALANCING OXIDATION-REDUCTION EQUATIONS

Many redox reactions that are normally encountered by chemists are not as simple as the example of zinc and copper(II) ions. There are two common methods suited to the task of completing and balancing redox reactions: the **oxidation number method** for molecular equations and the **ion-electron method** for ionic equations. The first method is easier and is therefore a good starting point, while the latter method is more useful, since it omits "spectator" ions.

We know from experimentation that potassium permanganate reacts with iron(II) sulfate in an acid environment (sulfuric acid is suitable) to produce manganese(II) sulfate, iron(III) sulfate, potassium sulfate, and water. The equation,

$$KMnO_4(aq) + FeSO_4(aq) + H_2SO_4(aq) \rightarrow$$
$$MnSO_4(aq) + K_2SO_4(aq) + Fe_2(SO_4)_3(aq) + H_2O(\ell)$$

may be described as "molecular," since the spectator ions are included.

Some redox reactions are difficult to balance by inspection. The numerical coefficients that represent the molar ratios are often so varied that a "trial and error" method usually results in acute frustration.

Begin by looking for common oxidizing and reducing agents. $KMnO_4$ is quickly recognized as an oxidizing agent, but the iron(II) in $FeSO_4$ is not so easily identified. An alternative method for identifying oxidizing and reducing agents is based on three clues.

Clues Used to Identify Redox Participants

1. An element is in the **middle** of a ternary (three-element) compound on one side of the equation but not on the other side. Example:

KMnO$_4$ MnSO$_4$

in the middle of a ternary compound   not in the middle

2. An element is in a compound on one side of the equation but is in the "free" (uncombined) state on the other side. Example:

CuSO$_4$ Cu

combined with sulfate ion   free (uncombined)

3. A metallic ion may exhibit one oxidation number on one side of the equation and a different oxidation number on the other side. Example:

FeSO$_4$ Fe$_2$(SO$_4$)$_3$

iron(II)   iron(III)

Using the original equation, these clues may be applied to find the elements undergoing oxidation and reduction.

OXIDATION-REDUCTION AND AN INTRODUCTION
TO ELECTROCHEMISTRY

$$\text{K}\widehat{\text{Mn}}\text{O}_4 + \widehat{\text{Fe}}\text{SO}_4 + \text{H}_2\text{SO}_4 \rightarrow \widehat{\text{Mn}}\text{SO}_4 + \text{Fe}_2(\text{SO}_4)_3 + \text{K}_2\text{SO}_4 + \text{H}_2\text{O}$$

iron(II) ——————————— iron(III)

change in oxidation number   (clue #3)

| middle of ternary compound | not in the middle of ternary compound | (clue #1) |

The next steps are to write the oxidation-reduction **half-reactions** apart from the equation, to determine the electron loss and gain, to balance the half-reactions, and finally to replace the coefficients in the equation. The electron loss and gain require preliminary calculations.

1. Determine the oxidation number of manganese in the compound KMnO_4.
   (a) From the table of common oxidation numbers we find that the oxidation number of potassium is $+1$, and that of oxygen is -2. K^+, $\text{MnO}_4{}^-$
   (b) There is a total of four oxygen atoms in KMnO_4. Therefore, the *total* oxidation value for the oxygen is $4 \times (-2) = -8$.
   (c) Since the sum of the oxidation numbers of any compound must equal zero, the value for the Mn in the formula has to be $+7$.

   $\overset{+1 \quad +7 \quad -8\,=\,\text{zero}}{\text{KMnO}_4}$

2. The oxidation number of manganese in MnSO_4 is $+2$. This is indicated by the name of the compound, manganese(II) sulfate.
3. Iron(II) sulfate and iron(III) sulfate also indicate the oxidation numbers of the iron by their names.
4. Write the half-reactions:

   (a) $\overset{+7}{\text{Mn}}$ is reduced to $\overset{+2}{\text{Mn}}$. This requires a *gain* of five electrons:

   $$\overset{+7}{\text{KMnO}_4} + 5\text{e}^- \rightarrow \overset{+2}{\text{Mn}^{2+}}$$

   (b) Fe^{2+} is oxidized to Fe^{3+}, indicating a *loss* of one electron per ion. Since two ions of Fe^{3+} are produced, two ions of Fe^{2+} must be used. Two ions of Fe^{3+} undergoing oxidation will require a two-electron loss. Write the two half-reactions together for addition and balancing:

   $$\text{KMnO}_4 + 5\text{e}^- \rightarrow \text{Mn}^{2+} \text{ (reduction)}$$
   $$2\,\text{Fe}^{2+} \rightarrow 2\,\text{Fe}^{3+} + 2\text{e}^- \text{ (oxidation)}$$

5. The law of conservation requires a balanced loss and gain of electrons. To accomplish this, the half-reactions must be multiplied

through by the smallest possible number that will equalize the number of electrons gained and lost. This means that 10 moles of Fe^{2+} will lose 10 moles of electrons to produce a total of 10 moles of Fe^{3+}. Two moles of $KMnO_4$ will gain 10 moles of electrons to produce 10 moles of Mn^{2+}.

$$\begin{array}{l}
2\ (KMnO_4 + 5e^- \rightarrow Mn^{2+}) \\
\underline{5\ (2\ Fe^{2+} \rightarrow 2\ Fe^{3+} + 2e^-)} \\
2\ KMnO_4 + 10\ Fe^{2+} \rightarrow 2\ Mn^{2+} + 10\ Fe^{3+}
\end{array} \left.\vphantom{\begin{array}{l}X\\X\end{array}}\right\} \text{add}$$

6. Place the coefficients in the original equation (note that 10 moles of Fe^{3+} are in 5 moles of $Fe_2(SO_4)_3$):

$$2\ KMnO_4(aq) + \underline{10}\ FeSO_4(aq) + H_2SO_4(aq)$$

$$\underline{2}\ MnSO_4(aq) + \underline{5}\ Fe_2(SO_4)_3(aq) + K_2SO_4(aq) + H_2O(\ell)$$

7. The completion of the balancing is done by inspection. The following order of inspection is recommended, as before:
   (a) Balance the metals
   (b) Balance the nonmetals, including the polyatomic ions
   (c) Balance the hydrogen
   (d) Balance the oxygen

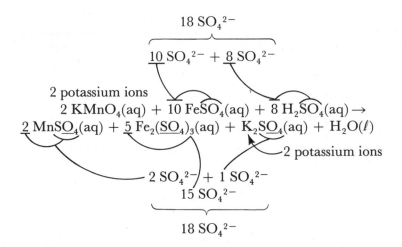

8. (a) $8\ H_2SO_4$ gives a total of 16 hydrogen ions on the left.
   (b) The final balancing requires $8\ H_2O$ to produce 16 hydrogen atoms on the right.
   (c) The total number of oxygen atoms is also balanced at this point.
   (d) The balanced equation is

$$2\ KMnO_4(aq) + 10\ FeSO_4(aq) + 8\ H_2SO_4(aq) \rightarrow$$

$$2\ MnSO_4(aq) + 5\ Fe_2(SO_4)_3(aq) + K_2SO_4(aq) + 8\ H_2O(\ell)$$

OXIDATION-REDUCTION AND AN INTRODUCTION
TO ELECTROCHEMISTRY

Another example, more concisely developed, should reinforce the previous rules and principles.

Example 12.2

Balance the following equation:

$$MnO_2(s) + HCl(aq) \rightarrow MnCl_2(aq) + Cl_2(g) + H_2O(l)$$

1. Identify the redox participants.
   (a) chloride ion changes to free chlorine gas (clu e #2)
   (b) manganese changes from Mn +4 in MnO_2 (manganese(IV) oxide) to Mn^{2+} in $MnCl_2$ (manganese(II) chloride) (clue #3)
2. Write the half-reactions.
   (a) balance the chlorine:

   $$\overset{-1}{2\,Cl^-} \rightarrow \overset{0}{Cl_2} + 2e^-$$

   Going from -1 *up* to zero is oxidation. One mole of chloride ion loses one mole of electrons. Two moles lose two moles of electrons. The Cl^- in $MnCl_2$ has not changed oxidation state. It is a spectator ion.
   (b) One mole of manganese(IV) atoms gains two moles of electrons in decreasing its oxidation number to manganese(II).

   $$\overset{+4}{MnO_2} + 2e^- \rightarrow \overset{+2}{Mn^{2+}}$$

3. Combine and add the half-reactions. Since the loss and gain of electrons are equal:

   $$\overset{-1}{2\,Cl^-} - 2e^- \rightarrow \overset{0}{Cl_2} \text{ (oxidation)}$$
   $$\rightarrow \overset{+4}{MnO_2} + 2e^- \rightarrow \overset{+2}{Mn^{2+}} \text{ (reduction)}$$
   $$\overline{\overset{+4}{MnO_2} + \overset{-1}{2\,Cl^-} \rightarrow \overset{0}{Cl_2} + \overset{+2}{Mn^{2+}}}$$

4. Replace the coefficients in the original equation:

   $$MnO_2 + \underline{2}\,HCl \rightarrow MnCl_2 + Cl_2 + H_2O$$

5. Balance the equation by adjusting the number of chlorine units:

   $$MnO_2 + 4\,HCl \rightarrow MnCl_2 + Cl_2 + H_2O$$

   4 chloride ions   2 chloride ions + 2 chlorine atoms

   4 chlorine units

6. Balance the hydrogen atoms and check the total number of oxygen atoms:

   2 oxygen atoms   2 oxygen atoms

   $$MnO_2(s) + 4\,HCl(aq) \rightarrow MnCl_2(aq) + Cl_2(g) + 2\,H_2O(l)$$

   4 hydrogen ions   4 hydrogen atoms

The equation is balanced.

Summary of Steps in the Oxidation Number Method of Balancing Molecular Equations

1. Identify the elements involved in the redox activity by use of the three clues.
2. Write the half-reactions, showing the loss and gain of electrons.
3. Balance the number of units undergoing oxidation or reduction if subscripts are in the formulas.
4. Add the balanced half-reactions and transfer the coefficients to the original equation.
5. Balance the metals, nonmetals, and polyatomic ions, and then the hydrogen, in that order.
6. Check the oxygens for final balancing.

Exercise 12.2

Balance the following equations by the oxidation number method.

(a) $Bi_2S_3(s) + HNO_3(aq) \rightarrow Bi(NO_3)_3(aq) + S(s) + NO(g) + H_2O(l)$

(b) $MnO(s) + PbO_2(s) + HNO_3(aq) \rightarrow HMnO_4(aq) + Pb(NO_3)_2(aq) + H_2O(l)$

(c) $K_2Cr_2O_7(aq) + HCl(aq) + FeCl_2(aq) \rightarrow CrCl_3(aq) + FeCl_3(aq) + KCl(aq) + H_2O(l)$

(d) $PbCrO_4(s) + KI(aq) + HCl(aq) \rightarrow PbCl_2(s) + CrCl_2(aq) + KCl(aq) + I_2(s) + H_2O(l)$

(e) $AuCl_3(aq) + H_2C_2O_4(aq) \rightarrow Au(s) + HCl(aq) + CO_2(g)$

(f) $KMnO_4(aq) + HBr(aq) \rightarrow MnBr_2(aq) + KBr(aq) + Br_2(l) + H_2O(l)$

(g) $(NH_4)_2Cr_2O_7(aq) \rightarrow N_2(g) + Cr_2O_3(s) + H_2O(l)$

(h) $KIO_3(aq) + HCl(aq) + H_2SO_3(aq) \rightarrow ICl(aq) + KCl(aq) + H_2SO_4(aq) + H_2O(l)$

(i) $Cu_2S(s) + HNO_3(aq) \rightarrow Cu(NO_3)_2(aq) + NO_2(g) + S(s) + H_2O(l)$

(j) $H_2S(g) + Br_2(l) + H_2O(l) \rightarrow HBr(aq) + H_2SO_4(aq)$

(k) $HgS(s) + HNO_3(aq) + HCl(aq) \rightarrow HgCl_2(aq) + S(s) + NO(g) + H_2O(l)$

(l) $KMnO_4(aq) + HCl(aq) \rightarrow MnCl_2(aq) + KCl(aq) + Cl_2(g) + H_2O(l)$

(m) $NaI(ag) + H_2SO_4(aq) \rightarrow Na_2SO_4(aq) + I_2(s) + H_2S(g) + H_2O(l)$

(n) $K_2Cr_2O_7(aq) + H_2S(g) + HCl(aq) \rightarrow CrCl_3(aq) + S(s) + KCl(aq) + H_2O(l)$

(o) $I_2(s) + Cl_2(g) + H_2O(l) \rightarrow HCl(aq) + HIO_3(aq)$

12.6 THE ION-ELECTRON METHOD

The ion-electron method is concerned with conservation of ionic charge rather than with the individual oxidation numbers of elements. The same reaction previously investigated looks different when stripped of "spectator" ions.

OXIDATION-REDUCTION AND AN INTRODUCTION TO ELECTROCHEMISTRY

$$KMnO_4(aq) + FeSO_4(aq) + H_2SO_4(aq) \rightarrow$$
$$MnSO_4(aq) + Fe_2(SO_4)_3(aq) + K_2SO_4(aq) + H_2O(\ell)$$

becomes:

$$MnO_4^-(aq) + Fe^{2+}(aq) + H^+(aq) \rightarrow Mn^{2+}(aq) + Fe^{3+}(aq) + H_2O(\ell)$$

The oxidizing and reducing agents and their products often can be identified immediately. As familiarity is gained in predicting reaction products, an equation can be completed when only the reactants are given.

1. MnO_4^- (permanganate ion) is identified as an oxidizing agent that is reduced to manganese(II) ion in acid medium (Table 12–1):

$$MnO_4^- \rightarrow Mn^{2+}$$

2. The four atoms of oxygen on the left are balanced by the oxygen contained in four molecules of water on the right:

$$MnO_4^- \rightarrow Mn^{2+} + 4H_2O$$

3. Since the reaction occurs in an acid medium, the eight hydrogen atoms on the right are balanced by eight $H^+(aq)$ on the left:

$$MnO_4^-(aq) + 8\,H^+(aq) \rightarrow Mn^{2+}(aq) + 4\,H_2O(\ell)$$

4. The total ionic charge on the left is $7+$ (one negative permanganate unit added to eight positive hydrogen ion units). The total ionic charge on the right is $2+$ (from the single unit of Mn^{2+}). This imbalance is adjusted by adding five electrons to the left or by subtracting five electrons from the right:

$$MnO_4^-(aq) + 8\,H^+(aq) + 5e^- \rightarrow Mn^{2+}(aq) + 4\,H_2O(\ell)$$

5. Fe^{2+} (iron(II) ion) is a reducing agent that becomes oxidized to Fe^{3+} (iron(III) ion):

$$Fe^{2+}(aq) \rightarrow Fe^{3+}(aq)$$

6. Balancing of the ionic charge is accomplished by adding one electron to the right side.

$$Fe^{2+}(aq) \rightarrow Fe^{3+}(aq) + e^-$$

7. Add the two-half reactions after balancing the loss and gain of electrons by multiplying the iron half-reaction by 5:

$$\begin{array}{ll} MnO_4^-(aq) + 8\ H^+(aq) + 5e^- & \rightarrow Mn^{2+}(aq) + 4\ H_2O(l) \\ 5\ Fe^{2+}(aq) - 5\ e^- & \rightarrow 5\ Fe^{3+}(aq) \end{array}$$

$$MnO_4^-(aq) + 5\ Fe^{2+}(aq) + 8\ H^+(aq) \rightarrow Mn^{2+}(aq) + 5\ Fe^{3+}(aq) + 4\ H_2O(l)$$

In conclusion, we can see that the ratio of five moles of Fe^{2+} reacting with one mole of MnO_4^- is exactly the same ratio as exhibited by the stoichiometry of the molecular equation:

$$\underline{2}\ KMnO_4 + \underline{10}\ FeSO_4$$
$$2 \qquad\qquad : 10 \qquad = 1{:}5$$

Another example of this method is illustrated in the reaction in which ethanol (ethyl alcohol) is oxidized to acetic acid. In more familiar terms, this is what happens when wine turns to vinegar. The laboratory reaction is observed when potassium dichromate (oxidizing agent) is added to ethanol (reducing agent) in an acidic medium provided by sulfuric acid.

Example 12.3

Balance the following ionic equation:

$$C_2H_5OH(l) + Cr_2O_7^{2-}(aq) + H^+\ (aq) \rightarrow C_2H_4O_2(aq) + Cr^{3+}(aq) + H_2O(l)$$

ethanol   acetic acid

1. The dichromate ion is the oxidizing agent which becomes reduced to chromium (III) ion:

$$Cr_2O_7^{2-}(aq) \rightarrow Cr^{3+}(aq)$$

2. The seven atoms of oxygen on the left are balanced by seven molecules of water on the right:

$$Cr_2O_7^{2-}(aq) \rightarrow Cr^{3+}(aq) + 7\ H_2O\ (l)$$

3. Since the reaction occurs in acid medium, the 14 hydrogen atoms on the right are balanced by 14 H^+ (aq) on the left:

$$Cr_2O_7^{2-}(aq) + 14\ H^+(aq) \rightarrow Cr^{3+}(aq) + 7\ H_2O\ (l)$$

OXIDATION-REDUCTION AND AN INTRODUCTION
TO ELECTROCHEMISTRY

4. The two chromium atoms on the left are balanced by placing the coefficient, 2, in front of Cr^{3+} on the right.

$$Cr_2O_7{}^{2-}(aq) + 14\ H^+ (aq) \rightarrow 2\ Cr^{3+} (aq) + 7\ H_2O\ (l)$$

5. The total ionic charge on the left is 12 + (2− for dichromate ion and 14+ for hydrogen ion). The total ionic charge on the right is 6+ from the two units of chromium(III) ion. The imbalance is adjusted by adding six electrons to the left side:

$$Cr_2O_7{}^{2-}(aq) + 14\ H^+(aq) + 6e^- \rightarrow 2\ Cr^{3+}(aq) + 7\ H_2O(l)$$

6. The oxidation of ethanol to acetic acid half-reaction is balanced by adding water to the left to obtain the extra oxygen atom needed and by adding the hydrogen ions to the right:

$$C_2H_5OH\ (l) + H_2O(l) \rightarrow C_2H_4O_2(aq) + 4\ H^+(aq)$$

7. The total ionic charge on the left is zero. The total of 4+ (due to 4 H+) is balanced by adding four electrons to the right side:

$$C_2H_5OH\ (l) + H_2O\ (l) \rightarrow C_2H_4O_2(aq) + 4\ H^+ (aq) + 4\ e^-$$

8. Balance the loss and gain of electrons and add the two half-reactions:

multiply by $2(Cr_2O_7{}^{2-} + 14\ H^+ + 6e^- \rightarrow 2\ Cr^{3+} + 7\ H_2O)$

multiply by $3(C_2H_5OH + H_2O \rightarrow C_2H_4O_2 + 4\ H^+ + 4e^-)$

add: $\begin{cases} 2\ Cr_2O_7{}^{2-} + 28\ H^+ + \cancel{12e^-} \rightarrow 4\ Cr^{3+} + 14\ H_2O \\ 3\ C_2H_5OH + 3\ H_2O \rightarrow 3\ C_2H_4O_2 + 12\ H^+ + \cancel{12e^-} \end{cases}$

$$2Cr_2O_7{}^{2-} + 3C_2H_5OH + 28H^+ + 3H_2O \rightarrow 4Cr^{3+} + 3C_2H_4O_2 + 12H^+ + 14H_2O$$

9. The equation should be simplified by subtracting 3 H_2O and 12 H^+ from both sides.

$\begin{array}{ll} 2Cr_2O_7{}^{2-} + 3C_2H_5OH + 28H^+ + 3\cancel{H_2O} & 4Cr^{3+} + 3C_2H_4O_2 + 12\cancel{H^+} + 14H_2O \\ -12H^+ - \cancel{3H_2O} & -\cancel{12H^+} - 3H_2O \end{array}$

$$2Cr_2O_7{}^{2-}(aq) + 3C_2H_5OH(l) + 16H^+(aq) \rightarrow 4Cr^{3+}(aq) + 3C_2H_4O_2(aq) + 11H_2O(l)$$

The previous examples of oxidation-reduction were in acidic media. An example of redox in a basic medium involves some modification.

Example 12.4

Permanganate ion reacts with cyanide ion (a strong base, since it is a vigorous proton acceptor) to produce manganese(IV) oxide and cyanate ion. Because of the CN⁻, the reaction occurs in a basic environment (otherwise, it would evolve poisonous HCN gas):

$$MnO_4{}^-(aq) + CN^- (aq) \rightarrow MnO_2(s) + OCN^- (aq)$$

1. The $MnO_4{}^-$ undergoes reduction to MnO_2:

$$MnO_4{}^- \rightarrow MnO_2$$

2. The four atoms of oxygen on the left compared to the two atoms on the right require water for balancing:

$$MnO_4^- \rightarrow MnO_2 + 2\ H_2O$$

3. The four hydrogen atoms on the right are balanced by 4 H$^+$. Just the same as in an acidic medium reaction!

$$MnO_4^- + 4\ H^+ \rightarrow MnO_2 + 2\ H_2O$$

4. Now add 4 OH$^-$ units to both sides of the equation in order to convert the 4 H$^+$ to water, since H$^+$ cannot remain as a reactant in a basic environment:

$$MnO_4^- + 4\ H^+ \rightarrow MnO_2 + 2\ H_2O$$
$$+ 4\ OH^- \qquad\qquad\qquad +4\ OH^-$$
$$MnO_4^- + 4\ H_2O \rightarrow MnO_2 + 2\ H_2O + 4\ OH^-$$

5. Simplify the half-reaction by subtracting 2 H$_2$O from both sides:

$$MnO_4^- + 4\ H_2O \rightarrow MnO_2 + 2\ \cancel{H_2O} + 4\ OH^-$$
$$- 2\ H_2O \qquad\qquad\qquad - \cancel{2 H_2O} -$$
$$MnO_4^- + 2\ H_2O \rightarrow MnO_2 + 4\ OH^-$$

6. Balance the electronic charge. Since the total on the left is 1$-$ and the total on the right side is 4$-$, add three electrons to the left side.

$$MnO_4^-(aq) + 2\ H_2O(l) + 3e^- \rightarrow MnO_2(s) + 4\ OH^-(aq)$$

7. The oxidation of cyanide ion to cyanate ion follows the same pattern:

$$CN^-(aq) \rightarrow OCN^-(aq)$$

8. Add water on the left to balance the oxygen and add H$^+$ on the right to balance the hydrogen in H$_2$O:

$$CN^- + H_2O \rightarrow OCN^- + 2\ H^+$$

9. Add 2 OH$^-$ to both sides of the equation in order to convert the H$^+$ to water:

$$CN^- + H_2O + 2\ OH^- \rightarrow OCN^- + 2\ H_2O$$

10. Simplify the half-reaction by subtracting one H$_2$O from both sides of the equation:

$$CN^- + 2\ OH^- \rightarrow OCN^- + H_2O$$

11. Balance the electronic charge by adding two electrons to the right side so that the net ionic charge is 3$-$ on both sides of the equation:

$$CN^-(aq) + 2\ OH^-(aq) \rightarrow OCN^-(aq) + H_2O(l) + 2e^-$$

12. The two half-reactions are then balanced with regard to loss and gain of electrons and finally are added after multiplying by the least common multiple:

multiply by 2($MnO_4^- + 2\ H_2O + 3e^- \rightarrow MnO_2 + 4\ OH^-$)

multiply by 3($CN^- + 2\ OH^- \rightarrow OCN^- + H_2O + 2e^-$)

$$2\ MnO_4^- + 4\ H_2O + 6e^- \rightarrow 2\ MnO_2 + 8\ OH^-$$

add: 3 $CN^- + 6\ OH^- \rightarrow 3\ OCN^- + 3\ H_2O + 6e^-$

$$\overline{2\ MnO_4^- + 3\ CN^- + 6\ OH^- + 4\ H_2O \rightarrow 2\ MnO_2 + 3\ OCN^- + 8\ OH^- + 3\ H_2O}$$

OXIDATION-REDUCTION AND AN INTRODUCTION
TO ELECTROCHEMISTRY

13. Simplify the final equation by subtracting 6 OH⁻ and 3 H₂O from both sides:

$$2\ MnO_4^- + 3\ CN^- + 6\ \cancel{OH^-} + 4\ H_2O$$
$$\underline{-\ \cancel{6}\ OH^- - 3\ H_2O}$$
$$2\ MnO_4^-(aq) + 3\ CN^-(aq) + H_2O\ (l) \rightarrow$$

$$2\ MnO_2 + 3\ OCN^- + 8\ OH^- + 3\ H_2\cancel{O}$$
$$\underline{-\ 6\ OH^- - \cancel{3}\ H_2O}$$
$$2\ MnO_2(s) + 3\ OCN^-(aq) + 2\ OH^-(aq)$$

Summary

The following checks should be applied to the balancing of redox equations when the ion-electron method is used.

1. The actual number of each species of atoms must be balanced. H^+ and H_2O will have to be added in acidic medium and OH^- and H_2O in basic medium.
2. Electrons must be added or subtracted to each half-reaction to correct for the difference in ionic charge.
3. The loss and gain of electrons must be balanced before the half-reactions are added.
4. The final equation must be simplified by removing excess H^+, OH^-, and H_2O.

Exercise 12.3

Balance the following equations by the ion-electron method.
(a) $Ag(s) + NO_3^-(aq) \rightarrow Ag^+(aq) + NO(g)$ (acid solution)
(b) $Mn^{2+} + Br\ (l) \rightarrow MnO_2(s) + Br^-\ (aq)$ (basic solution)
(c) $Cr(OH)_4^-(aq) + H_2O_2 \rightarrow CrO_4^{2-}(aq) + H_2O\ (l)$ (basic solution)
(d) $Cr_2O_7^{2-}(aq) + I^-(aq) \rightarrow Cr^{3+}(aq) + I_2(s)$ (acid solution)
(e) $PbO_2(s) + Cl^-(aq) \rightarrow Pb^{2+}(aq) + Cl_2(g)$ (acid solution)
(f) $MnO_4^-\ (aq) + H_2S\ (g) \rightarrow Mn^{2+}\ (aq) + S(s)$ (acid solution)
(g) $Cr_2O_7^{2-}(aq) + HNO_2(aq) \rightarrow Cr^{3+}(aq) + NO_3^-(aq)$ (acid solution)
(h) $BrO^-(aq) + Cr(OH)_4^-\ (aq) \rightarrow Br^-(aq) + CrO_4^{2-}(aq)$ (basic solution)
(i) $MnO_4^-(aq) + NO_2^-\ (aq) \rightarrow MnO_2(s) + NO_3^-\ (aq)$ (basic solution)
(j) $H_2C_2O_4(aq) + MnO_2(s) \rightarrow Mn^{2+}(aq) + CO_2(g)$ (acid solution)
(k) $MnO_4^-\ (aq) + C_2H_4\ (g) \rightarrow Mn^{2+}\ (aq) + CO_2(g)$ (acid solution)
(l) $Br_2(aq) \rightarrow BrO_3^-(aq) + Br^-(aq)$ (basic solution)
(m) $MnO_2(s) + Cl^-(aq) \rightarrow Mn^{2+}\ (aq) + Cl_2(g)$ (acid solution)
(n) $Cr_2O_7^{2-}(aq) + Fe^{2+}(aq) \rightarrow Cr^{3+}\ (aq) + Fe^{3+}\ (aq)$ (acid solution)
(o) $H_2S\ (g) + I_2(s) \rightarrow S(s) + I^-(aq)$ (acid solution)

When an oxidation-reduction reaction is known to be a complete (stoichiometric) conversion of reactants to products, the methods for performing useful calculations are essentially the same as those employed for nonredox reactions.

Example 12.5

Calculate the mass of sodium iodide required for the production of 5.0 g of iodine in the following reaction:

$$NaI(aq) + H_2SO_4(aq) \rightarrow Na_2SO_4(aq) + I_2(s) + H_2S(g) + H_2O\ (l)$$

1. Balance the equation by the oxidation number method:

$$8\ NaI(aq) + 5\ H_2SO_4(aq) \rightarrow 4\ Na_2SO_4(aq) + 4\ I_2(s) + H_2S(g) + 4\ H_2O\ (l)$$

2. Solve by the factor-label method :

$$\text{mass of } I_2 \rightarrow \text{moles } I_2 \rightarrow \text{moles } NaI \rightarrow \text{mass } NaI$$

$$5.0\ g\ I_2 \times \frac{1\ \text{mole } I_2}{254\ g\ I_2} \times \frac{8\ \text{moles } NaI}{4\ \text{moles } I_2} \times \frac{150\ g\ NaI}{1\ \text{mole } NaI} = \text{mass of } NaI$$

$$\boxed{\text{mass} = 5.9\ g\ of\ NaI}$$

Example 12.6

What mass of solid $K_2Cr_2O_7$ must be added to a solution of oxalic acid to produce 500 mL of CO_2 gas when measured at STP? The ionic equation is:

$$H^+(aq) + H_2C_2O_4\ (aq) + Cr_2O_7{}^{2-}(aq) \rightarrow Cr^{3+}(aq) + CO_2(g) + H_2O\ (l)$$

1. Balance the equation by the ion-electron method:

$$8\ H^+(aq) + 3\ H_2C_2O_4(aq) + Cr_2O_7{}^{2-}(aq) \rightarrow 2\ Cr^{3+} + 6\ CO_2(g)$$
$$+ 7\ H_2O\ (l)$$

2. Solve by the factor-label method:

$$\text{volume of } CO_2 \rightarrow \text{moles } CO_2 \rightarrow \text{moles } K_2Cr_2O_7 \rightarrow \text{mass } K_2Cr_2O_7$$

$$\text{mass} = 0.500\ L\ CO_2 \times \frac{1\ \text{mole } CO_2}{22.4\ L\ CO_2} \times \frac{1\ \text{mole } K_2Cr_2O_7}{6\ \text{moles } CO_2} \times \frac{294\ g\ K_2Cr_2O_7}{1\ \text{mole } K_2Cr_2O_7}$$

$$\boxed{\text{mass} = 1.09\ g\ of\ K_2Cr_2O_7}$$

OXIDATION-REDUCTION AND AN INTRODUCTION
TO ELECTROCHEMISTRY

12.8 ELECTROCHEMICAL CELLS

Earlier in this chapter we used the example of the reaction between zinc metal and the copper ion (in a copper sulfate solution) as an illustration of oxidation-reduction. Look at Figure 12–2 again and review the explanation, the essence of which is that the atomic structure of zinc gives zinc atoms the chemical property of a *tendency* to "lose" electrons (i.e., to undergo oxidation) to many atoms or ions having a tendency to "gain" electrons in the process of reduction. Figure 12–3 represents a model for this exchange.

In the oxidation of metallic zinc and the reduction of copper(II) ion, the transfer of electrons occurs *directly* between the two species. As long as the strip of zinc metal is immersed in the copper ion solution, the reaction is direct and vigorous. This is an example of the phenomenon we investigated in Chapter 5 in terms of the "activity series" (see

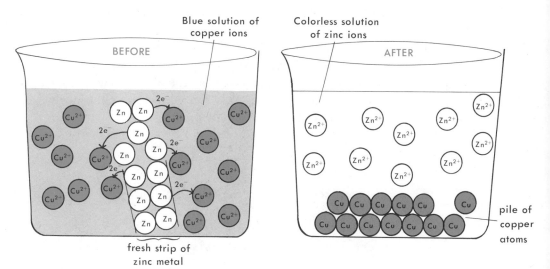

Figure 12–3   A model representing the oxidation of zinc metal and the reduction of copper ion.

Table 5–1). On the basis of the relative positions of zinc and copper in the activity series we would have predicted the simple replacement (in reality, oxidation-reduction reaction) of copper ion by zinc metal:

$$Zn(s) + CuSO_4(aq) \rightarrow ZnSO_4(aq) + Cu(s)$$

or, more to the point,

$$Zn\ (s) + \underset{\text{(blue)}}{Cu^{2+}(aq)} \rightarrow \underset{\text{(colorless)}}{Zn^{2+}(aq)} + Cu\ (s)$$

In this chapter we shall revisit the activity series under a different name and attempt to provide a better insight into what it is and how it works. Our starting point is the consideration of **electrochemical** cells. An electrochemical cell is an apparatus in which either chemical reactions are produced by an electrical current or in which an electrical current is produced by chemical reactions. The process of effecting a chemical change by passing an electrical current through a substance is known as *electrolysis,* and the cell in which an electrolysis occurs is known as an *electrolytic* cell. The device in which an electric current is produced by a chemical reaction is called a *voltaic* (or galvanic) cell. Reactions in voltaic cells are spontaneous reactions, but those that occur in electrolytic cells must be induced by using a direct current of electricity. The voltage of the direct current must be sufficiently high to remove electrons from negative ions (oxidation) and to add electrons to positive ions (reduction).

The reactions that occur in both types of cell are oxidation-reduction reactions. In both the electrolytic cell and the voltaic cell the anode is the electrode at which oxidation takes place, and the cathode is the electrode where reduction occurs. Let us examine what happens when an aqueous solution of copper(II) chloride is electrolyzed (Fig. 12-4).

In the electrolysis of $CuCl_2(aq)$, direct current passes through the solution, with the result that copper "plates out" at the cathode and elemental chlorine gas escapes at the anode. The total equation for the electrolysis indicates that the copper(II) chloride has been decomposed into its constituent elements:

$$CuCl_2(aq) \xrightarrow{\text{elect}} Cu(s) + Cl_2(g)$$

Now consider separately the reactions that occur at each of the two inert electrodes.

Metallic copper forms at the cathode because the Cu^{2+} ions attract electrons and are reduced.

$$\text{at the cathode: } Cu^{2+} + 2e^- \rightarrow Cu(s) \text{ (reduction)}$$

OXIDATION-REDUCTION AND AN INTRODUCTION
TO ELECTROCHEMISTRY

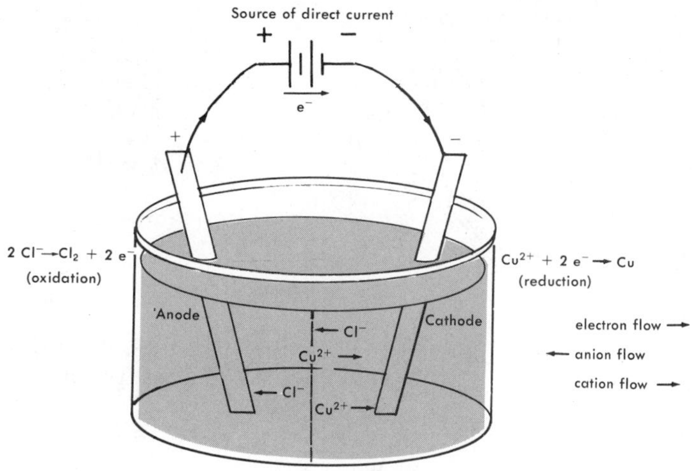

Figure 12–4   An electrolytic cell showing electrolysis of $CuCl_2(aq)$.

$2\ Cl^- \rightarrow Cl_2 + 2\ e^-$
(oxidation)

$Cu^{2+} + 2\ e^- \rightarrow Cu$
(reduction)

Anode

Cathode

Cl^-

Cu^{2+}

Cl^-

Cu^{2+}

electron flow →
← anion flow
cation flow →

Source of direct current
+   −
e^-

Simultaneously, the chloride ions move to the anode, where two chloride ions give up two electrons and are thereby oxidized, forming a molecule of $Cl_2(g)$.

at the anode: $2\ Cl^- \rightarrow Cl_2(g) + 2e^-$ (oxidation)

In all oxidation-reduction processes, the number of electrons gained in reduction must be equal to those lost in oxidation. Therefore,

cathode reaction: $Cu^{2+} + 2e^- \rightarrow Cu(s)$
anode reaction: $2\ Cl^- \rightarrow Cl_2 + 2e^-$

The net equation $Cu^{2+}(aq) + 2\ Cl^-(aq) \rightarrow Cl_2(g) + Cu(s)$

Electrolytic cells are used extensively in industry for the production of numerous chemicals such as chlorine, sodium hydroxide, hydrogen, copper, aluminum, and fluorine. The familiar plating operations for silver-plating of jewelry and chrome-plating of auto bumpers are performed in electrolytic cells.

The electrochemical cell is based on the *physical separation* of oxidizing and reducing agents in a system that allows the free migration of ions between the two and an external flow of electrons. For example, when we construct a physical arrangement that promotes the electron transfer tendencies between zinc and copper ion *indirectly through an external metal wire,* an electricity-producing phenomenon will result. Such a system is called an electrochemical cell.

If direct contact occurs between the oxidizing and reducing agents, a primitive "short circuit" will result. The separation can be made with a "salt bridge" or by using a porous barrier similar to a clay flowerpot. The salt bridge permits the migration of ions between the two separated halves of the cell in order to keep the charges

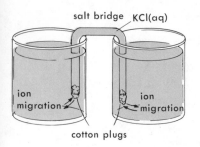

salt bridge KCl(aq)

porous cup

ion migration

ion migration

ion migration

cotton plugs

Figure 12–5   Simplified diagrams of the salt bridge and porous cup arrangements for construction of an electrochemical cell.

balanced and to produce a complete circuit. Potassium chloride solution in a length of glass tubing is a commonly used salt bridge for two good reasons.

1. KCl(aq) forms an ionic junction between the half-cells, and the bridge allows the free migration of ions between the separated halves.
2. $K^+(aq)$ and $Cl^-(aq)$ have less of a tendency to undergo oxidation or reduction than do the components of the constructed cell.

The porous barrier allows the free migration of ions between the separated halves. Examples of the salt bridge and the porous barrier are shown in Figure 12–5.

Now, let us consider the zinc-copper system as it is arranged in an electrochemical cell. In Figure 12–6 you will observe that the zinc strip is separated from the copper ion by a porous barrier. Since the ions in the solutions of $ZnSO_4(aq)$ and $CuSO_4$ are free to migrate through the porous barrier, oxidation and reduction occur until one **electrode** (zinc strip) accumulates as many electrons as it can, while the other electrode (copper strip) has lost an equal number of electrons. At this point, where the *difference* in electron "pressure" is at a maximum, we say that an emf **(electromotive force)** has been developed. Synonyms for emf are **voltage, potential,** and **potential difference.**

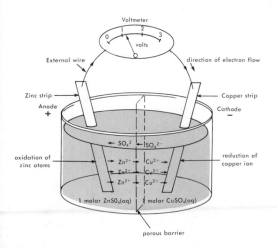

Figure 12–6   A diagrammatic representation of a zinc-copper electrochemical cell.

OXIDATION-REDUCTION AND AN INTRODUCTION TO ELECTROCHEMISTRY

Electromotive force is an appropriately descriptive term. It suggests that the difference in electron "pressure" creates a tendency for the electrons to *move* from the electrode of high electron density *toward* the electrode of low electron density. This can happen *if we connect the two electrodes with an external wire* (a good electron conductor). A voltmeter measures the potential difference. Under standard conditions (25°C and 1 molar ion concentration), the observed voltage is 1.10 volts, written $\mathscr{E}^0{}_{cell} = +1.10$ volts. The superscript of zero to the upper right of the symbol $(\mathscr{E}^0)$ indicates standard conditions. The oxidation-reduction reaction is thus free to continue until the oxidation of zinc or the reduction of copper ion is complete. When this happens, the electrochemical cell no longer functions; such a cell is commonly described as a "dead battery."

The increase in the number of electrons on the zinc strip and the simultaneous decrease in the number of electrons on the copper strip owing to the reactions,

$$\text{electrons lost: } Zn - 2e^- \rightarrow Zn^{2+}$$

$$\text{electrons gained: } Cu^{2+} + 2e^- \rightarrow Cu$$

is illustrated in Figure 12–7, which symbolizes the gradual disappearance of the zinc strip and an increase in the amount of solid copper.

The term **anode** and **cathode** are assigned according to the direction of ion flow. The anode attracts anions, and the cathode attracts cations. Since the negatively charged ions (anions) are shown (Fig. 12–6) as $SO_4{}^{2-}(aq)$ moving toward the zinc strip, the zinc is properly called the anode. Hence, the term **anodic oxidation** enters the vocabulary of electrochemical cell technology. Conversely, the positively charged ions (cations) are shown (Fig. 12–7) as the metallic ions moving toward the copper strip, which makes the copper the cathode. Now we have the other term, **cathodic reduction.** Figure 12–8 is a representation of the electrode labels.

It should be mentioned that although copper is the site of the reduction of copper ion when zinc is in the system, we would expect a reversal in behavior when copper is coupled in an electrochemical cell with some other metal ion that is more easily reduced. If we look again at the activity series, we can predict that copper metal will replace silver from silver nitrate solution:

$$Cu\ (s) + 2\ Ag^+(aq) \rightarrow Cu^{2+}(aq) + 2\ Ag\ (s)$$

Indeed, setting up a standard copper-silver cell would show that anodic oxidation occurs at the copper electrode and that the silver strip is the site of cathodic reduction. The cell voltage, as illustrated in Figure 12–9, is found to be 0.46 volts or $\mathscr{E}^0{}_{cell} = +0.46V$.

Up to this point we have seen how the order of elements on the activity series can be related to the construction of electrochemical

---

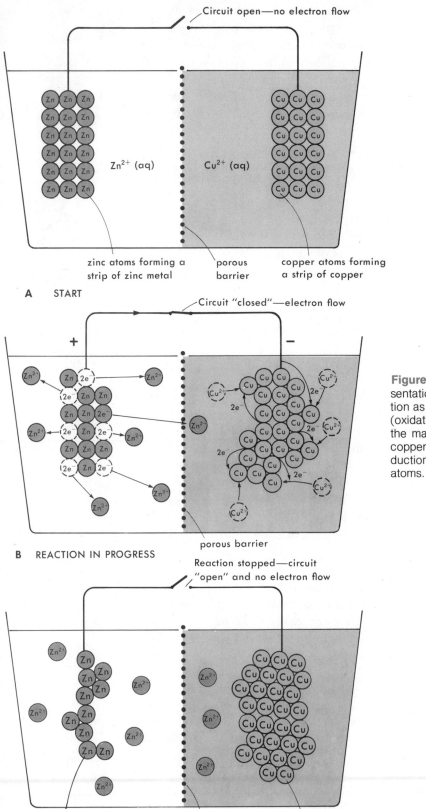

Figure 12–7 Symbolic representation of zinc strip disintegration as zinc atoms lose electrons (oxidation) and the increase in the mass of the copper strip as copper ions gain electrons (reduction) in the change to copper atoms.

A START

B REACTION IN PROGRESS

C END OF REACTION

OXIDATION-REDUCTION AND AN INTRODUCTION
TO ELECTROCHEMISTRY

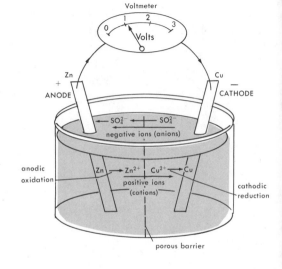

Figure 12–8 The labeling of the electrodes depends on the oxidation and reduction sites and the resultant direction of ion flow.

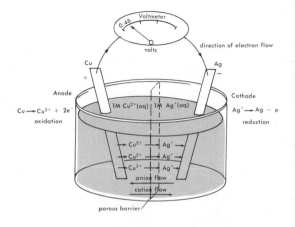

Figure 12–9 Diagrammatic representation of the arrangement of a copper-silver voltaic cell.

12.8 ELECTROCHEMICAL CELLS

353

cells. But now it's time to look at some wider applications that can be made if we evolve the activity series into a table of **standard electrode potentials,** which provides the information we need in order to make useful predictions about the tendency of various oxidation-reduction reactions to occur spontaneously under standard conditions.

12.9 STANDARD ELECTRODE POTENTIALS

A standard electrode potential of a half-cell is the voltage produced when the particular half-cell is coupled with a hydrogen electrode. The hydrogen electrode is arbitrarily assigned standard oxidation or reduction potentials of zero. If the direction of the electron flow is from the second half-cell to the hydrogen half-cell, it is an indication that the second half-cell is undergoing an oxidation reaction and that the hydrogen half-cell is undergoing reduction. The voltage produced would be directly due to the oxidation reaction, since the hydrogen is assigned a value of 0.000 volts.* For example, the electrode potential for a zinc-zinc ion half-cell is observed according to the diagram in Figure 12–10. In the voltaic cell in that diagram, zinc is oxidized to Zn^{2+}, and $H^+(aq)$ is reduced to $H_2(g)$.

$$Zn \rightarrow Zn^{2+} + 2e^- \qquad \mathscr{E}^0{}_{ox} = +0.76V$$

$$2\,H^+ + 2e^- \rightarrow H_2 \qquad \mathscr{E}^0{}_{red} = 0.00V$$

The $\mathscr{E}^0{}_{cell}$ is the sum of the standard oxidation and reduction potentials. Since the $\mathscr{E}^0{}_{cell}$ is directly related to the zinc-zinc ion half-cell ($Zn/Zn^{2+}(aq)$), this value is recorded as the standard oxidation potential, $\mathscr{E}^0{}_{ox}$. Because Zn is oxidized, the sign of $\mathscr{E}^0{}_{ox}$ is positive. If zinc ion were reduced, a negative sign would indicate the reduction potential, $-0.76V$.

An example of a negative oxidation potential is provided by a copper-copper ion half-cell ($Cu/Cu^{2+}(aq)$, in which reduction takes place more easily than in the $H_2/2\,H^+(aq)$ half-cell (Fig. 12–11). In this cell, the $Cu^{2+}(aq)$ is reduced to free copper and the $H_2(g)$ is oxidized to $2\,H^+(aq)$:

$$Cu^{2+} \rightarrow Cu + 2e^- \qquad \mathscr{E}^0{}_{red} = +0.34\,V$$

$$H_2 \rightarrow 2\,H^+ + 2e^- \qquad \mathscr{E}^0{}_{ox} = +0.00\,V$$

---

*By international convention, the hydrogen electrode is accepted as the standard by which all other half-cells are measured and it is called the standard reference electrode.

OXIDATION-REDUCTION AND AN INTRODUCTION TO ELECTROCHEMISTRY

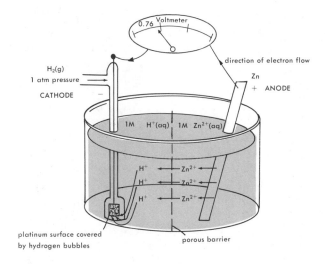

Figure 12–10   Diagrammatic representation of a zinc-hydrogen standard cell.

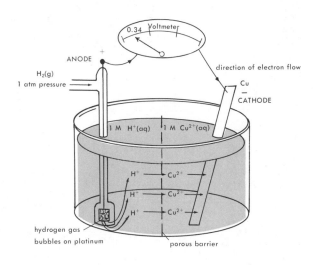

Figure 12–11   Diagrammatic representation of a copper-hydrogen standard cell.

Since the voltage is developed by the reduction of $Cu^{2+}(aq)$ to Cu the cell voltage is known as the *standard reduction potential* of the Cu/Cu^{2+} half-cell and is written Cu^{2+}/Cu.

Consider again the voltaic cell made of a Zn/Zn^{2+} half-cell and a Cu/Cu^{2+} half-cell that produced a cell voltage of 1.10 volts. The tendency of zinc to undergo oxidation coupled with the tendency of Cu^{2+} to undergo reduction made a cell that vigorously and spontaneously

TABLE 12-2 Standard Reduction Potentials

Reduction Half-Reaction		Standard Reduction Potential (volts) $\mathscr{E}^0_{red}$
$Li^+ + e^-$	$\rightarrow Li(s)$	-3.05
$K^+ + e^-$	$\rightarrow K(s)$	-2.93
$Ba^{2+} + 2e^-$	$\rightarrow Ba(s)$	-2.90
$Ca^{2+} + 2e^-$	$\rightarrow Ca(s)$	-2.87
$Na^+ + e^-$	$\rightarrow Na(s)$	-2.71
$Mg^{2+} + 2e^-$	$\rightarrow Mg(s)$	-2.37
$Al^{3+} + 3e^-$	$\rightarrow Al(s)$	-1.66
$Zn(OH_4)^{2-} + 2e^-$	$\rightarrow Zn(s) + 4\ OH^-$	-1.22
$Mn^{2+} + 2e^-$	$\rightarrow Mn(s)$	-1.18
$Fe(OH)_2(s) + 2e^-$	$\rightarrow Fe(s) + 2\ OH^-$	-0.88
$2H_2O + 2e^-$	$\rightarrow H_2(g) + 2\ OH^-$	-0.83
$Zn^{2+} + 2e^-$	$\rightarrow Zn(s)$	-0.76
$Cr^{3+} + 3e^-$	$\rightarrow Cr(s)$	-0.74
$Fe(OH)_3(s) + e^-$	$\rightarrow Fe(OH)_2(s) + OH^-$	-0.56
$S(s) + 2e^-$	$\rightarrow S^{2-}$	-0.48
$Fe^{2+} + 2e^-$	$\rightarrow Fe(s)$	-0.44
$Cr^{3+} + e^-$	$\rightarrow Cr^{2+}$	-0.41
$Cd^{2+} + 2e^-$	$\rightarrow Cd(s)$	-0.40
$PbSO_4(s) + 2e^-$	$\rightarrow Pb(s) + SO_4^{2-}$	-0.36
$Cu(OH)_2(s) + 2e^-$	$\rightarrow Cu(s) + 2\ OH^-$	-0.36
$Tl^+ + e^-$	$\rightarrow Tl(s)$	-0.34
$Co^{2+} + 2e^-$	$\rightarrow Co(s)$	-0.28
$Ni^{2+} + 2e^-$	$\rightarrow Ni(s)$	-0.25
$AgI(s) + e^-$	$\rightarrow Ag(s) + I^-$	-0.15
$Sn^{2+} + 2e^-$	$\rightarrow Sn(s)$	-0.14
$Pb^{2+} + 2e^-$	$\rightarrow Pb(s)$	-0.13
$CrO_4^{2-} + 4H_2O + 3e^-$	$\rightarrow Cr(OH)_3(s) + 5\ OH^-$	-0.12
$2H^+ + 2e^-$	$\rightarrow H_2(g)$	0.00
$NO_3^- + H_2O + 2e^-$	$\rightarrow NO_2^- + 2\ OH^-$	0.01
$AgBr(s) + e^-$	$\rightarrow Ag(s) + Br^-$	0.10
$S(s) + 2H^+ + 2e^-$	$\rightarrow H_2S$	0.14
$Sn^{4+} + 2e^-$	$\rightarrow Sn^{2+}$	0.15
$Cu^{2+} + e^-$	$\rightarrow Cu^+$	0.15
$SO_4^{2-} + 4H^+ + 2e^-$	$\rightarrow SO_2(g) + 2H_2O$	0.20
$Cu^{2+} + 2e^-$	$\rightarrow Cu(s)$	0.34
$Ag_2O(s) + H_2O + 2e^-$	$\rightarrow 2Ag(s) + 2\ OH^-$	0.34
$ClO_4^- + H_2O + 2e^-$	$\rightarrow ClO_3^- + 2\ OH^-$	0.36
$O_2(g) + 2H_2O + 4e^-$	$\rightarrow 4OH^-$	0.40
$Cu^+ + e^-$	$\rightarrow Cu(s)$	0.52
$I_2(s) + 2e^-$	$\rightarrow 2I^-$	0.53
$ClO_3^- + 3H_2O + 6e^-$	$\rightarrow Cl^- + 6\ OH^-$	0.62
$Fe^{3+} + e^-$	$\rightarrow Fe^{2+}$	0.77
$Hg_2^{2+} + 2e^-$	$\rightarrow 2Hg(l)$	0.79
$Ag^+ + e^-$	$\rightarrow Ag(s)$	0.80
$ClO^- + H_2O + 2e^-$	$\rightarrow Cl^- + 2\ OH^-$	0.89
$2Hg^{2+} + 2e^-$	$\rightarrow Hg_2^{2+}$	0.92
$NO_3^- + 4H^+ + 3e^-$	$\rightarrow NO(g) + 2H_2O$	0.96
$AuCl_4^- + 3e^-$	$\rightarrow Au(s) + 4Cl^-$	1.00
$Br_2(l) + 2e^-$	$\rightarrow 2Br^-$	1.07
$O_2(g) + 4H^+ + 4e^-$	$\rightarrow 2H_2O$	1.23
$MnO_2(s) + 4H^+ + 2e^-$	$\rightarrow Mn^{2+} + 2H_2O$	1.23
$Cr_2O_7^{2-} + 14H^+ + 6e^-$	$\rightarrow 2Cr^{3+} + 7H_2O$	1.33
$Cl_2(g) + 2e^-$	$\rightarrow 2\ Cl^-$	1.36
$ClO_3^- + 6H^+ + 5e^-$	$\rightarrow \frac{1}{2}Cl_2(g) + 3H_2O$	1.47
$Au^{3+} + 3e^-$	$\rightarrow Au(s)$	1.50
$MnO_4^- + 8H^+ + 5e^-$	$\rightarrow Mn^{2+} + 4H_2O$	1.52
$H_2O_2 + 2H^+ + 2e^-$	$\rightarrow 2H_2O$	1.77
$Co^{3+} + e^-$	$\rightarrow Co^{2+}$	1.82
$F_2(g) + 2e^-$	$\rightarrow 2F^-$	2.87

OXIDATION-REDUCTION AND AN INTRODUCTION
TO ELECTROCHEMISTRY

produced a fairly high voltage. If the sum of the $\mathscr{E}^0{}_{ox}$ of the zinc half-cell is *added* to the $\mathscr{E}^0{}_{red}$ of the copper half-cell, the actual $\mathscr{E}^0{}_{cell}$ is observed.

$$
\begin{array}{lll}
& Zn \rightarrow Zn^{2+} + 2e^- & \mathscr{E}^0{}_{ox} = +0.76 \text{ V} \\
\text{add} & Cu^{2+} \rightarrow Cu - 2e^- & \mathscr{E}^0{}_{red} = +0.34 \text{ V} \\
\hline
& Zn + Cu^{2+} \rightarrow Cu + Zn^{2+} & \mathscr{E}^0{}_{cell} = +1.10 \text{ V}
\end{array}
$$

Whenever a redox reaction occurs spontaneously, or if there is a *tendency* for the reaction to occur, the cell voltage will have a *positive* sign. If two half-cells (one oxidation and one reduction) produce a negative $\mathscr{E}^0{}_{cell}$ when added, the prediction can be made that this reaction will not have a tendency to proceed spontaneously. This does not mean that the reaction cannot be forced to proceed. Remember, there is a phenomenon known as an endothermic (energy-absorbing) reaction. Many oxidation-reduction reactions vital to the energy-absorbing needs of living organisms have negative cell potentials, but they do occur. The use of standard electrode potentials allows predictions of tendencies, but such predictions do not represent inflexible laws.

12.10 PREDICTING TENDENCIES OF REACTIONS

Given a pair of half-cell reactions, it is possible to predict the reaction that is most likely to occur by arranging them so that the sum of the standard electrode potentials is a positive value. These data can be obtained from a table of standard reduction potentials (Table 12–2). The table could be arranged as one of standard oxidation potentials, but most of the world's scientific community prefers the former arrangement.

Example 12.7

Given two reduction half-reactions, write the equation having the greatest tendency to proceed spontaneously.

$$
\begin{array}{ll}
Cr^{3+} + 3e^- \rightarrow Cr & \mathscr{E}^0{}_{red} = -0.74 \text{ V} \\
Sn^{4+} + 2e^- \rightarrow Sn^{2+} & \mathscr{E}^0{}_{red} = +0.15 \text{ V}
\end{array}
$$

1. The only half-reductions that can possibly yield a positive value for the sum of electrode potentials is the oxidation of Cr^0 to Cr^{3+} and the relatively easy reduction of Sn^{4+} to Sn^{2+}. The ease of reduction is indicated by the positive sign of the standard reduction potential. (Note that the voltage is the same; only the sign is reversed.)

---

2. Arrange the appropriate half-cells for addition

$$Cr \rightarrow Cr^{3+} + 3e^- \qquad \mathscr{E}^0_{ox} = +0.74 \text{ V}$$

$$Sn^{4+} + 2e^- \rightarrow Sn^{2+} \qquad \mathscr{E}^0_{red} = +0.15 \text{ V}$$

3. The loss and gain of electrons must be balanced:

$$2(Cr \rightarrow Cr^{3+} + 3e^-)$$

$$3(Sn^{4+} + 2e^- \rightarrow Sn^{2+})$$

This becomes

add:
$2\ Cr \rightarrow 2\ Cr^{3+} + 6e^-$	$\mathscr{E}^0_{ox} = +0.74 \text{ V}$
$3\ Sn^{4+} + 6e^- \rightarrow 3\ Sn^{2+}$	$\mathscr{E}^0_{red} = +0.15 \text{ V}$
$2\ Cr(s) + 3\ Sn^{4+}(aq) \rightarrow 2\ Cr^{3+}(aq) + 3\ Sn^{2+}(aq)$	$\mathscr{E}^0_{cell} = +0.89 \text{ V}$

Note: The half-reactions are multiplied, but the standard electrode potentials are not. This is due to the fact that the potential is not a function of the number of electrons, but rather it is a measure of the vigor with which oxidation and reduction occur.

Example 12.8

Would the following reaction be expected to occur spontaneously under standard conditions?

$$5\ Fe^{2+}(aq) + MnO_4^- + 8\ H^+(aq) \rightarrow Mn^{2+}(aq) + 4\ Fe^{3+}(aq) + 5\ H_2O(l)$$

1. Arrange the reaction into oxidation and reduction half-cells and add the electrode potentials:

$5\ Fe^{2+} \rightarrow 5\ Fe^{3+} + 5e^-$	$\mathscr{E}^0_{ox} = -0.77 \text{ V}$
$MnO_4^- + 8\ H^+ + 5e^- \rightarrow Mn^{2+} + 4\ H_2O$	$\mathscr{E}^0_{red} = +1.52 \text{ V}$
	$\mathscr{E}^0_{cell} = +0.75 \text{ V}$

2. The cell potential is positive, which results in a prediction that the reaction does have a tendency to occur spontaneously.

Example 12.9

Predict the probable tendency for the following reaction to occur:

$$2\ Au(s) + Cr_2O_7^{2-}(aq) + 14\ H^+(aq) \rightarrow 2\ Au^{3+}(aq) + 2\ Cr^{3+}(aq) + 7\ H_2O\ (l)$$

1. Add the two indicated half-cell reaction potentials:

$2\ Au \rightarrow 2\ Au^{3+} + 6e^-$	$\mathscr{E}^0_{ox} = -1.50 \text{ V}$
$Cr_2O_7^{2-} + 14\ H^+ + 6e^- \rightarrow 2\ Cr^{3+} + 7\ H_2O$	$\mathscr{E}^0_{red} = +1.33 \text{ V}$
	$\mathscr{E}^0_{cell} = -0.17 \text{ V}$

2. The conclusion drawn from the negative cell voltage is that the dichromate oxidation of gold is *not* likely to occur spontaneously.

OXIDATION-REDUCTION AND AN INTRODUCTION
TO ELECTROCHEMISTRY

Later on in your study of chemistry you will see how cell voltages are related to energy changes. An understanding of energy changes in an oxidation-reduction reaction gives us a means of predicting more accurately the feasibility of a spontaneous reaction. As energy change is further related to the concept of chemical equilibrium, you will learn how to alter conditions, as we learned in the case of LeChatelier's principle, so that reactions that would not be expected to "go" under standard conditions might indeed work if the conditions of temperature, ion concentration, and pressure are purposefully controlled.

Exercise 12.5

1. An electrochemical cell is composed of one half-cell having a strip of nickel in a 1.0 M solution of $Ni^{2+}(aq)$ and the other half-cell having a strip of lead in a 1.0 M solution of $Pb^{2+}(aq)$.
   (a) Write the equation for the reaction.
   (b) Diagram and label the cell.
   (c) Calculate the cell voltage.
2. Given the following pairs of half-cell reactions and their standard electrode potentials, write the equations for the reactions most likely to proceed spontaneously, and calculate the cell voltages.
   (a) $Co^{2+}(aq) + 2e^- \rightarrow Co(s)$ $\qquad \mathcal{E}^0_{red} = -0.28$ V
   $\qquad SO_4^{2-}(aq) + 4 H^+(aq) + 2e^- \rightarrow SO_2(g) + 2 H_2O(l)$
   $\qquad\qquad\qquad\qquad\qquad\qquad\qquad\qquad \mathcal{E}^0_{red} = +0.20$ V
   (b) $CrO_4^{2-}(aq) + 4 H_2O(l) + 3e^- \rightarrow Cr(OH)_3(s) + 5 OH^-(aq)$
   $\qquad\qquad\qquad\qquad\qquad\qquad\qquad\qquad \mathcal{E}^0_{red} = -0.12$ V
   $\qquad Cu^{2+}(aq) + 2e^- \rightarrow Cu(s)$ $\qquad \mathcal{E}^0_{red} = +0.34$ V
3. Predict the tendency of the following reactions to occur spontaneously:
   (a) $MnO_4^-(aq) + 2 H^+(aq) + \frac{1}{2} Cl_2(g) \rightarrow ClO_3^-(aq) + Mn^{2+}(aq)$
   $\qquad\qquad\qquad\qquad\qquad\qquad\qquad\qquad\qquad + H_2O(l)$
   (b) $Zn(s) + Ag_2O(s) + 2 OH^-(aq) \rightarrow Zn(OH)_4^{2-}(aq) + 2 Ag^0(s)$
   $\qquad\qquad\qquad\qquad\qquad\qquad\qquad\qquad\qquad + H_2O(l)$

QUESTIONS & PROBLEMS

12.1   Define oxidation and reduction in terms of loss and gain of electrons and change in oxidation number.

12.2   What is meant by oxidizing and reducing agents? Give illustrations of each.

12.3   Label the following changes as oxidation or reduction:
(a) $Al^{3+} \rightarrow Al$
(b) $Au \rightarrow AuCl_4^-$
(c) $SO_3^{2-} \rightarrow SO_4^{2-}$
(d) $Br^- \rightarrow Br_2$
(e) $Cl_2 \rightarrow ClO_3^-$

Complete the blanks below each equation. Equation 12.4 is given as an illustration.

12.4 $MnO_2(s) + HCl(aq) \rightarrow Cl_2(g) + MnCl_2(aq) + H_2O(l)$

Ox. agt. ___MnO_2___   Species reduced ___Mn___   e− gained ___2___

Red. agt. ___Cl^-___   Species oxidized ___Cl^-___   e− lost ___1___

12.5 $KIO_3(aq) + KI(aq) + HOAc(aq) \rightarrow KOAc(aq) + H_2O(l) + I_2(s)$

Ox. agt. _____   Species reduced _____   e− gained _____

Red. agt. _____   Species oxidized _____   e− lost _____

12.6 $K_2Cr_2O_7(aq) + H_2SO_4(aq) + HI(aq) \rightarrow$
                           $KHSO_4(aq) + Cr_2(SO_4)_3(aq) + H_2O(l) + I_2(s)$

Ox. agt. _____   Species reduced _____   e− gained _____

Red. agt. _____   Species oxidized _____   e− lost _____

12.7 $Ag(s) + HNO_3(aq) \rightarrow AgNO_3(aq) + NO(g) + H_2O(l)$

Ox. agt. _____   Species reduced _____   e− gained _____

Red. agt. _____   Species oxidized _____   e− lost _____

12.8 $FeSO_4(aq) + KMnO_4(aq) + H_2SO_4(aq) \rightarrow$
                    $Fe_2(SO_4)_3(aq) + MnSO_4(aq) + K_2SO_4(aq) + H_2O(l)$

Ox. agt. _____   Species reduced _____   e− gained _____

Red. agt. _____   Species oxidized _____   e− lost _____

12.9 $CrCl_3(aq) + Na_2O_2(s) + NaOH(aq) \rightarrow$
                          $Na_2CrO_4(aq) + NaCl(aq) + H_2O(l)$

Ox. agt. _____   Species reduced _____   e− gained _____

Red. agt. _____   Species oxidized _____   e− lost _____

12.10 $KClO_3(aq) + SnCl_2(aq) + HCl(aq) \rightarrow$
                          $KCl(aq) + SnCl_4(aq) + H_2O(l)$

Ox. agt. _____   Species reduced _____   e− gained _____

Red. agt. _____   Species oxidized _____   e− lost _____

12.11 $CuS(s) + HNO_3(aq) \rightarrow Cu(NO_3)_2(aq) + NO(g) + S(s)$
                                       $+ H_2O(l)$

Ox. agt. _____   Species reduced _____   e− gained _____

Red. agt. _____   Species oxidized _____   e− lost _____

12.12 $H_3AsO_4(aq) + H_2S(g) \rightarrow H_2O(l) + S(s) + H_3AsO_3(aq)$

Ox. agt. _____   Species reduced _____   e− gained _____

Red. agt. _____   Species oxidized _____   e− lost _____

12.13 $H_2SO_4(aq) + HI(aq) \rightarrow H_2O(l) + I_2(s) + H_2S(g)$

Ox. agt. _____   Species reduced _____   e− gained _____

Red. agt. _____   Species oxidized _____   e− lost _____

OXIDATION-REDUCTION AND AN INTRODUCTION
TO ELECTROCHEMISTRY

12.14 $K_2Cr_2O_7(aq) + H_2S(g) + HCl(aq) \rightarrow$
$$S(s) + CrCl_3(aq) + KCl(aq) + H_2O(l)$$
Ox. agt. ———   Species reduced ———   e⁻ gained ———
Red. agt. ———   Species oxidized ———   e⁻ lost ———

12.15 $NaOH(aq) + Cl_2(g) \rightarrow NaClO_3(aq) + NaCl(aq) + H_2O(l)$
Ox. agt. ———   Species reduced ———   e⁻ gained ———
Red. agt. ———   Species oxidized ———   e⁻ lost ———

12.16   Balance the following equations by the oxidation number method:
(a) $HI(aq) + H_2SO_4(aq) \rightarrow$
$$H_2S(g) + I_2(s) + H_2O(l)$$
(b) $H_2SO_4(aq) + HBr(aq) \rightarrow$
$$SO_2(g) + Br_2(l) + H_2O(l)$$
(c) $MnO_2(s) + HCl(aq) \rightarrow$
$$MnCl_2(aq) + Cl_2(g) + H_2O(l)$$
(d) $NaNO_3(aq) + Fe(s) \rightarrow$
$$Fe_2O_3(s) + NaNO_2(aq)$$
(e) $CuO(s) + NH_3(g) \rightarrow$
$$Cu(s) + N_2(g) + H_2O(l)$$
(f) $As_4(s) + HNO_3(aq) + H_2O(l) \rightarrow$
$$H_3AsO_4(aq) + NO(g)$$
(g) $H_2O_2(aq) + H_2S(g) \rightarrow S(s) + H_2O(l)$
(h) $Mo_2O_3(s) + KMnO_4(aq) + H_2SO_4(aq) \rightarrow$
$$MoO_3(s) + K_2SO_4(aq) + MnSO_4(aq) + H_2O(l)$$
(i) $V_2O_2(s) + H_2SO_4(aq) + KMnO_4(aq) \rightarrow$
$$V_2O_5(s) + MnSO_4(aq) + K_2SO_4(aq) + H_2O(l)$$
(j) $AuCl_3(aq) + H_2S(aq) + H_2O(l) \rightarrow$
$$Au(s) + HCl(aq) + H_2SO_4(aq)$$

12.17   Write balanced net ionic equations for any of these reactions that occur spontaneously in 1 M aqueous solution:
(a) $I^- + Br_2(l)$
(b) $Ag^+ + Fe(s)$
(c) $Ag^+ + Fe^{2+}$
(d) $Fe(s) + Fe^{3+}$
(e) $Cu(s) + Cl_2(g)$
(f) $Mn(s) + Rb^+$
(g) $Cu^{2+} + Br^-$
(h) $Hg(l) + H_3O^+$
(i) $Mg(s) + Sn^{2+}$
(j) $Cu(s) + Ag^+$
(k) $Al(s) + H_3O^+$
(l) $Cu^{2+} + Fe(s)$

(m) $MnO_4^- + Br^-$
(n) $Fe^{3+} + I^-$

12.18   Write balanced net ionic equations for any of these reactions that occur spontaneously in 1 M aqueous solution:
(a) $Fe + H_2SO_4(aq)$
(b) $Na + H_2O(l)$
(c) $Al + HCl(aq)$
(d) $Fe + ZnSO_4(aq)$
(e) $Mg + H_3PO_4(aq)$
(f) $Pb + CuSO_4(aq)$
(g) $Ag + Pb(NO_3)_2(aq)$
(h) $Zn + Mg(NO_3)_2(aq)$
(i) $Fe + CuSO_4(aq)$
(j) $Ca + H_2O(l)$
(k) $Mg + NiSO_4(aq)$
(l) $Sn + ZnSO_4(aq)$
(m) $Cr + MgCl_2(aq)$
(n) $Ba + K_2SO_4(aq)$
(o) $Cu + Al_2(SO_4)_3(aq)$

12.19   Balance these equations, all of which take place in aqueous solution:
(a) $P(s) + HNO_3 + H_2O \rightarrow H_3PO_4 + NO(g)$
(b) $MnO_4^- + H_2S(g) + H^+ \rightarrow$
$$Mn^{2+} + S(s) + H_2O$$
(c) $BrO_3^- + I^- + H^+ \rightarrow Br^- + I_2(s) + H_2O$
(d) $Fe^{2+} + MnO_4^- + H^+ \rightarrow$
$$Fe^{3+} + Mn^{2+} + H_2O$$
(e) $NO_3^- + Cl^- + H^+ \rightarrow$
$$NO(g) + Cl_2(g) + H_2O$$
(f) $HBr + H_2SO_4 \rightarrow$
$$SO_2(g) + Br_2(l) + H_2O$$
(g) $Cu(s) + NO_3^- + H^+ \rightarrow$
$$Cu^{2+} + NO(g) + H_2O$$
(h) $Cr_2O_7^{2-} + Fe^{2+} + H^+ \rightarrow$
$$Cr^{3+} + Fe^{3+} + H_2O$$

12.20 Balance the following equations by the ion-electron method, all of which take place in aqueous solution:

(a) $MnO_4^- + Br^- \rightarrow$
$\qquad Br_2(l) + MnO_2(s)$ (basic solution)

(b) $Ag_2O + CH_2O \rightarrow$
$\qquad Ag(s) + HCO_2^-$ (basic solution)

(c) $H_2SO_3 + Fe^{3+} \rightarrow$
$\qquad Fe^{2+} + SO_4^{2-}$ (acid solution)

(d) $Sn^{2+} + Cr_2O_7^{2-} \rightarrow$
$\qquad Sn^{4+} + Cr^{3+}$ (acid solution)

(e) $MnO_4^- + Sn^{2+} \rightarrow$
$\qquad Mn^{2+} + Sn^{4+}$ (acid solution)

(f) $CrO_4^{2-} + HSnO_2^- \rightarrow$
$\qquad CrO_2^- + HSnO_3^-$ (basic solution)

(g) $BaO_2 + Cl^- \rightarrow$
$\qquad Ba^{2+} + Cl_2(g)$ (acid solution)

(h) $C_2H_4O + NO_3^- \rightarrow$
$\qquad C_2H_4O_2 + NO(g)$ (acid solution)

(i) $CrO_2^- + ClO^- \rightarrow$
$\qquad Cl^- + CrO_4^{2-}$ (basic solution)

(j) $H_3AsO_4 + I^- \rightarrow$
$\qquad H_3AsO_3 + I_2(s)$ (acid solution)

12.21 How many grams of copper can be obtained when 1.6 g of CuO is allowed to react with an excess of ammonia? See Equation 12.16(e).

12.22 What volume of SO_2 gas (STP) can be obtained when 0.030 mole of HBr reacts completely with H_2SO_4? See Equation 12.16(b).

12.23 How many milliliters of 0.022 M HCl are required to produce 72.0 mL (STP) of chlorine gas when they react with MnO_2? See Equation 12.16(c).

12.24 What is the molar concentration of dichromate ion if 34.0 mL are needed to oxidize completely the iron(II) ion in 8.0 g of $FeCl_2$? The skeleton equation is: $Cr_2O_7^{2-} + Fe^{2+} \rightarrow Cr^{3+} + Fe^{3+}$ (acid solution).

12.25 A voltaic cell uses the following reaction: $2 Al(s) + 3 Cu^{2+}(aq) \rightarrow 2 Al^{3+}(aq) + 3 Cu(s)$. Assuming that standard conditions prevail,
(a) diagram and label the cell
(b) calculate the cell voltage

12.26 Calculate the cell voltages and write the equations for the most feasible reactions, given the following pairs of standard half-cell reactions:

(a) $NO_3^-(aq) + H_2O(l) + 2e^- \rightarrow$
$\qquad NO_2^-(aq) + 2 OH^-(aq)$
$\qquad \mathscr{E}^0_{red} = +0.01$ V

$\quad Zn(OH)_4^{2-}(aq) + 2e^- \rightarrow$
$\qquad Zn(s) + 4 OH^-(aq)$
$\qquad \mathscr{E}^0_{red} = -1.22$ V

(b) $AgBr(s) + e^- \rightarrow Ag(s) + Br^-(aq)$
$\qquad \mathscr{E}^0_{red} = +0.10$ V
$\quad MnO_2(s) + 4 H^+(aq) + 2e^- \rightarrow$
$\qquad Mn^{2+}(aq) + 2 H_2O(l)$
$\qquad \mathscr{E}^0_{red} = +1.23$ V

12.27 Predict the tendency of the following reactions to occur spontaneously.
(a) $3 Sn^{2+}(aq) + 2 Cr^{3+}(aq) \rightarrow$
$\qquad 3 Sn^{4+}(aq) + 2 Cr(s)$
(b) $F_2(g) + Cu(s) \rightarrow 2 F^-(s) + Cu^{2+}(s)$

12.28 Define anode and cathode in terms of oxidation and reduction.

12.29 Diagram an apparatus to show how

OXIDATION-REDUCTION AND AN INTRODUCTION
TO ELECTROCHEMISTRY

metallic zinc and metallic copper and their salts could be assembled to make a voltaic cell. Identify the anode and cathode, and write equations for the anode reaction, cathode reaction, and overall reaction. Also show the flow of anions, cations, and electrons.

12.30   Sketch an apparatus that could be used to electrolyze aqueous copper(II) bromide. Identify the anode and cathode, and show direction of electron flow and ion migration. Write equations for reactions at each electrode and describe the overall reaction.

CHAPTER THIRTEEN

LEARNING OBJECTIVES:

At the completion of this chapter, you should be able to:

1. Define the following terms: *organic chemistry, hydrocarbon, hydrocarbon derivative, alkane, alkene, alkyne, functional group, general formula, homologous series, homolog, isomer, structural formula.*

2. Write general formulas for the *aliphatic hydrocarbons* and draw structural formulas for simple aliphatic hydrocarbons and their isomers.

3. Give common names and IUPAC names for the first 10 members of the *alkanes.*

4. Give names of the first 10 straight chain *alkyl* groups.

5. Write structural formulas and give IUPAC names for all isomers of the first five alkanes.

6. Give common names for the first two *alkenes* and *alkynes* and IUPAC names for simple alkenes and alkynes.

7. Draw the structural formula when given the name of a simple alkene or alkyne.

8. Explain the difference between *aliphatic* and *aromatic* hydrocarbons.

9. Give examples and illustrations of seven different functional groups.

10. Give IUPAC and common names for simple *monohydric alcohols.*

11. Draw structural formulas for simple monohydric alcohols when given the name.

12. Give structural formulas and names of three primary, secondary, and tertiary alcohols.

13. Write equations illustrating the "gentle" oxidation of primary and secondary alcohols.

14. Give IUPAC names and common names for simple aldehydes and ketones.

15. Draw structural formulas for simple aldehydes and ketones when given the name.

16. Give common names for simple *ethers,* and draw their structural formulas when given the name.

17. Give IUPAC names and common names for simple *carboxylic acids* and *esters* and draw their structural formulas when given the name.

18. Write balanced equations for the following:
    (a) the complete oxidation or combustion of any hydrocarbon or any compounds containing only carbon, hydrogen, and oxygen atoms.
    (b) the addition reaction resulting when Cl_2 or Br_2 is added to an alkene or alkyne.
    (c) the substitution reaction occurring when a halogen is mixed with an alkane to form an alkyl halide.
    (d) the polymerization of ethylene to form polyethylene.
    (e) the reaction of a carboxylic acid with a simple alcohol to form an ester.
    (f) the alkaline hydrolysis of the ester of a fatty acid to form a soap.

INTRODUCTION TO ORGANIC CHEMISTRY

INTRODUCTION TO ORGANIC CHEMISTRY

13.1 WHAT IS ORGANIC CHEMISTRY?

To this point we have been studying materials, concepts, and reactions dealing principally with inorganic chemistry. In the early part of the nineteenth century it was recognized that fats, sugars, and foods are the results of living processes and that clay, ores, and rocks are obtained directly from the earth. Materials that occur in (or that may be prepared from) lifeless matter were known as *inorganic substances,* and those associated with living matter were termed *organic substances.* For example, coal and petroleum were classified as organic materials because they are the products of the decay of plant life, and urea was considered organic because it is a waste product in the metabolism of mammals. Some chemists subdivided organic materials even further, according to the sources of the living matter. Animal chemistry dealt with such substances as blood, saliva, and body fluids, and vegetable chemistry was concerned with sugars, gums, acids, and rubber. It was later realized that the one factor that all these compounds have in common is the element *carbon.*

Early in the nineteenth century it was found that organic compounds can be made synthetically in the laboratory from substances that are not associated with life processes at all. About 150 years ago the German chemist Friedrich Wöhler accidentally prepared the "organic" compound urea, $CO(NH_2)_2$, through the thermal decomposition of "inorganic" ammonium cyanate, NH_4CNO. As a result of his work, the separation of chemistry into divisions of organic and inorganic lost its meaning. Urea is excreted by a normal human adult at a rate of approximately 25 grams per day. When Wöhler wrote to his teacher, Berzelius, "I must tell you that I can prepare urea without requiring a kidney of an animal, either man or dog," a milestone was reached in chemistry, and the name "organic chemistry" became a historical relic. Today, *organic chemistry is defined as the chemistry of carbon compounds.* More than a million organic compounds have now been identified, a majority of which are made synthetically and many of which are unknown in nature.

Despite these indistinct boundaries, chemistry is such a broad subject that we find it a great convenience to subdivide it into smaller areas. There is ample justification for treating organic chemistry as a special branch of chemistry. One important consideration is that carbon forms four covalent bonds. This characteristic of the carbon atom results in the formation of large and complex organic molecules with a variety of structural arrangements. Also, there are many more compounds containing the carbon atom than there are noncarbon compounds. The versatility of the carbon atom is demonstrated by the following characteristics:

1. Carbon atoms can combine to form straight chains thousands of atoms long. The term *straight-chain* means nonbranched; it should not be interpreted to mean that the molecules form a rigid body with the carbon atoms in line.

$$-\overset{|}{\underset{|}{C}}-\overset{|}{\underset{|}{C}}-\overset{|}{\underset{|}{C}}-\overset{|}{\underset{|}{C}}-\overset{|}{\underset{|}{C}}-\overset{|}{\underset{|}{C}}-\overset{|}{\underset{|}{C}}-$$

2. In addition to the straight carbon chains, carbon atoms can form compounds in branches and rings.

Branched Chain | Straight Chain | Ring (or cyclic)

3. Other atoms, notably H, O, N, S, P, and the halogens, can be attached to the carbon atoms in these chains, branches, or rings to yield an enormous number of derivatives.

$$-\overset{|}{\underset{|}{C}}-\overset{|}{\underset{|}{C}}-\overset{|}{\underset{|}{C}}-Cl$$

4. The bonds between adjacent carbon atoms may be single, double, or triple, or combinations of all of these.

Single   Double   Double   Triple   Cyclic & Double

Each different arrangement of these combinations of bonds and atoms corresponds to a separate compound having its own unique properties. The result is that there are over two million known organic compounds, and millions more are theoretically possible.

The principles and concepts learned in inorganic chemistry apply equally to organic chemistry, but there are some important differences between organic and inorganic compounds. Organic compounds are more numerous; they always contain the carbon atom; they are usually complex in structure; and often they are of high molecular mass.

There are marked differences in the solubilities of organic and inorganic compounds. Many inorganic compounds are made of ions and are water-soluble, whereas organic compounds are covalent and generally insoluble in water. Organic compounds are often easily soluble in organic solvents such as ether, alcohol, benzene, and carbon tetrachloride. The chemical reactions of inorganic substances are more rapid because of the presence of ions. While ions tend to interact vigorously and directly, the reactions of organic compounds involve the slower process of breaking a covalent bond and forming a new one

TABLE 13–1 **Some Properties of Organic and Inorganic Compounds**

Number of Known Compounds	Inorganic Compounds ~ 50,000	Organic Compounds > 1 million
Structure	Generally a simple combination of a few atoms. Nonvolatile solids of low molecular mass. Isomers are rare. Shapes are of some importance.	Often complex, having a large number of atoms in branches, chains, and/or loops. Some volatile solids, but most are liquids or gases. Many isomers, therefore shapes and structures are of critical importance.
Bonding	Mostly ionic or polar-covalent.	Mostly covalent and polar-covalent.
Melting and boiling points	Generally high.	Generally low.
Solubility	Most are soluble in water; insoluble in organic solvents.	Most are insoluble in water; soluble in organic solvents.
Conductivity of aqueous solutions	Most are electrolytes.	Most are nonelectrolytes.
Speed of reactions	Fast, even at room temperature.	Very low, even at high temperatures. Many reactions require a catalyst.
Flammability	Few are flammable.	Most are flammable.
Decomposition by heat	Resistant to all but very high temperatures.	Decompose at relatively low temperatures (300°C). None can survive a dull-red heat.

in its place. These and other properties and differences of organic and inorganic compounds are summarized in Table 13–1.

13.3 ISOMERISM

Molecular structure is a valuable concept in the study of inorganic chemistry; its understanding is critical in the consideration of organic compounds. Let us see why this is so.

Ethyl alcohol is a liquid at room temperature, and it has a boiling point of 78°C. It is completely miscible with water. The gas called dimethyl ether has a boiling point of -24°C, is only slightly soluble in water, and has other properties to show that it is quite different from the alcohol. Yet each of these compounds has a molecular mass of 46.0 amu and an empirical formula of C_2H_6O. How can we have two different compounds with the same empirical formula and molecular mass? Upon investigation we find that the compounds differ in the manner in which the atoms are bonded to each other; they differ in *molecular structure.*

Compounds that have the same molecular formula but which differ in their *structural formulas* are called **isomers.** See Figure 13–1.

13.4 THE HYDROCARBONS

The large family of organic compounds containing only the elements carbon and hydrogen is appropriately named the **hydrocarbons.** Hydrocarbons are of importance because they may be considered as the basis of all the organic compounds. From the viewpoint of structure, it is useful to divide organic compounds into two broad classes, **aliphatic** (fatty) and **aromatic** (fragrant). The original applications

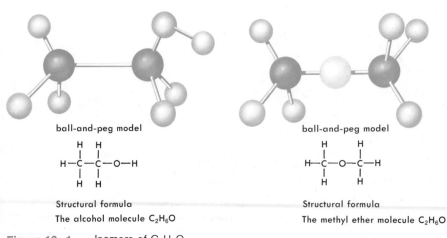

ball-and-peg model

Structural formula

The alcohol molecule C_2H_6O

ball-and-peg model

Structural formula

The methyl ether molecule C_2H_6O

Figure 13–1   Isomers of C_2H_6O.

INTRODUCTION TO ORGANIC CHEMISTRY

TABLE 13–2 Families of Hydrocarbons

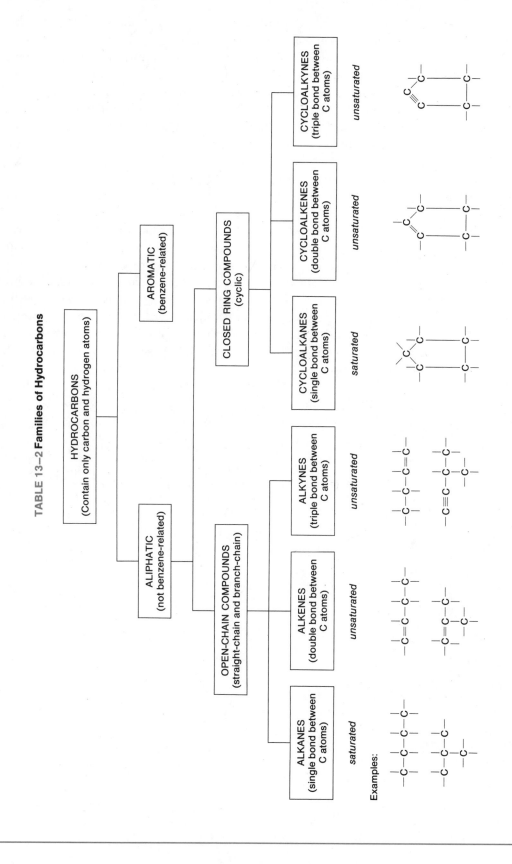

of the terms "fragrant" and "fatty" are no longer significant, but the terms aliphatic and aromatic persist.

Aliphatic hydrocarbons are divided into the categories **saturated** and **unsaturated** and into the subdivisions **alkanes, alkenes,** and **alkynes.** These divisions and subdivisions reflect the type of bond between the carbon atoms. See Table 13–2. We will confine most of our discussion to the aliphatic hydrocarbons and their alkane (saturated) derivatives.

13.5 THE ALKANES

The alkanes (formerly called **paraffins**) contain only carbon and hydrogen atoms joined by single covalent bonds. Each carbon atom is flanked by four covalent bonds, and each hydrogen atom shares one pair of electrons with a carbon atom. Methane is the first and simplest member of the alkanes, most of which occur in petroleum. All have the same general structure and are similar in chemical properties. They show a regular change in their physical properties, as shown in Table 13–4. The alkanes are colorless, practically odorless, insoluble in water, and readily soluble in nonpolar solvents such as benzene, ether, and carbon tetrachloride. Here are the names and structural formulas for the first four alkanes:

$$
\begin{array}{c}
\text{H} \\
| \\
\text{H}-\text{C}-\text{H} \\
| \\
\text{H}
\end{array}
\qquad
\begin{array}{c}
\text{H} \quad \text{H} \\
| \quad\ | \\
\text{H}-\text{C}-\text{C}-\text{H} \\
| \quad\ | \\
\text{H} \quad \text{H}
\end{array}
\qquad
\begin{array}{c}
\text{H} \quad \text{H} \quad \text{H} \\
| \quad\ | \quad\ | \\
\text{H}-\text{C}-\text{C}-\text{C}-\text{H} \\
| \quad\ | \quad\ | \\
\text{H} \quad \text{H} \quad \text{H}
\end{array}
\qquad
\begin{array}{c}
\text{H} \quad \text{H} \quad \text{H} \quad \text{H} \\
| \quad\ | \quad\ | \quad\ | \\
\text{H}-\text{C}-\text{C}-\text{C}-\text{C}-\text{H} \\
| \quad\ | \quad\ | \quad\ | \\
\text{H} \quad \text{H} \quad \text{H} \quad \text{H}
\end{array}
$$

methane, CH_4 ethane, C_2H_6 propane, C_3H_8 butane, C_4H_{10}

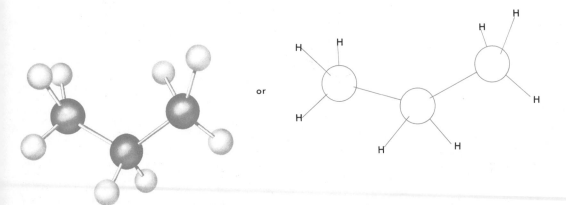

or

Figure 13–2   The propane molecule.

                INTRODUCTION TO ORGANIC CHEMISTRY

The bond angles and bond distances of the alkanes are almost all the same as in methane. You will recall that in methane the bond angles are 109°28′, and the carbon-hydrogen bond lengths are about 1.09 Å. Studies show that the C—H bond length in ethane is 1.54 Å. With very slight variations, these values are characteristic of the bond lengths and bond angles in all of the alkanes.

We will be using structural formulas often in our study of organic molecules. Let's examine the types of formulas that chemists find convenient in their illustrations and explanations. In the compounds propane and butane shown above, the carbon atoms are not in a straight line and the molecule is not in the plane of the page. The propane molecule is more accurately illustrated in Figure 13–2.

But these illustrations take up a great deal of space and are inconvenient to reproduce, so graphic formulas are used to show a two-dimensional structure, with zig-zags between carbon atoms.

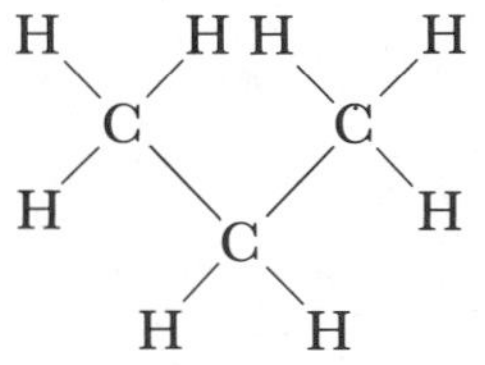

Although not as correct as the zig-zag structure, it is convenient to show the carbon atoms in a straight line, with the hydrogen atoms extending at angles of 90° to one another. Even this is often too cumbersome, so a further condensation of the structure formula shows only the bonds between carbon atoms and omits those between carbon and hydrogen. These and other types of shortcut formulas commonly used to describe organic structures are summarized in Table 13–3.

Because of the large number of compounds involved and the overlapping characteristics of compounds, hydrocarbon nomenclature has always been inexact. Many names used in the early era of organic chemistry arose from the origins of the materials, from some of their properties, or from the whims of the scientists who first prepared the compounds. This is reflected in the names of the first four members of the alkanes. All alkanes are named by attaching the suffix -ane to a prefix signifying the number of carbon atoms in the compound. The prefixes *meth-*, *eth-*, *prop-*, or *but-* in the first four members are trivial (common) prefixes of historical origin which have survived because of long usage. Except for these first four alkanes, the names of all others

TABLE 13–3 **Types of Formulas Used in Organic Chemistry**

1. Structural or graphic formula (shows all bonds and angles):

2. Two-dimensional structural formula (shows bonds but not angles). This is the most frequently used structural formula:

3. Simplified structural formula. Shows only carbon-to-carbon bonds. For clarity and further simplification, these are often used with hydrogen atoms not shown:

4. Condensed formula (shows bonds between *groups* of atoms):

$$CH_3-CH_2-CH_2-CH_3$$

5. Simplified condensed formula (shows only the bonding groups but suggests the order of linkage). The condensed formula is commonly used by organic chemists:

$$CH_3CH_2CH_2CH_3$$

6. Molecular formula. Shows total number of all atoms without regard for bonding or structure. Sometimes used in equation writing:

$$C_4H_{10}$$

contain the Greek prefix indicating the number of carbon atoms in the molecule.

prefix	number of carbon atoms
meth-	one
eth-	two
prop-	three
but-	four
pent-	five
hex-	six
hept-	seven
oct-	eight
non-	nine
dec-	ten

Each successive member of the alkane family differs by one CH_2 increment. Propane contains one more carbon atom and two more hydrogen atoms than methane. A series of compounds in which each member differs from the next by a constant amount of atoms is known as a **homologous series,** and the series members are termed **homologs.** The number of carbon and hydrogen atoms in each alkane is related by the *general formula* C_nH_{2n+2}, where n is an integral number (e.g., if n = 4, $C_4H_{2(4)+2}$, or C_4H_{10}).

The names and properties of some of the common alkanes are given in Table 13–4. Observe that the boiling points and melting points rise with increasing molecular mass. The first four compounds are gaseous at room temperature, and pentane is a low-boiling liquid. Starting with butane, C_4H_{10}, more than one arrangement of atoms (isomers) is possible. As the number of atoms in a compound increases, so does the number of possible isomers. It is apparent that some systematic method of chemical nomenclature is needed if we are to name this multitude of isomers in a clear and unambiguous manner.

13.6 ALKYL GROUPS

We have found charged groups such as SO_4^{2-} and NH_4^+ helpful in naming inorganic compounds and in writing equations. The need for modifying groups in the naming of organic compounds is even greater than in inorganic nomenclature. The **alkyl groups** are covalent groups obtained by dropping the *-ane* suffix from an alkane and replacing it with *-yl.* In so doing, butane (C_4H_{10}) becomes the *butyl group* (C_4H_9—), and hexane (C_6H_{14}) becomes the hexyl group (C_6H_{13}—). Inasmuch as the alkane general formula is C_nH_{2n+2}, the alkyl groups have the general formula C_nH_{2n+1}. Alkyl groups are not individual

TABLE 13–4 **Names and Properties of Alkanes**

Name	Formula	Melting Point, °C	Boiling Point, °C	Density g/mL at 20°C	Number of Isomers
Methane	CH_4	−183	−162		1
Ethane	C_2H_6	−172	−88		1
Propane	C_3H_8	−187	−42		1
n-Butane	C_4H_{10}	−138	0		2
n-Pentane	C_5H_{12}	−130	36	0.626	3
n-Hexane	C_6H_{14}	−95	69	0.659	5
n-Heptane	C_7H_{16}	−90	98	0.684	9
n-Octane	C_8H_{18}	−57	126	0.703	18
n-Nonane	C_9H_{20}	−54	151	0.718	35
n-Decane	$C_{10}H_{22}$	−30	174	0.730	75
n-Octadecane	$C_{18}H_{38}$	28	308	0.785	60523
n-Eicosane	$C_{20}H_{42}$	36		0.778	366319
Isobutane	C_4H_{10}	159	−12	0.600	
Isopentane	C_5H_{12}	−160	28	0.620	
Neopentane	C_5H_{12}	−17	9.5	0.610	

TABLE 13–5 Some Alkyl Groups

Alkane (R—H)		Alkyl Group (R—)	
H−C−H with H above and below (H on left)	CH_4 methane	H−C− with H above and below (H on left)	CH_3- methyl group
H−C−C−H (H above and below each C)	CH_3CH_3 ethane	H−C−C− (H above and below each C)	CH_3CH_2- ethyl group
H−C−C−C−H (H above and below each C)	$CH_3CH_2CH_3$ propane	H−C−C−C− (H above and below each C)	$CH_3-CH_2-CH_2-$ propyl group
		H−C−C−C−H (H above and below middle)	$CH_3-CH-CH_3$ isopropyl group

species. They are always associated with a particular organic molecule. Table 13–5 illustrates the relationship of these groups to their parent alkanes.

Condensed formulas (CH_3-, C_2H_5-, C_3H_7-, etc.) are usually used. R— is the general symbol for any alkyl group.

The prefix *n-* (meaning normal) indicates that all carbon atoms in an alkyl group are in a continuous chain. Except in the case of isopropyl, the prefix *iso-* shows that an alkyl group is present as a branch

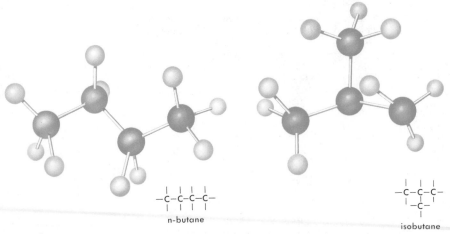

−C−C−C−C−
n-butane

−C−C−C−
 −C−
isobutane

Figure 13–3   Structural models and structural formulas for n-butane and isobutane.

INTRODUCTION TO ORGANIC CHEMISTRY

chain. Structural models and structural formulas for *n*-butane and isobutane are shown in Figure 13–3. For clarity and simplification the hydrogen atoms are left off the structural formula.

Let's investigate the structures of a few alkanes and try to name them properly.

The first three members of the group—methane, CH_4, ethane, C_2H_6, and propane, C_3H_8, have no isomers. The fourth member, butane, C_4H_{10}, has the two isomers mentioned above. When we consider the five-carbon isomer, pentane, C_5H_{12}, we find that there are three isomers possible:

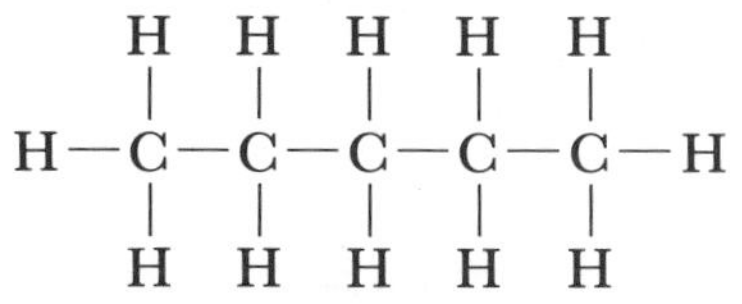

n-pentane

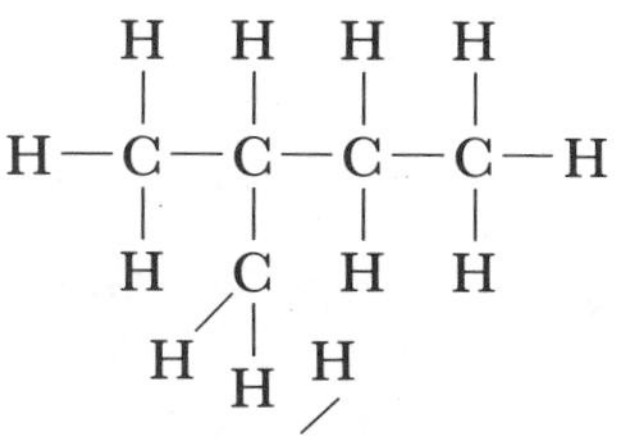

isopentane

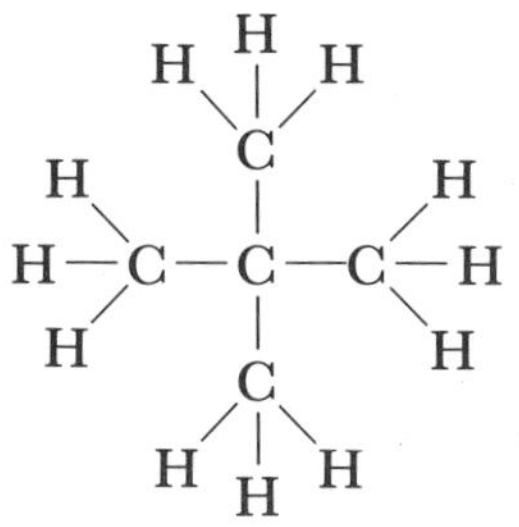

neopentane

But beyond pentane, the *iso*-prefix becomes useless in alkane nomenclature, and we must resort to the IUPAC system for naming compounds.

13.7 SYSTEMATIC NOMENCLATURE OF ORGANIC COMPOUNDS

The International Union of Pure and Applied Chemistry **(IUPAC)** system of nomenclature is applicable to the naming of all families of organic compounds. As applied to the naming of alkanes, the rules are:

1. The stem of the name of the alkane is derived from the longest continuous chain of carbon atoms. This is the parent structure. The alkyl groups replacing the hydrogen atoms are listed as modifying

groups to the parent structure, much as adjectives are used to modify nouns. Thus:

$$-\overset{|}{\underset{|}{C}}-\overset{|}{\underset{|}{C}}-\overset{|}{\underset{|}{C}}-\overset{|}{\underset{|}{C}}-$$
$$\underset{|}{-C-}$$

is called methylbutane, and

$$-\overset{|}{\underset{|}{C}}-\overset{|}{\underset{|}{C}}-\overset{|}{\underset{|}{C}}-\overset{|}{\underset{|}{C}}-\overset{|}{\underset{|}{C}}-\overset{|}{\underset{|}{C}}-\overset{|}{\underset{|}{C}}-\overset{|}{\underset{|}{C}}-$$
$$-\overset{|}{\underset{|}{C}}-$$
$$-\underset{|}{C}-$$

is ethyloctane (but this name is incomplete; see explanation below)

2. When it is necessary to indicate the position of an attached group (e.g., the alkyl group in the case of ethyloctane above), the carbon atoms in the longest continuous chain are numbered from the end that results in the lowest numbers.

$$-\overset{|}{\underset{|}{Ⓒ}}-\overset{|}{\underset{|}{Ⓒ}}-\overset{|}{\underset{|}{Ⓒ}}-\overset{|}{\underset{|}{Ⓒ}}-\overset{|}{\underset{|}{Ⓒ}}-\overset{|}{\underset{|}{Ⓒ}}-\overset{|}{\underset{|}{Ⓒ}}-\overset{|}{\underset{|}{Ⓒ}}-$$

correct

3-ethyloctane

$$-\overset{|}{\underset{|}{Ⓒ}}-\overset{|}{\underset{|}{Ⓒ}}-\overset{|}{\underset{|}{Ⓒ}}-\overset{|}{\underset{|}{Ⓒ}}-\overset{|}{\underset{|}{Ⓒ}}-\overset{|}{\underset{|}{Ⓒ}}-\overset{|}{\underset{|}{Ⓒ}}-\overset{|}{\underset{|}{Ⓒ}}-$$

incorrect

6-ethyloctane

3. If the same alkyl group appears more than once as a side-chain modifying group, this is indicated by the prefixes *di-, tri-, tetra-,* and the like. For example,

$$-\overset{|}{\underset{|}{Ⓒ}}-\overset{|}{\underset{|}{Ⓒ}}-\overset{|}{\underset{|}{Ⓒ}}-\overset{|}{\underset{|}{Ⓒ}}-\overset{|}{\underset{|}{Ⓒ}}-\overset{|}{\underset{|}{Ⓒ}}-\overset{|}{\underset{|}{Ⓒ}}-\overset{|}{\underset{|}{Ⓒ}}-$$

$CH_3 \qquad CH_3$

longest chain is 8 carbons ∴ *octane*

{ methyl groups are on #3 and #5 carbon atoms

is 3-methyl-5-methyloctane, which is shortened to 3,5-dimethyloctane.

$$-\overset{|}{\underset{|}{Ⓒ}}-\overset{|}{\underset{|}{Ⓒ}}-\overset{|}{\underset{|}{Ⓒ}}-\overset{|}{\underset{|}{Ⓒ}}-$$
$$-\underset{|}{C}-\underset{}{C}-$$

is 2,3-dimethylbutane, and

$$\underset{|}{-C-}$$
$$-\overset{|}{\underset{|}{Ⓒ}}-\overset{|}{\underset{|}{Ⓒ}}-\overset{|}{\underset{|}{Ⓒ}}-\overset{|}{\underset{|}{Ⓒ}}-$$
$$-\underset{|}{C}-$$

is 2,2-dimethylbutane.

INTRODUCTION TO ORGANIC CHEMISTRY

There are additional rules and conventions, but the above will suffice for our level of discussion.

Reactions of Saturated Hydrocarbons

The alkane or paraffin hydrocarbons tend to be chemically unreactive. The term *paraffin* means "little affinity" in Greek and refers to the low reactivity of these compounds. The reactions that they do undergo are called halogenation, combustion, and thermal decomposition or "cracking."

1. **Halogenation.** The halogens, particularly chlorine and bromine, react with the alkanes to replace a hydrogen atom by a halogen atom. For example, methane can be converted to methyl bromide (bromomethane) if it is allowed to react with bromine.

$$CH_4 + Br_2 \rightarrow CH_3Br + HBr$$

methane   methyl bromide

In this reaction, a hydrogen atom is removed from CH_4 and a bromine atom substituted in its place. This is referred to as a *substitution reaction*. The four hydrogen atoms of methane can be replaced successively by bromine atoms or chlorine atoms to form the mono-, di-, tri-, and tetra-halogenated alkanes by such substitution reactions. In the sequence where chlorine is added to methane, the products are CH_3Cl (chloromethane), CH_2Cl_2 (dichloromethane), $CHCl_3$ (trichloromethane, or chloroform), and CCl_4 (carbon tetrachloride), respectively.

2. **Combustion.** All compounds containing only carbon and hydrogen atoms can be completely burned in oxygen to form carbon dioxide gas and water. Because this process is highly exothermic (heat-evolving), hydrocarbons are frequently used for fuel. In the case of methane,

$$CH_4(g) + 2\ O_2(g) \rightarrow CO_2(g) + 2\ H_2O\ (\ell) + 213\ kcal$$

3. **Cracking.** This is the process of breaking large hydrocarbon molecules into smaller ones. A simple cracking reaction is that of propane cracking into methane and ethylene, an alkene discussed in Section 13.8.

$$C_3H_8 \xrightarrow[\text{heat}]{\text{catalyst}} CH_4 + C_2H_4\ \text{(ethylene)}$$

13.8 UNSATURATED ALIPHATIC HYDROCARBONS—ALKENES

Another homologous series of hydrocarbons has the general formula C_nH_{2n} and is called the **alkene** series. It is sometimes called the

ethylene series because ethylene is the first member of the series. The alkenes are hydrocarbons containing two fewer hydrogen atoms than the corresponding member of the alkane series. Because of a double bond, hydrocarbons containing four fewer hydrogen atoms than their corresponding alkanes are named **alkynes.** They have the general formula C_nH_{2n-2}. Since the alkenes and the alkynes have less than the maximum number of hydrogen atoms, they are considered **unsaturated** hydrocarbons. We will discuss the alkenes now and take up the alkynes in Section 13.9.

The chief industrial source of alkene hydrocarbons is the cracking or thermal decomposition of the long-chain alkanes from petroleum. The alkenes are made by the removal of two hydrogen atoms (dehydrogenation) from adjacent carbon atoms of an alkane molecule. The resulting structure and distinguishing feature of all alkene structures is the existence of a double bond between two adjacent carbon atoms.

$$
\underset{\text{ethane}}{
\begin{array}{c}
\text{H} \quad \text{H} \\
| \quad\; | \\
\text{H}-\text{C}-\text{C}-\text{H} \\
| \quad\; | \\
\text{H} \quad \text{H}
\end{array}}
\quad
\xrightarrow[\text{(dehydrogenation)}]{\substack{\text{loses two} \\ \text{H atoms}}}
\quad
\underset{\text{ethylene}}{
\begin{array}{c}
\text{H} \qquad\quad \text{H} \\
\diagdown \qquad\; \diagup \\
\text{C}=\text{C} \\
\diagup \qquad\; \diagdown \\
\text{H} \qquad\quad \text{H}
\end{array}}
$$

The bond distance between the double-bonded carbons is less than that between single-bonded carbon atoms. The carbon-to-carbon distance in ethylene (C_2H_4) is 1.34 Å, compared with a distance of 1.54 Å between carbon atoms in ethane. All atoms are in the same plane, and bond angles are 120°. See Figure 13–4.

The first two alkenes, ethylene, C_2H_4, and propylene, C_3H_6, have no isomers, but we find three* possible structural arrangements for butylene, C_4H_8. Two of them are straight-chained and differ in the position of the double bond in the chain. The other is a branch-chained structure. These compounds are given the names 1-butene, 2-butene, and isobutylene, respectively.

---

*Actually, there are four possible arrangements, but we will omit discussion of the *cis-trans* alkene isomers.

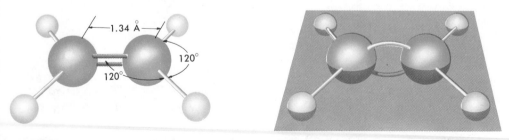

Figure 13–4   Structure of the ethylene molecule, C_2H_4. All atoms are in the same plane, and bond angles are 120°.

$$C=C-C-C \qquad C-C=C-C \qquad \begin{array}{c} C=C-C \\ | \\ C \end{array}$$

| 1-butene | 2-butene | 2-methylpropene (isobutylene) |

If we try to name the higher-carbon alkenes, we find once again that we are in trouble, but the IUPAC system provides us with unambiguous names for the multitude of higher-carbon isomers. IUPAC rules for naming the alkenes are:

1. The *-ane* ending of the corresponding alkane is replaced by *-ene,* making the parent structure ethene, propene, butene, etc., depending upon the longest continuous chain containing the carbon-to-carbon double bond.
2. The carbons of the parent compound are numbered starting from the end of the chain nearest the double bond. The position of the double bond is indicated by the smaller of the two numbers of the carbon atoms to which the double bond is attached. For example:

$$CH_3-CH_2-CH=CH_2 \qquad CH_3-CH_2-CH=CH-CH_3$$

1 butene   2-pentene

3. The names and positions of side-chains are handled in the same manner as in the naming of the alkanes. The numbered positions are determined by the double bond:

$$\underset{\displaystyle CH_3}{CH_3-CH_2-\underset{|}{C}=CH-CH_3} \qquad \underset{\displaystyle CH_3}{CH_3-\underset{|}{CH}-CH=CH_2}$$

3-methyl-2 pentene   3-methyl-1 butene

$$\begin{array}{c} -C-C=C-C-C-C- \\ | \quad\quad | \\ -C- \\ | \end{array} \text{ is 3-methyl-2-hexene}$$

$$\begin{array}{c} -C-C=C- \\ | \\ -C- \\ | \end{array} \text{ is methylpropene (isobutylene)}$$

$$\begin{array}{c} -C- \\ | \\ -C-C=C-C-C- \\ | \quad\quad | \quad | \\ -C- \quad -C- \\ | \quad\quad | \end{array} \text{ is 2,4,4-trimethyl-2 pentene}$$

TABLE 13–6 **Names and Properties of Alkenes**

Name	Formula	Melting Point, °C	Boiling Point, °C	Density g/mL at 20°C
Ethylene (ethene)	C_2H_4	−169	−102	
Propylene (propene)	C_3H_6	−185	− 48	
1-Butene	C_4H_8	−185	− 6.5	
1-Pentene	C_5H_{10}	−138	30	0.643
1-Hexene	C_6H_{12}	−140	63.5	0.675
1-Heptene	C_7H_{14}	−119	93	0.698
1-Octene	C_8H_{16}	−104	122.5	0.716
1-Nonene	C_9H_{18}	− 81	146	0.731
1-Decene	$C_{10}H_{20}$	− 66	171	0.743
3-Methyl-1-butene	C_5H_{10}	−135	25	0.648
2-Methyl-1-butene	C_5H_{10}	−123	39	0.660

The boiling points, melting points, and other physical properties of the alkenes are similar to those of the corresponding alkanes. See Table 13–6.

In contrast to their saturated-hydrocarbon counterparts, the alkenes are quite reactive chemically. They undergo *addition* reactions form saturated compounds that contain only single bonds between the carbon atoms. In other words, an alkene may react with hydrogen to yield an alkane. For example, ethylene adds a molecule of hydrogen to produce ethane:

$$H-\underset{\underset{H}{|}}{\overset{\overset{}{}}{C}}=\underset{\underset{H}{|}}{\overset{\overset{}{}}{C}}-H \qquad + H_2 \xrightarrow{Pt} \qquad H-\underset{\underset{H}{|}}{\overset{\overset{H}{|}}{C}}-\underset{\underset{H}{|}}{\overset{\overset{H}{|}}{C}}-H$$

ethylene	(hydrogenation)	ethane
(unsaturated)		(saturated)

In an addition reaction a halogen is added to ethylene to form an alkyl halide. If chlorine is used, the product is 1,2-dichloroethane:

$$H-\underset{\underset{H}{|}}{\overset{\overset{}{}}{C}}=\underset{\underset{H}{|}}{\overset{\overset{}{}}{C}}-H \qquad + Cl_2 \rightarrow \qquad H-\underset{\underset{H}{|}}{\overset{\overset{Cl}{|}}{C}}-\underset{\underset{H}{|}}{\overset{\overset{Cl}{|}}{C}}-H$$

	(chlorination)	1,2-dichloroethane
		(an alkyl halide)

Alkene molecules can unite to form large molecules by a mechanism called *addition polymerization*. **Polymers** are compounds of high molecular mass that result from the chemical combination of a large number of identical simple molecules. Polymerization reactions are important in the formation of plastic materials such as polyethylene, the familiar polymer used for protective coatings and as a packaging material.

$$n(CH_2 = CH_2) \rightarrow (-CH_2-CH_2-CH_2-CH_2-CH_2-CH_2)_n$$

n units of ethylene *monomer* → one unit of polyethylene *polymer*

Ethylene is also used as an intermediate in the manufacture of other organic materials. When mixed with oxygen, ethylene is a good general anesthetic. The artificial ripening of citrus fruit is hastened by exposure to a low concentration of ethylene in the air.

13.9 THE ALKYNES

The third homologous series of hydrocarbons we shall study is the family of compounds containing two fewer hydrogen atoms than their corresponding alkenes. These have the general formula C_nH_{2n-2} and are called *alkynes*. The alkyne structure has a triple bond between adjacent carbon atoms; hence alkynes are even more unsaturated (and more reactive) then their alkene counterparts. The simplest and by far the most useful compound in this series is ethyne, often called by its common name, acetylene (Fig. 13–5). It is a linear molecule having bond angles of 180°, a carbon-to-carbon distance of 1.20 Å (this is the triple bond), and a carbon-to-carbon distance of 1.06 Å. The name acetylene is unfortunate because its *-ene* suffix gives the impression that it is an alkene.

Acetylene is a colorless, flammable gas with an unpleasant odor. It is made by the reaction of calcium carbide with water.

$$CaC_2(s) + 2\ H_2O(\ell) \rightarrow Ca(OH)_2(s) + HC\equiv CH(g)$$

About 10 per cent of the acetylene made commercially is used in the oxyacetylene torch for cutting and welding steel. Most of it is used as a raw material intermediate in the synthesis and manufacture of important organic plastics, fibers, resins, and other materials. Table 13–7 lists the properties of some common alkynes.

Figure 13–5 Structure of the acetylene molecule, C_2H_2.

TABLE 13–7 Alkynes

Name	Formula	Melting Point, °C	Boiling Point, °C	Density g/mL at 20°C
Acetylene (ethyne)	C_2H_2	−82	sublimes	
Propyne	C_3H_4	−101	−23	
1-Butyne	C_4H_6	−122	8	
1-Pentyne	C_5H_8	−98	40	0.695
1-Hexyne	C_6H_{10}	−124	72	0.719
2-Butyne	C_4H_6	−24	27	0.694
2-Hexyne	C_6H_{10}	−92	84	0.730
3-Hexyne	C_6H_{10}	−51	81	0.725

The same IUPAC rules of nomenclature apply to alkynes as to the alkenes except that *-yne* replaces the *-ene* of the alkenes.

The simpler alkynes are often named as derivatives of acetylene. The names and structures of the first three members of the series illustrate this characteristic:

$$H—C\equiv C—H \qquad H—C\equiv C—\underset{\underset{H}{|}}{\overset{\overset{H}{|}}{C}}—H \qquad H—C\equiv C—\underset{\underset{H}{|}}{\overset{\overset{H}{|}}{C}}—\underset{\underset{H}{|}}{\overset{\overset{H}{|}}{C}}—H$$

ethyne	propyne	1-butyne
(acetylene)	(methylacetylene)	(ethylacetylene)

The alkynes undergo addition reactions with hydrogen bromine and hydrogen bromide. In the case of addition reactions with bromine, the first step is the formation of an alkene dibromide. Upon further addition of bromine, a tetrabromoalkane is produced.

$$CH_3C\equiv CH + Br_2 \rightarrow CH_3—\underset{\underset{Br}{|}}{C}=\underset{\underset{Br}{|}}{C}H \qquad \text{(First step)}$$

propyne   1,2-dibromopropene

$$CH_3—\underset{\underset{Br}{|}}{C}=\underset{\underset{Br}{|}}{C}H + Br_2 \rightarrow CH_3—\underset{\underset{Br}{|}}{\overset{\overset{Br}{|}}{C}}—\underset{\underset{Br}{|}}{\overset{\overset{Br}{|}}{C}} \qquad \text{(Final step)}$$

1,1,2,2,-tetrabromopropane

13.10 CYCLIC ALIPHATIC HYDROCARBONS

Petroleum crude oil coming from the western United States contains a high percentage of closed-chain or cyclic hydrocarbons called naphthenes. California and Texas crude contains appreciable quantities of cyclohexanes and cyclopentanes, and gasoline has cycloalkanes and cycloalkenes as ingredients. Cycloalkanes are also found in lubricating oils.

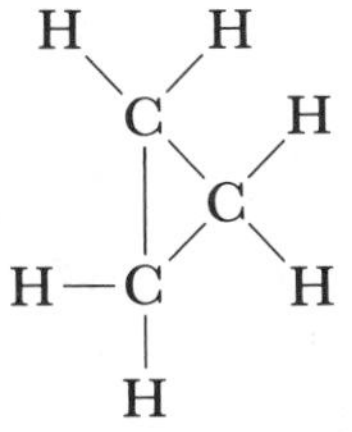

C_3H_6
cyclopropane

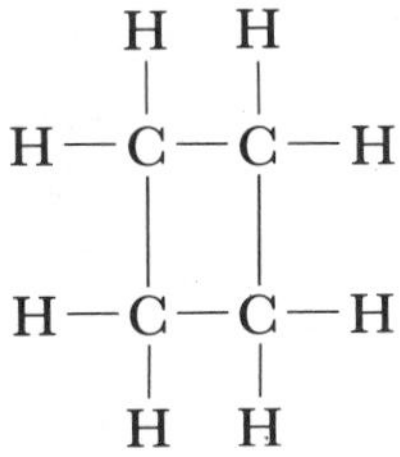

C_4H_8
cyclobutane

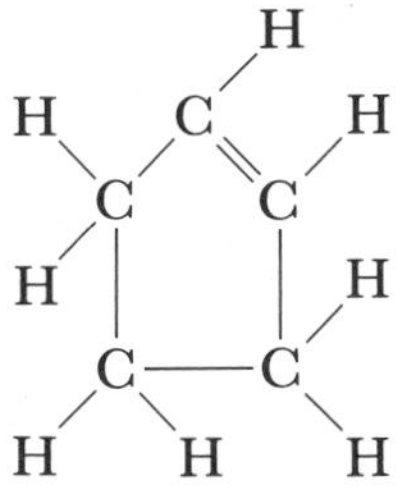

C_5H_8
cyclopentene

The names of cyclic aliphatic hydrocarbons are formed by adding the prefix *cyclo-* to the name of the alkane or alkene having the same number of carbon atoms. The simplest possible cycloalkane is cyclopropane, C_3H_6, a sweet-smelling gas used as a general anesthetic. The general formula for a cycloalkane is C_nH_{2n}, and for a cycloalkene it is C_nH_{2n-2}.

Cyclopropane and cyclobutane are colorless gases, and cyclopentane is a colorless liquid with a low boiling point. The boiling points of the cycloalkanes are usually 10° to 20°C higher than those of the corresponding alkanes. See Table 13–8. In general, the properties of the cyclic compounds resemble those of open-chain aliphatic compounds of the same number of carbon atoms.

With few exceptions, cyclic aliphatic compounds undergo the same reactions as their open-chain analogs.

TABLE 13–8 **Cyclic Aliphatic Hydrocarbons**

Name	Melting Point, °C	Boiling Point, °C	Density g/mL at 20°C
Cyclopropane	−127	− 33	
Cyclobutane	− 80	13	
Cyclopentane	− 94	49	0.746
Cyclohexane	6.5	81	.778
Cycloheptane	− 12	118	.810
Cyclooctane	14	149	.830
Cyclopentene	− 93	46	.774
Cyclohexene	−104	83	.810

The families of hydrocarbons discussed to this point are the open-chain compounds and those cyclic compounds resembling open-chain compounds in their behavior. We now come to the final major class of hydrocarbons, termed **aromatic** hydrocarbons. These are cyclic compounds such as benzene, C_6H_6, and the other compounds that behave like benzene in their chemical reactions. The name "aromatic" is based upon the fact that derivatives of these hydrocarbons have distinctive pleasant odors. This name is no longer meaningful. You are probably familiar with the pleasant (but toxic) odor of benzene, which is the solvent for rubber cement.

Benzene is the first member of this homologous series of hydrocarbons with the general formula C_nH_{2n-6}. Benzene and its homologs, toluene, C_7H_8, and the three isomeric xylenes, C_8H_{10}, are colorless, highly flammable liquids at room temperature. See Table 13–9.

Despite the high degree of unsaturation suggested by the formula C_6H_6, benzene and its derivatives are found to be much more stable than ordinary unsaturated compounds. Benzene behaves chemically more like a saturated hydrocarbon. Experiments show that the characteristic reaction of aromatic compounds is *substitution* rather than addition:

$$C_6H_6 \ + \ Br_2 \xrightarrow{\text{Fe}} \ C_6H_5Br \ + \ HBr$$

benzene   bromobenzene

The molecular structure of aromatic compounds is characterized by the six-member ring of carbon atoms known as the *benzene ring*. Studies indicate that the benzene molecule is made up of six carbon atoms linked together in the form of a hexagonal ring, all lying in a single plane. One hydrogen atom is attached to each of the carbon atoms. All bond angles are $120°$.

TABLE 13–9 **Aromatic Hydrocarbons**

Formula	Name	State at 20°C	Melting Point, °C	Boiling Point, °C	Density g/mL at 20°C
C_6H_6	Benzene	liquid	5.5	80	0.879
C_7H_8	Toluene	liquid	−95	111	0.866
C_8H_{10}	o-Xylene	liquid	−25	144	0.880
C_8H_{10}	m-Xylene	liquid	−48	139	0.864
C_8H_{10}	p-Xylene	liquid	13	138	0.861

H
|
H. C H
C
|
C. C.
H C H
|
H

But this structure accounts for only three of the four bonds of each carbon atom. Where does the fourth bond go, and where are the remaining six electrons? In answer to these questions, a benzene molecule having alternate single and double bonds was proposed.

H   H
| |
C   C
H—C   C—H   H—C   C—H
H—C   C—H   and H—C 120° C—H   or   [hexagon]   and   [hexagon]
C   120° C 120°
| |
H   H

Further study revealed that the carbon-to-carbon bond lengths were neither double nor single but that they were all identical, 1.40 Å. This length is intermediate between that of an aliphatic single bond, 1.54 Å, and a double bond, 1.34 Å. Since it is impossible to write a single structure that can explain the properties and structural arrangement of benzene, it is conventionally represented by a circle within a hexagon or simply by a hexagon with three lines resembling double bonds.

[hexagon with circle]   or   [hexagon with three lines]

This simplified representation serves to emphasize the flat, hexagonal structure of the benzene molecule and its six identical carbon-to-carbon bonds, which are equivalent in all respects. One electron from each carbon atom is **delocalized.** The concept of electron delocalization is meant to suggest that the "delocalized" electrons belong to the molecule as a whole rather than to any specific atoms. This concept will be studied more fully in your organic chemistry course.

Aromatic compounds are extremely useful as raw materials for the chemical synthesis of other organic compounds. Benzene is used in the manufacture of plastics, synthetic rubbers, nylon, pesticides, explosives, pharmaceuticals, and a host of other important products. Benzene and toluene and their derivatives are used as solvents in the paint and varnish industry.

During most of the nineteenth century petroleum was relatively scarce, and the major source of aromatic hydrocarbons was the destructive distillation of bituminous coal. When bituminous coal is heated to about 1200°C in the absence of air, coal gas and coal tar are obtained, and a residue of coke remains. The coal gas contains benzene (not to be confused with *benzine*) and toluene, and the coal tar contains other valuable aromatic compounds. Many of the aromatic hydrocarbons are now obtained from petroleum through a process of **catalytic reforming.**

13.12 THE CONCEPT AND MEANING OF FUNCTIONAL GROUPS

The physical properties and chemical behavior of an alkene are attributed to the activity of its carbon-to-carbon double bond. We find that any compound having a carbon-to-carbon triple bond has properties related to the reactivity of that bond. In the next section we shall see that the —OH group is the one structural feature common to all alcohols. The —OH group defines the structure and determines the

"WHAT I'D LIKE TO DO IS TRANSMUTE SOME CARBON, SULFUR, OIL AND CLAY INTO POLYSTYRENE."

TABLE 13–10 **Important Categories of Hydrocarbon Derivatives**

Class of Compound	Type Formula	Functional Group	Example	Common Name	Systematic Name
Alkene	$*RC\!=\!CR'$ with H H	$\backslash C\!=\!C /$	$HC\!=\!CH$ with H H	ethylene	eth*ene*
Alkyne	$*RC\!\equiv\!CR'$	$-C\!\equiv\!C-$	$HC\!\equiv\!CH$	acetylene	eth*yne*
Alcohol	$R-OH$	$-OH$ hydroxyl	C_2H_5-OH	ethyl alcohol	ethan*ol*
Acid	$*R-\overset{O}{\overset{\|}{C}}-OH$	$-\overset{O}{\overset{\|}{C}}-OH$ carboxyl	$CH_3-\overset{O}{\overset{\|}{C}}-OH$	acetic acid	ethan*oic acid*
Aldehyde	$*R-\overset{O}{\overset{\|}{C}}-H$	$-\overset{O}{\overset{\|}{C}}-H$ formyl	$CH_3-\overset{O}{\overset{\|}{C}}-H$	acetaldehyde	ethan*al*
Ketone	$R-\underset{O}{\overset{\|}{C}}-R'$	$-\underset{O}{\overset{\|}{C}}-$ carbonyl	$CH_3-\underset{O}{\overset{\|}{C}}-CH_3$	dimethyl ketone (acetone)	propan*one*
Ether	$R-O-R'$	$-O-$ oxy	CH_3-O-CH_3	dimethyl ether	methoxymethane
Ester	$*R-\overset{O}{\overset{\|}{C}}-OR'$	$-\overset{O}{\overset{\|}{C}}-OR'$	$CH_3-\overset{O}{\overset{\|}{C}}-O-C_2H_5$	ethyl acetate	ethyl ethan*oate*

*R and R' may be identical or different groups, or *R may be an H atom (except in ethers and ketones) or an aromatic group.

properties and activities of the alcohols. In fact, all organic compounds can be divided into a relatively small number of systems based upon the functional groups they contain. A functional group is an atom or a group of atoms that imparts to all compounds of a family its particular characteristics. In nearly all reactions, except those in which the hydrocarbon chain is broken (in combustion, for example), it is the functional group that reacts. The hydrocarbon portion of the molecule is relatively inert and usually does not react. Since we can associate a set of properties with a particular functional group in a compound, the understanding of organic chemistry can be resolved into a study of the chemistry of the various functional groups. Some important functional groups and their structural arrangements are listed in Table 13–10.

13.13 FAMILIES OF ALIPHATIC COMPOUNDS

The hydrogen atoms of a hydrocarbon may be replaced by a wide variety of atoms or groups of atoms to form other series of organic compounds. These compounds are considered to be derived from their parent hydrocarbons; hence the name *hydrocarbon derivatives.* The compounds we shall study and their general formulas are as follows:

Compound	General Formula
alcohols	R—OH
ethers	R—O—R′
carboxylic acids	*R—COOH
aldehydes	*R—CHO
ketones	R—CO—R′
esters	*R—COO—R′

(Except in ethers and ketones, R′ and *R may be identical or different groups and *R may be a hydrogen atom or an aromatic group.)

13.14 ALCOHOLS

Alcohols may be thought of as being composed of an —OH (hydroxyl) group substituted for one or more of the hydrogens in an aliphatic hydrocarbon.† In this sense, alcohols can be considered as being derived from their corresponding hydrocarbons. It should be noted that the —OH functional group is not the same as the hydroxide ion and that alcohols are molecular and not ionic in nature. There are saturated, unsaturated, aromatic, and aliphatic alcohols, some with one or more —OH groups attached. Some of the common alcohols and their types are listed in Table 13–11. At our level of discussion we will confine ourselves primarily to the saturated monohydric (one —OH) alcohols.

The simplest alcohols are those derived from the alkanes, with only one hydroxyl group per molecule. They have the general formula R—OH, where R represents an alkyl group. The first two members, methyl alcohol, CH_3OH, and ethyl alcohol, C_2H_5OH, have no isomers.

$$
\begin{array}{cc}
\quad\ \ H & \quad\ \ H \quad\ H \\
\quad\ \ | & \quad\ \ | \quad\ \ | \\
H-C-OH & H-C-C-OH \\
\quad\ \ | & \quad\ \ | \quad\ \ | \\
\quad\ \ H & \quad\ \ H \quad\ H \\
\text{methanol} & \text{ethanol} \\
\text{(methyl alcohol)} & \text{(ethyl alcohol)}
\end{array}
$$

However, isomers do occur in the third member of the family, propyl alcohol, C_3H_7OH. There is an n-propyl alcohol in which a terminal (end) hydrogen atom is replaced by the —OH group, and there is also

---

†Aromatic compounds in which the hydroxyl group is directly attached to the ring structure differ markedly in properties from alcohols. They are called *phenols.*

INTRODUCTION TO ORGANIC CHEMISTRY

TABLE 13–11 **Examples of Alcohols**

Formula	Name (common name in parentheses)	Type
CH_3OH	methanol (methyl alcohol)	saturated monohydric
CH_3-CH_2OH	ethanol (ethyl alcohol)	
$CH_3-CH_2-CH_2OH$	1-propanol (n-propyl alcohol)	
$CH_3-\overset{\displaystyle OH}{\underset{\displaystyle \vert}{CH}}-CH_3$	2-propanol (isopropyl alcohol)	
$\underset{\displaystyle CH_2-CH_2}{\overset{\displaystyle OH \quad OH}{\vert \quad \vert}}$	1,2-ethanediol (ethylene glycol)	saturated dihydric
$\overset{\displaystyle CH_2-CH-CH_2}{\underset{\displaystyle OH \quad OH \quad OH}{\vert \quad \vert \quad \vert}}$	1,2,3-propanetriol (glycerol)	saturated trihydric
$CH_3-CH_2-CH_2-CH_3-CH_2OH$	1-pentanol (n = amyl alcohol)	saturated monohydric
$CH_3-CH_2-CH_2-\underset{\displaystyle OH}{\underset{\displaystyle \vert}{CH}}-CH_2-CH_3$	3-hexanol	
	cyclopentanol	cyclic saturated monohydric
$CH_2{=}CH-CH_2OH$	1-propene-3-ol (allyl alcohol)	unsaturated monohydric
$-CH_2OH$	1-phenylmethanol (benzyl alcohol)	aromatic monohydric

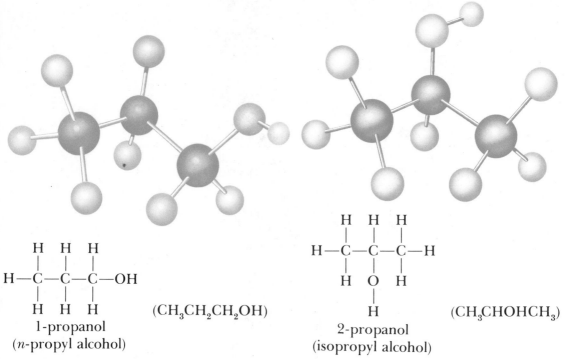

H H H
| | |
H—C—C—C—OH
| | |
H H H   (CH₃CH₂CH₂OH)

1-propanol
(*n*-propyl alcohol)

H H H
| | |
H—C—C—C—H
| | |
H O H   (CH₃CHOHCH₃)
|
H

2-propanol
(isopropyl alcohol)

Figure 13–6   Structural models of n-propyl and isopropyl alcohol.

an isopropyl alcohol isomer having the —OH group attached to a center carbon. See Figure 13–6.

Table 13–12 shows some saturated monohydric alcohols and their properties.

Monohydric alcohols are classified as primary, secondary, or tertiary, according to the nature of the carbon atom to which the —OH group is attached. A primary carbon atom is bonded to only one other carbon atom; a secondary carbon atom is bonded to two other carbon atoms; and a tertiary carbon atom is bonded to three other carbon atoms.

TABLE 13–12 **Saturated Monohydric Alcohols**

Name	Formula	Melting Point, °C	Boiling Point, °C	Density g/mL at 20°C	Solubility g/100g H₂O
Methanol	CH₃OH	− 97	65	0.793	∞
Ethanol	CH₃CH₂OH	−115	78	0.789	∞
1-Propanol	CH₃CH₂CH₂OH	−126	97	0.804	∞
1-Butanol	CH₃(CH₂)₂CH₂OH	− 89	118	0.810	7.9
1-Pentanol	CH₃(CH₂)₃CH₂OH	− 78	138	0.817	2.3
2-Propanol	CH₃CHOHCH₃	− 86	82	0.789	∞
1-Hexanol	C₆H₁₃OH	− 52	156	0.819	0.2

$$R-CH_2OH \qquad \begin{array}{c} R \overset{\displaystyle H}{\underset{\displaystyle |}{\diagdown}} R \\ C \\ | \\ OH \end{array} \qquad \begin{array}{c} R \overset{\displaystyle R}{\underset{\displaystyle |}{\diagdown}} R \\ C \\ | \\ OH \end{array}$$

primary   secondary   tertiary

This classification will be illustrated by considering the four isomers of the fourth alcohol, butyl alcohol, C_4H_9OH:

$$\begin{array}{cccc} | & | & | & | \\ -C- & C- & C- & C-OH \\ | & | & | & | \end{array} \qquad \begin{array}{c} -C- \\ | \\ \begin{array}{ccc} | & | & | \\ -C- & C- & C-OH \\ | & | & | \end{array} \end{array}$$

n-butyl alcohol
or 1-butanol
(a primary alcohol)

2-methyl-1-propanol
(a primary alcohol)

$$\begin{array}{cccc} | & | & | & | \\ -C- & C- & C- & C- \\ | & | & | & | \\ & & & O \\ & & & H \end{array} \qquad \begin{array}{c} -C- \\ | \\ \begin{array}{ccc} | & | & | \\ -C- & C- & C- \\ | & | & | \\ & O & \\ & H & \end{array} \end{array}$$

sec-butyl alcohol
or 2-butanol
(a secondary alcohol)

tert-butyl alcohol
or 2-methyl-2-propanol
(a tertiary alcohol)

The lower alcohols (those having one to four carbon atoms) have common names made up of the prefix of the appropriate alkyl group followed by the word "alcohol." For the more complex alcohols the IUPAC system is used:

1. Select the parent alkane having the longest continuous carbon chain that contains the $-OH$ group. Replace the terminal *-e* by *-ol*.
2. Indicate by the lowest number possible the position of the $-OH$ group in the parent chain, retaining number sequence for naming other substituted groups. Some examples:

$$CH_3CHCH_2CH_2CH_2OH \qquad CH_3CH_2-\overset{\displaystyle CH_3}{\underset{\displaystyle OH}{\overset{\displaystyle |}{\underset{\displaystyle |}{C}}}}-CH_3$$
$$\overset{\displaystyle |}{\underset{\displaystyle CH_3}{}}$$

4-methyl-1-pentanol   2-methyl-2-butanol

$$CH_3CHCHCH_3 \qquad CH_3CH_2CHCH_2OH$$
$$\overset{CH_3}{|} \qquad\qquad\qquad \overset{CH_3}{|}$$
$$\underset{OH}{|}$$

3-methyl-2-butanol   2-methyl-1-butanol

The alcohols produced in the largest quantities are methanol ("wood alcohol"), ethanol ("grain alcohol"), and 2-propanol. Methanol, or methyl alcohol, formerly was made exclusively by the destructive distillation of hardwood; that is, by the heating of hardwood in the absence of air—hence the name "wood alcohol." Today, most methanol is made synthetically by the reaction of hydrogen with carbon monoxide. It is an extremely toxic liquid, and for that reason it is sometimes added to grain alcohol (as a **denaturant**) to make it unfit for drinking. Methanol is used industrially as a paint and lacquer solvent, in automotive antifreezes, and in the synthesis of plastics and other compounds.

Ethanol, or ethyl alcohol, is the second member of the saturated monohydric alcohols. It is known as the beverage "grain alcohol" and has long been produced by the bacterial fermentation of starch or molasses. As was mentioned, denatured ethanol contains methanol or some other toxic ingredient, added so that it can be sold tax-free. The strength of alcoholic beverages is given in **proof,** 100 proof whiskey being 50 per cent pure ethanol by *mass.* Solutions using ethanol as the solvent are called **tinctures** (as in tincture of iodine). Ethanol is used extensively in the manufacture of dyes, perfumes, and medicines.

Isopropyl alcohol, an isomer of propanol, is the familiar "rubbing alcohol." It is synthesized from propene and used in large quantities in the manufacture of acetone. There are two important saturated polyhydric alcohols, ethylene glycol and glycerol. Ethylene glycol is a saturated dihydric (two —OH groups) alcohol prepared from ethylene. It is used as an automotive antifreeze ("Prestone") and as a solvent in inks, paints, and plastics. Glycerol, or glycerine, is a saturated trihydric alcohol obtained as a by-product in the manufacture of soap. It is used in large quantities in the formulation of cosmetics and pharmaceuticals, and in inks, paints, and explosives.

$$
\begin{array}{cc}
\text{CH}_2-\text{CH}_2 & \text{CH}_2-\text{CH}-\text{CH}_2 \\
|\quad\ \ | & |\quad\ \ |\quad\ \ | \\
\text{OH}\ \ \ \text{OH} & \text{OH}\ \ \text{OH}\ \ \text{OH} \\
\text{(ethylene glycol)} & \text{(glycerol)} \\
\text{1,2-ethanediol} & \text{1,2,3-propanetriol}
\end{array}
$$

The melting point, boiling point, and solubility in water of the saturated monohydric alcohols are generally higher than those of their parent alkanes. Although methane, ethane, and propane are insoluble in water, methanol, ethanol, and both propanols are miscible with water in all proportions. The higher melting and boiling points and greater solubility of the lower alcohols are explained by the ability of the —OH group of the alcohol to form hydrogen bonds with water or with other alcohol molecules. In a sense, alcohols are organic analogs of water in which a hydrogen atom is replaced by an alkyl group. Thus it is understandable that the solubility of an alcohol in water

INTRODUCTION TO ORGANIC CHEMISTRY

decreases as the length of the hydrocarbon portion of the molecule increases and the influence of the —OH group becomes comparatively smaller. (See Table 13–12.)

The alcohols are much more reactive than their alkane counterparts. The lower alcohols react with sodium metal to produce hydrogen gas and a salt, much like the reaction of water with metallic sodium.

$$CH_3OH + Na \rightarrow \tfrac{1}{2}H_2 + CH_3ONa \text{ (sodium methoxide)}$$

$$HOH + Na \rightarrow \tfrac{1}{2}H_2 + HONa \quad \text{(sodium hydroxide)}$$

Alcohols are versatile compounds that are used to make nearly every other kind of aliphatic compound, such as alkenes, ethers, aldehydes, ketones, acids, and esters.

Primary alcohols have two hydrogen atoms attached to the carbon atom bearing the hydroxyl group. They can be *gently* oxidized to produce aldehydes.

$$R-CH_2OH \xrightarrow{\text{Oxidation}} R-\overset{\overset{\displaystyle O}{\|}}{C}-H$$

primary   an aldehyde
alcohol

A secondary alcohol has only one hydrogen atom attached to the alcohol carbon. It can be *gently* oxidized to form a ketone.

$$R-\overset{\overset{\displaystyle H}{|}}{\underset{\underset{\displaystyle OH}{|}}{C}}-R \xrightarrow{\text{Oxidation}} R-\overset{}{\underset{\underset{\displaystyle O}{\|}}{C}}-R$$

secondary   a ketone
alcohol

Note the emphasis on the word *gentle* oxidation. An alcohol or any other organic compound containing only carbon, hydrogen, and oxygen atoms may be *completely* oxidized to yield carbon dioxide and water:

$$CH_3OH + O_2 \rightarrow CO_2 + H_2O$$

13.15 ETHERS

There are two general types of reactions by which small molecules may be used to build larger ones. One is the addition reaction in which two molecules add to each other, as in the addition of bromine to acetylene (p. 382). The other is the condensation or dehydration reaction in which the combination of two molecules is accompanied by the elimination of a molecule of water. Condensation reactions that take place between molecules of alcohols produce **ethers,** whereas those involving molecules of an alcohol and an acid result in the formation of **esters** (Section 13.19). In both cases a molecule of water is eliminated and a

$$-\overset{|}{\underset{|}{C}}-O-\overset{|}{\underset{|}{C}}-$$ linkage is formed.

$$
\underset{\substack{\displaystyle \\ \text{diethyl ether} \\ \text{(a simple ether)}}}{\text{H}-\overset{\overset{\displaystyle H}{|}}{\underset{\underset{\displaystyle H}{|}}{C}}-\overset{\overset{\displaystyle H}{|}}{\underset{\underset{\displaystyle H}{|}}{C}}-O-\overset{\overset{\displaystyle H}{|}}{\underset{\underset{\displaystyle H}{|}}{C}}-\overset{\overset{\displaystyle H}{|}}{\underset{\underset{\displaystyle H}{|}}{C}}-\text{H}}
\qquad
\underset{\substack{\displaystyle \\ \text{ethyl acetate} \\ \text{(an organic ester)}}}{\text{H}-\overset{\overset{\displaystyle H}{|}}{\underset{\underset{\displaystyle H}{|}}{C}}-\overset{\overset{\displaystyle }{|}}{\underset{\underset{\displaystyle O}{||}}{C}}-O-\overset{\overset{\displaystyle H}{|}}{\underset{\underset{\displaystyle H}{|}}{C}}-\overset{\overset{\displaystyle H}{|}}{\underset{\underset{\displaystyle H}{|}}{C}}-\text{H}}
$$

Ethers are often referred to as organic oxides. They are made up of two alkyl groups joined to an oxygen atom, and they have the general formula R—O—R′. If both alkyl groups are the same, the ether is called symmetrical or simple; if the R's are different, the compound is a mixed ether. Ethers are commonly named by giving the names of the alkyl groups attached to the oxygen, followed by the word *ether.* If the ether is symmetrical, it is only necessary to name the alkyl group once.

> $CH_3—O—CH_3$ is a simple ether called methyl ether.(The name "dimethyl ether" is correct, but methyl ether is adequate.)
>
> $CH_3—O—C_2H_5$ is methyl ethyl ether (a mixed ether)
>
> $C_2H_5—O—C_2H_5$ is ethyl ether (a simple ether)

According to the IUPAC system of nomenclature, the *-yl* of the alkyl group is replaced by *-oxy,* and the ether is named as though it were a substituted alkane. For example, the systematic name for ethyl ether is ethoxyethane.

Simple ethers are prepared by the elimination of water from two molecules of alcohol.

$$
\underset{\text{alcohol}}{\underset{\substack{\displaystyle ROH \\ \text{ether}}}{ROH} \rightarrow ROR + HOH}
$$

Ethyl ether may be prepared by the careful condensation of two molecules of ethanol in the presence of concentrated sulfuric acid.

$$\text{C}_2\text{H}_5\text{OH} \atop \text{C}_2\text{H}_5\text{OH} \xrightarrow[\text{H}_2\text{SO}_4]{} \text{C}_2\text{H}_5-\text{O}-\text{C}_2\text{H}_5 + \text{H}_2\text{O}$$

Ethers and alcohols may be **functional isomers.** They have the same molecular formulas but contain different functional groups. Unlike the alcohols, ethers have no hydroxyl group and cannot undergo hydrogen bonding, so the boiling point of an ether is well below that of an alcohol with the same number of carbon atoms. For example, ethanol and methyl ether are functional isomers with the formula $\text{C}_2\text{H}_6\text{O}$, yet the boiling point of ethanol is $78°\text{C}$, which is $102°$ higher than the boiling point of methyl ether, $-24°\text{C}$. Another effect of the difference between the alcohol and ether functional groups is illustrated by the fact that the lower alcohols are miscible with water in all proportions, whereas ethers have a negligible water solubility. Ethers are much more volatile and much less chemically active than the isomeric alcohols.

In addition to their industrial importance as solvents for fats, waxes, and oils, certain ethers have valuable medical uses. Ethyl ether, commonly referred to as "ether," was first used as a general anesthetic more than 100 years ago, and it remains an important general anesthetic today, despite its flammability and undesirable side-effects.

13.16 ALDEHYDES AND KETONES

Aldehydes and ketones (as well as the carboxylic acids that we shall study in the next section) are oxidation products of the alcohols. Each of these types of compounds contains the carbonyl group, $\text{C}=\text{O}$.

	Aldehyde	Ketone	Carboxylic acid
general formula	$\text{R}-\overset{\overset{\text{O}}{\|\|}}{\text{C}}-\text{H}$	$\text{R}-\overset{\overset{\text{O}}{\|\|}}{\text{C}}-\text{R}'$	$\text{R}-\overset{\overset{\text{O}}{\|\|}}{\text{C}}-\text{OH}$
condensed structural formula	$\text{CH}_3-\overset{\overset{\text{O}}{\|\|}}{\text{C}}-\text{H}$	$\text{CH}_3-\overset{\overset{\text{O}}{\|\|}}{\text{C}}-\text{CH}_3$	$\text{CH}_3-\overset{\overset{\text{O}}{\|\|}}{\text{C}}-\text{OH}$
name	ethanal (acetaldehyde)	propanone (dimethyl ketone) (acetone)	ethanoic acid (acetic acid)

Since both aldehydes and ketones are carbonyl compounds, it is not surprising that their reactions are similar. Often the only difference is in the rate of their reactions, ketone reactions being slower.

The name "aldehyde" is a contraction of the term "*alcohol dehy-*drogenation," indicating that two hydrogen atoms are removed from an end carbon when aldehydes are prepared from primary alcohols. With the exception of formaldehyde, R in the general formula R—CHO can be any alkyl group. In formaldehyde, HCHO, R is an H atom. The general formula shows that the aldehydes differ from their parent hydrocarbon molecules in that the two hydrogen atoms on the end carbon have been replaced by a single oxygen atom with a double bond to a primary carbon atom.

Inasmuch as the hydroxyl group in a primary alcohol is at the end of a carbon chain, the aldehyde functional group, —CHO, is *always* located at the end of a carbon chain.

$$\underset{\substack{\text{any primary}\\\text{alcohol}}}{R-\overset{\displaystyle H}{\underset{\displaystyle H}{C}}-OH} \xrightarrow[\text{oxidation}]{\text{gentle}} \underset{\text{an aldehyde}}{R-\overset{\displaystyle H}{C}=O}$$

oxygen from a mild
oxidizing agent

$$\underset{\substack{\text{ethanol}\\\text{(a primary alcohol)}}}{H-\overset{\displaystyle H}{\underset{\displaystyle H}{C}}-\overset{\displaystyle H}{C}-O\boxed{H + O}} \rightarrow \underset{\substack{\text{ethanal (acetaldehyde)}\\\text{(an aldehyde)}}}{H-\overset{\displaystyle H}{\underset{\displaystyle H}{C}}-\overset{\displaystyle H}{C}=O} + HOH$$

The common name of any aldehyde is determined by the name of the acid that it yields on oxidation. Formaldehyde is so named because it may be oxidized to make formic acid; acetaldehyde can be oxidized to produce acetic acid.

$$\underset{\text{(an aldehyde)}}{R-\overset{\displaystyle H}{C}=O} \xrightarrow{[O]} \underset{\text{(a carboxylic acid)}}{R-\overset{\displaystyle OH}{C}=O}$$

$$\underset{\text{formaldehyde}}{H-\overset{\displaystyle O}{\overset{\|}{C}}-H} \xrightarrow{[O]} \underset{\text{formic acid}}{H-\overset{\displaystyle O}{\overset{\|}{C}}-OH}$$

$$\underset{\text{acetaldehyde}}{CH_3-\overset{\displaystyle O}{\overset{\|}{C}}-H} \xrightarrow{[O]} \underset{\text{acetic acid}}{CH_3-\overset{\displaystyle O}{\overset{\|}{C}}-OH}$$

(The above type of abbreviated equation is a shortcut often used by organic chemists. The [O] means that oxygen is obtained from some suitable but unspecified oxidizing agent.)

Formaldehyde is the simplest aldehyde. It can be prepared by the mild oxidation of methanol.

$$CH_3OH \xrightarrow{[O]} H-\overset{\displaystyle O}{\overset{\displaystyle \|}{C}}-H$$

The next higher homolog is acetaldehyde, CH_3CHO, which can be made by gently oxidizing ethyl alcohol. If care is not exercised in the oxidation, however, the aldehyde formed will be oxidized to the corresponding carboxylic acid. This ease of oxidation means that aldehydes are excellent reducing agents. The first three members of the aldehydes derived from the alkanes are:

$$H-\overset{\displaystyle O}{\overset{\displaystyle \|}{C}}-H \qquad CH_3-\overset{\displaystyle O}{\overset{\displaystyle \|}{C}}-H \qquad CH_3CH_2-\overset{\displaystyle O}{\overset{\displaystyle \|}{C}}-H$$

methanal   ethanal   propanal
(formaldehyde)   (acetaldehyde)   (propionaldehyde)

Systematic naming of the aldehydes involves replacing the *-e* of the corresponding hydrocarbon with *-al*. The parent structure is considered to be the longest chain carrying the $-CHO$ functional group. However, the common names shown above in parentheses are almost always used for the first three aldehydes.

Formaldehyde is a gas at room temperature. It is used as a fungicide and germicide, as a preservative of biological specimens, in embalming fluids, and in the manufacture of plastics. It is widely marketed as a preservative and fumigant when packaged as a 37 per cent water solution called **formalin.** Acetaldehyde is used chiefly to make acetic acid. Formaldehyde and acetaldehyde undergo a variety of addition and condensation reactions, and both are useful in the production of many other organic compounds.

Ketones contain a carbonyl group attached to two carbon atoms and have the general formula RCOR′. If both alkyl groups are the same, it is termed a simple ketone; if the groups are not the same, it is a mixed ketone. The common name of a ketone consists of the names of its two alkyl groups followed by the word **ketone.** According to the systematic method, the *-e* of the parent alkane is replaced with the ending *-one.* The simplest and most important aliphatic ketone is dimethyl ketone (acetone).

$$\begin{array}{ccc}
\overset{\displaystyle O}{\underset{\displaystyle \|}{}} & \overset{\displaystyle O}{\underset{\displaystyle \|}{}} & \overset{\displaystyle O}{\underset{\displaystyle \|}{}} \\
CH_3-C-CH_3 & CH_3-C-CH_2CH_3 & CH_3CH_2-C-CH_2CH_3
\end{array}$$

propanone   butanone   3-pentanone
(dimethyl ketone)   (methyl ethyl ketone)   (diethyl ketone)
(acetone)   (MEK)

Ketones may be made by the mild oxidation of a secondary alcohol. For example, acetone can be prepared by the gentle oxidation of isopropyl alcohol:

$$\begin{array}{ccc}
\overset{\displaystyle H}{\underset{\displaystyle |}{}} & & \\
R-C-R & \xrightarrow{\text{oxidation}} & R-C-R' \\
\underset{\displaystyle |}{} & & \underset{\displaystyle \|}{} \\
OH & & O
\end{array}$$

any secondary   a ketone
alcohol

isopropyl alcohol   acetone
(a secondary alcohol)   (a ketone)

Because of the position of the carbonyl group in ketones, they are not as reactive as the aldehydes. Ketones do not polymerize readily; they are more resistant to oxidation; and they undergo addition reactions less readily than do aldehydes of comparable molecular mass. Acetone is a colorless liquid that is completely miscible with water and many organic solvents. It is extensively used as a starting material in the manufacture of many other chemicals. Both acetone (an important laboratory solvent) and methyl ethyl ketone (MEK) are sold in large quantities as industrial solvents for paints and surface coating materials.

13.17   CARBOXYLIC ACIDS

The continued gentle oxidation of aldehydes or primary alcohols results in the formation of another group of important compounds known as carboxylic acids, defined as organic acids containing one or more carboxyl groups in the molecule. These acids contain a carbon atom bonded to both a **carbonyl** group ($-C=O$) and a hydro*xyl* group ($-OH$). The combination of these two groups is called a

carboxyl group (a contraction of the two words), which is the characteristic functional group of the carboxylic acids and has the structural formula

$$-\overset{\displaystyle O}{\underset{\displaystyle \|}{C}}-OH \quad \text{or} \quad -COOH$$

Carboxylic acids are so named to distinguish them from other acids, such as hydrochloric and sulfuric acids, which do not contain the —COOH group. The carboxylic acids may be viewed as hydrocarbon derivatives in which one or more of the hydrogen atoms has been replaced by a —COOH group. *Aliphatic monocarboxylic* acids are those compounds having only one carboxyl group per molecule. *Saturated aliphatic monocarboxylic* acids have the general formula R—COOH, where R is an alkyl group except in the case of formic acid (HCOOH), where R is a hydrogen atom. Acids of the R—COOH type may be prepared by the mild oxidation of the appropriate aldehyde or primary alcohol.

$$R-CH_2OH \xrightarrow{[O]} R-CHO \xrightarrow{[O]} R-COOH$$

$$CH_3CH_2OH \xrightarrow{[O]} CH_3CHO \xrightarrow{[O]} CH_3COOH$$

ethanol	ethanal	ethanoic acid
(primary alcohol)	(aldehyde)	(carboxylic acid)

The first four members of the homologous series of the saturated aliphatic monocarboxylic acids are as follows (common names are shown in parentheses):

$$H-\overset{\displaystyle O}{\underset{\displaystyle \|}{C}}-OH \qquad CH_3\overset{\displaystyle O}{\underset{\displaystyle \|}{C}}-OH \qquad CH_3CH_2\overset{\displaystyle O}{\underset{\displaystyle \|}{C}}-OH \qquad CH_3CH_2CH_2\overset{\displaystyle O}{\underset{\displaystyle \|}{C}}-OH$$

HCOOH	CH_3COOH	CH_3CH_2COOH	$CH_3CH_2CH_2COOH$
methanoic acid	ethanoic acid	propanoic acid	butanoic acid
(formic acid)	(acetic acid)	(propionic acid)	(butyric acid)

There are organic acids that contain two, three, or more —COOH groups per molecule. Oxalic acid is the simplest *dicarboxylic* acid, and citric acid is a familiar *tricarboxylic* acid.

$$CH_3-COOH$$

$$\underset{\displaystyle COOH}{\overset{\displaystyle COOH}{|}}$$

$$\begin{array}{c} CH_2-COOH \\ | \\ OH-C-COOH \\ | \\ CH_2-COOH \end{array}$$

(acetic acid)	oxalic acid	citric acid
(a *mono*carboxylic acid)	(a *di*carboxylic acid)	(a *tricarboxylic monohydric* acid)

Organic acids are widely distributed in nature (especially in food-stuffs) either as free acid or as acid derivatives. Citric acid is found in lemons, oranges, and other citrus fruits. Oxalic acid occurs in fruits and vegetables. Lactic acid gives buttermilk its sour taste, and the fatty acids (principally stearic, palmitic, and oleic acids) are part of animal and vegetable fats.

Acids were among the earliest compounds investigated, and their common names were often taken from Latin words that refer to their sources rather than to their chemical structures.

The simple carboxylic acids are still referred to by their common names, especially those having one to four carbon atoms. IUPAC names for the aliphatic monocarboxylic acids are derived by replacing the -e of the corresponding alkane with -oic and adding the word acid. Since the —COOH group is always a terminal group, no number is needed to designate its position in the carbon chain. The parent alkane is the longest continuous chain attached to the —COOH group. A list of some important and familiar carboxylic acids is given in Table 13–13.

Formic acid is the lowest member of the saturated aliphatic mono-carboxylic acids. It is a colorless liquid with a sharp, biting odor, and it produces painful blisters when placed in contact with the skin. Formic acid occurs freely in nature in many plants, such as the nettle, and in many insects. Its name comes from the Latin word meaning "ant." It is very corrosive to living tissue and causes intense pain when it is injected under the skin by the sting of a bee, the bite of a red ant, or contact with the stinging nettle plant.

The only ionizable hydrogen atom in a carboxylic acid is the one contained in the carboxyl group, and the acidity in water is due to the slight ionization of that atom.

TABLE 13–13 **Carboxylic Acids**

Formula	Common Name	IUPAC Name	Occurrence	Solubility, g/100 g H_2O at 20°C
H—COOH	formic acid	methanoic acid	red ants	∞
CH_2—COOH	acetic acid	ethanoic acid	vinegar	∞
CH_3—CH_2—COOH	propionic acid	propanoic acid		∞
$CH_3(CH_2)_2COOH$	n-butyric acid	butanoic acid	rancid butter	∞
$CH_3(CH_2)_3COOH$	n-valeric acid	pentanoic acid		3.7
$CH_3(CH_2)_4COOH$	n-caproic acid	hexanoic acid	goat-milk fat	1.0
$CH_3(CH_2)_6COOH$	caprylic acid	octanoic acid	goat-milk fat	0.7
$CH_3(CH_2)_8COOH$	capric acid	decanoic acid	goat-milk fat	0.2
$CH_3(CH_2)_{10}COOH$	lauric acid	dodecanoic acid	coconut oil	
$CH_3(CH_2)_{12}COOH$	myristic acid	tetradecanoic acid	coconut oil, nutmeg oil	insol
$CH_3(CH_2)_{14}COOH$	palmitic acid	hexadecanoic acid	palm oil	
$CH_3(CH_2)_{16}COOH$	stearic acid	octadecanoic acid	lard, tallow	
$CH_3(CH_2)_7CH=CH(CH_2)_7COOH$	oleic acid	9-octadecanoic acid	olive oil	

$$R \quad -COOH + HOH \rightarrow R-COO^- + H_3O^+$$
$$CH_3-COOH + HOH \rightarrow CH_3-COO^- + H_3O^+$$

acetic acid   acetate ion

Generally, carboxylic acids are weak; formic acid is the strongest of the group, and it ionizes only about 5 per cent in dilute solution. The aliphatic acids show much the same solubility characteristics as the alcohols, the lower four being miscible with water; the five-carbon acid is partly soluble, and the higher ones are essentially insoluble (see Table 13–13). As is true of the alcohols, the long-chain acids are insoluble in water because they behave more like hydrocarbons as the carbon chain gets longer. The lower acids are liquids with disagreeable odors, but as the number of carbon atoms in the chain increases, the volatility decreases, so the acids with more than 10 carbon atoms are solids with little odor. The aliphatic acids with 12 to 18 carbons are important in the manufacture of soaps and candles.

By far the most important and familiar of the monocarboxylic acids is acetic acid, known for centuries as a component of vinegar. Vinegar is made through the fermentation (oxidation) of ethyl alcohol obtained from molasses, or cider or other fruit juices.

$$CH_3CH_2OH + O_2 \xrightarrow[\text{fermentation}]{\text{bacterial}} CH_3COOH + HOH$$

Vinegar produced in the manner contains about 4 to 5 per cent acetic acid and is used almost exclusively as a condiment and food preservative. Commercially, acetic acid is made by the oxidation of ethyl alcohol or acetaldehyde obtained through the catalytic oxidation of ethylene. Glacial acetic acid (100 per cent acetic acid) is a colorless liquid with a sharp, penetrating odor. It freezes at 16.7°C to form a solid that looks like ice. Large quantities of acetic acid are used for making drugs and medicines, in the dyeing of textiles, in the tanning of leather, and in the preparation of solvents.

Long-chain carboxylic acids found in animal and vegetable fats and oils are often referred to as *fatty acids.* Some common fatty acids are lauric (dodecanoic), myristic (tetradecanoic), palmitic, and stearic acids. Oleic acid is a naturally occurring unsaturated acid (see Table 13–13). It is an interesting fact that the naturally occurring fatty acids have an even number of carbon atoms in a continuous chain ranging from 6 to 18 carbons. The fatty acids exist as their glyceryl esters and are found mostly in animal fats; oleic acid is found as its glyceryl ester in liquid fats (called oils) such as olive oil and cottonseed oil.

There are two types of chemical reactions of acids that we shall consider. The first is the reaction of an acid with a base to form a salt. The second type of reaction occurs between an alcohol and an acid to form a compound called an *ester* (an organic salt). Esterification reactions will be taken up in Section 13.18.

In a manner analogous to inorganic acids, carboxylic acids react with bases to form salts:

$$HCOOH + KOH \rightarrow HCOOK + HOH$$

methanoic   potassium
acid   methanoate
(formic acid) (potassium formate)

The name of the salt is formed by first naming the cation, followed by the name of the acid, in which the *-ic* ending is changed to *-ate*. As we shall see in Section 13.18, the sodium and potassium salts of the higher carboxylic acids, especially of palmitic, oleic, and stearic acids, are called *soaps*.

13.18   ESTERS

As was mentioned earlier, the reaction of an organic or inorganic acid with an alcohol results in the formation of an *ester*. If the acid is a saturated aliphatic monocarboxylic acid, an organic ester is formed, having a general formula $R-COO-R'$, where R and R' may be identical or different alkyl groups (or R may be an H).

$$HBr + CH_3CH_2OH \rightarrow CH_3CH_2Br + HOH$$

ethyl bromide (an ester;
also called an alkyl halide)

$$HCOOH + CH_2CH_2OH \rightarrow HCOOCH_2CH_3 + HOH$$

formic acid   ethyl alcohol   ethyl formate (an ester)

The general formula and name of an ester may lead one to believe that the carboxyl hydrogen atom of the acid has been replaced by an alkyl group. Indeed, esters are named as if they were the alkyl salts of carboxylic acids, but the resemblance of esters to salts is superficial. Salts are ionic and esters are covalent, and it has been determined experimentally that in esterification reactions the water is formed from the combination of the OH of the acid and the H from the alcohol, unlike the mechanism occurring in neutralization reactions.

$$R-\overset{\overset{\textstyle O}{\|}}{C}-OH + H-OR' \rightarrow R-\overset{\overset{\textstyle O}{\|}}{C}-OR' + HOH$$

acid   alcohol   ester

The name of the ester is formed by changing the *-ic* ending of the acid to *-ate*, preceded by the name of the alkyl group of the alcohol. Thus, methyl acetate is the ester obtained from the reaction of methyl alcohol and acetic acid, and ethyl formate results from the reaction of ethyl alcohol with formic acid. In most esterification reactions some concentrated sulfuric acid is mixed with the reactants to catalyze the reaction and shift the equilibrium to favor the formation of the ester.

$$
\begin{array}{c}
\underset{\substack{| \\ H}}{\overset{\substack{H \quad O \\ | \quad \parallel}}{H-C-C-}}\overline{OH + H}O-\underset{\substack{| \\ H}}{\overset{\substack{H \\ |}}{C}}-H \xrightarrow{H_2SO_4} \underset{\substack{| \\ H}}{\overset{\substack{H \quad O \\ | \quad \parallel}}{H-C-C-}}O-\underset{\substack{| \\ H}}{\overset{\substack{H \\ |}}{C}}-H + HOH
\end{array}
$$

acid part   alcohol part

$$
\underset{\substack{\text{ethanoic acid} \\ \text{(acetic acid)}}}{CH_3COOH} + \underset{\substack{\text{methanol} \\ \text{(methyl alcohol)}}}{CH_3OH} \xrightarrow{H_2SO_4} \underset{\substack{\text{methyl ethanoate} \\ \text{(methyl acetate)}}}{CH_3COOCH_3 + HOH}
$$

The lower-carbon esters have pleasant odors and are found in the fragrances of flowers and ripe fruits. (See Table 13–14.)

The important esters occur in animal and vegetable fats and oils (liquid fats) such as corn oil, tallow, coconut oil, bacon grease, and butter. These fats are composed of mixtures of glycerol esters of carboxylic acids and are called **glycerides.** Because glycerol has three hydroxyl groups per molecule, there may be one, two, or three different acids forming a single ester molecule. If the acids in the molecule are of the same kind, the fat is called a *simple glyceride;* if they are different, the fat is called a *mixed glyceride.*

TABLE 13–14 **Examples of Common Esters**

Formula	Common Name (characteristic odor)	IUPAC Name
$\overset{\substack{O \\ \parallel}}{HC}-O-CH_2CH_3$	ethyl formate (rum)	ethyl methanoate
$\overset{\substack{O \\ \parallel}}{HC}-O-CH_2-\overset{\substack{CH_3 \\ \mid}}{CH}-CH_3$	isobutyl formate (raspberries)	2-methyl propyl methanoate
$CH_3-\overset{\substack{O \\ \parallel}}{C}-O-(CH_2)_4-CH_3$	n-amyl acetate (bananas)	pentyl ethanoate
$CH_3-\overset{\substack{O \\ \parallel}}{C}-O-(CH_2)_7-CH_3$	n-octyl acetate (oranges)	octyl ethanoate
$CH_3-CH_2-CH_2-\overset{\substack{O \\ \parallel}}{C}-O-CH_2-CH_3$	ethyl butyrate (pineapples)	ethyl butanoate

"I'm so glad you like it. Actually it's just sodium acid pyrophosphate, erythorbate, and glucono delta lactone with some meat flavoring."

$$CH_2-O-\overset{\overset{\displaystyle O}{\|}}{C}-C_{17}H_{33}$$

$$CH-O-\overset{\overset{\displaystyle O}{\|}}{C}-C_{17}H_{33}$$

$$CH_2-O-\overset{\overset{\displaystyle O}{\|}}{C}-C_{17}H_{33}$$

glycerol trioleate

(a simple glyceride)

$$CH_2-O-\overset{\overset{\displaystyle O}{\|}}{C}-C_3H_7$$

$$CH-O-\overset{\overset{\displaystyle O}{\|}}{C}-C_{11}H_{23}$$

$$CH_2-O-\overset{\overset{\displaystyle O}{\|}}{C}-C_{15}H_{31}$$

glyceryl butyro-
lauro-palmitate
(a mixed glyceride)

Natural fats and oils are glycerides of long-chain fatty acids containing an even number of carbon atoms. Animal fats consist mainly of glycerides of stearic and palmitic acids. A few of the more than 40 acids found in nature are listed in Table 13–15.

A common and characteristic reaction of esters is the alkaline hydrolysis with a strong base to produce glycerol and salts of the acids

TABLE 13–15 **Common Fatty Acids**

Name	Formula	Some Sources
Butyric	C_3H_7COOH	Butter
Lauric	$C_{11}H_{23}COOH$	Coconut oil
Myristic	$C_{13}H_{27}COOH$	Coconut oil, nutmeg butter
Palmitic	$C_{15}H_{31}COOH$	Animal and vegetable fats
Stearic	$C_{17}H_{35}COOH$	Animal and vegetable fats
Oleic	$C_{17}H_{33}COOH$	Olive oil

of which the ester was originally constituted. The mixture of fatty acid salts produced is called *soap,* and the process is termed *saponification.*

$$
\begin{array}{c}
CH_3(CH_2)_{14}\!-\!\overset{\displaystyle O}{\overset{\|}{C}}\!-\!O\!-\!CH_2 \\[1em]
CH_3(CH_2)_{14}\!-\!\overset{\displaystyle O}{\overset{\|}{C}}\!-\!O\!-\!CH \;+\; 3NaOH \;\rightarrow\; 3CH_3(CH_2)_{14}\!-\!C\!\!\begin{array}{l}{\diagup O}\\[-0.3em]{\diagdown ONa}\end{array} \;+\; \begin{array}{l}CH_2OH\\ |\\ CHOH\\ |\\ CH_2OH\end{array} \\[1em]
CH_3(CH_2)_{14}\!-\!\overset{\displaystyle O}{\overset{\|}{C}}\!-\!O\!-\!CH_2
\end{array}
$$

glyceryl palmitate		sodium palmitate	

natural fat	+ base →	soap	+ glycerol

Most soaps are mixtures of the sodium salts of fatty acids, especially sodium stearate and sodium palmitate. The softer and more soluble potassium salts are soaps used in shaving creams and other products calling for liquid soaps. The fatty-acid chain in ordinary soaps is 10 to 18 carbon atoms long. If the chain has less than 10 carbon atoms, the soap is unable to emulsify in water; if more than 18 carbons, the soap is not sufficiently soluble in water to be an effective detergent.

Soap-making is one of the earliest chemical syntheses, dating back to the tribesmen of Caesar's time and later to the American pioneers, who made soap by boiling animal grease with lye (K_2CO_3)* obtained by soaking wood ashes in water. The term "saponification" originally pertained specifically to soap-making, but now it is applied more generally to the alkaline hydrolysis of any ester.

---

*Lye is commonly used as a synonym for sodium hydroxide. Actually, it is the common name for any substance producing an alkaline solution.

13.1   Draw structural formulas for these compounds:
(a) 2,2,4-trimethylpentane
(b) 3,3-diethylpentane
(c) dimethylpropane

13.2   Draw two different structural formulas for dibromoethane and name each according to the IUPAC system.

13.3   Draw structural formulas for these compounds:
(a) propyne
(b) 3-methyl-1-butene

13.4   Write the chemical reaction that takes place between one mole of acetylene and two moles of bromine. Show structures and name the product formed.

13.5   Draw structural formulas and give names of:
(a) two possible compounds having the molecular formula C_3H_4
(b) the product resulting from the mild oxidation of 1-propanol; of 3-methyl-2-butanol
(c) the product of the reaction of propanoic acid with ethanol
(d) a compound that can be oxidized to form methyl ethyl ketone

13.6   Draw structural formulas and give names of:
(a) a dihydric alcohol
(b) a tertiary alcohol
(c) an eight-carbon branch-chained alkane
(d) a cycloalkene

13.7   Which of the following is (are) incorrect statement(s):
(a) ethyl propanoate is $C_3H_7COOC_2H_5$
(b) —COOH is called the carbonyl group
(c)

is 2, 4-dimethyl pentanol

(d) 2-methyl-2-pentanol is a secondary alcohol
(e) 2-pentyne is a saturated hydrocarbon

13.8   Give IUPAC names for:
(a) $CH_3-CH_2-CH-CH-CH_3$
with CH_3 CH_3 substituents

(b) $CH_3-CH_2-CH-CH=CH_2$
with CH_3 substituent

(c) $H-C-C-H$ with Br Br on top and Br Br on bottom

13.9   How many isomeric butyl alcohols (C_4H_9OH) are possible? Draw their structural formulas and give IUPAC names.

13.10   To which class of organic compounds does each of the following belong?
(a) $H_2C\underset{}{\overset{CH_2}{\diagup\diagdown}}CH_2$

(b) CH_3OH

(c) CH_3-C-H with $\|$ CH_2

(d) $CH_3CH_2CH_3$

(e) $CH_3-CH-CH_2-CH_3$
with CH_3 substituent

13.11   Give IUPAC names for:
(a) $CH_2=CH-CH_2-CH-CH_3$
with CH_3 substituent

(b) $CH_3-CH_2-CH-CH-CH_2-CH_2-CH_3$
with CH_3 CH_3 substituents

13.12   List some differences and similarities of organic and inorganic compounds.

13.13   Why are structural formulas so important in organic chemistry?

13.14   Of what significance was Wöhler's production of urea in 1828?

13.15   Explain why carbon forms so many different types of compounds; why is it so versatile?

13.16   How are hydrocarbons classified?

13.17   What are isomers? Draw structures and give names for the isomers of butane.

13.18   How do the reactions of saturated hydrocarbons differ from those of unsaturated hydrocarbons?

13.19   What is the general formula for the hydrocarbons?

13.20   Write the structural formula and explain the meaning of the prefixes, suffixes, branch groups, and parent structure in the following compound: 2-bromo-3,3 dimethylbutanoic acid.

13.21   List the prefixes for the first ten hydrocarbons.

13.22   List the suffixes for the alcohols, acids, aldehydes, and ketones.

13.23   Explain the meaning and give illustrations of:
(a) homologs
(b) isomers
(c) polymers
(d) esters
(e) a branch-chained aliphatic hydrocarbon
(f) a cyclic alkane
(g) an addition reaction
(h) a substitution reaction
(i) saponification

13.24   Match the letter of the most closely related answer in the space provided. The same letter may be used to answer more than one question, but only one answer is required for each question.

_____ saturated hydrocarbon
_____ ketone
_____ contains triple bond
_____ isomer of pentanal
_____ isomer of ethanol
_____ alkyl halide
_____ $HCOOH + CH_3OH$
_____ soap
_____ alcoholic solution
_____ undergoes addition

A. C_nH_{2n-6}
B. ethyl acetate
C. tetrachloroethane
D. 3-pentanone
E. 2-pentyne
F. alkane series
G. propanal
H. methyl ethyl ether
I. a salt
J. an ester
K. a tincture
L. C_3H_7Br
M. RCOOH
N. nothing matches

13.25   Which are incorrect?
(a) methyl-ethyl ketone is 2-propanone

(b) 2-pentyne is $-\overset{|}{\underset{|}{C}}-\overset{|}{\underset{|}{C}}-\overset{|}{C}=\overset{|}{C}-\overset{|}{\underset{|}{C}}-$

(c) 2-methyl,2-butanol is a secondary alcohol

(d)   3,4-dimethyl pentanal is

$-\overset{|}{\underset{|}{C}}-\overset{|}{\underset{|}{C}}-\overset{|}{\underset{|}{C}}-\overset{}{\underset{\underset{C_2H_5}{|}}{C}}-C\overset{H}{\underset{O}{\diagup\!\diagdown}}$

(e) ethyl propanoate is $C_3H_7COOCH_3$

CHAPTER FOURTEEN

LEARNING OBJECTIVES

At the completion of this chapter, you should be able to:

1. Define *radioactivity*.

2. Distinguish between ordinary chemical reactions and nuclear reactions.

3. Describe the three principal types of *ionizing radiation*.

4. Distinguish between natural and induced radioactivity.

5. Describe a radioactive disintegration series.

6. List and write the symbols for the primary particles and radiation associated with nuclear reactions.

7. Complete and balance equations for nuclear reactions.

8. Explain briefly the relationship between nuclear stability and the neutron-to-proton ratio.

9. Describe roughly the structure and function of a representative nuclear reactor.

10. Explain how a Geiger-Müller counter works.

11. List and define the units of radiation that relate to biological hazards.

12. Explain the meaning of *specific activity*.

13. Define *half-life* as it applies to radioisotopes.

14. Solve problems dealing with the rate of nuclear decay.

15. Explain what is meant by *mass decrement* (defect) of a nucleus.

16. Explain what is meant by *nuclear binding energy* and how it relates to *mass decrement*.

17. Distinguish between the phenomena of nuclear *fission* and *fusion*.

18. Calculate the energy changes in fission and fusion reactions.

19. List at least three practical applications of nuclear reactions.

NUCLEAR CHEMISTRY

In this chapter we shall consider a variety of phenomena that relate to the nucleus of the atom.

For the most part, however, our study of the atom has been concerned with electronic configuration. The concept of valence and the behavior of atoms in "ordinary" chemical reactions is necessarily the primary area of interest to the chemist. Ordinary chemical reactions can be described and explained in terms of intractions between ions, atoms, or molecules in which nuclear behavior is quite irrelevant. For example, molecules of CO_2 may contain various isotopes of carbon, such as ^{12}C and ^{14}C. But in terms of the physical and chemical properties of the CO_2 there is no significant difference. We would expect the ^{12}C and ^{14}C atoms to combine with oxygen in the same way to produce CO_2 molecules that for some, but not all, purposes would be practically identical. Hence, we arrive at the crux of the issue, because the essential difference between $^{12}CO_2$ and $^{14}CO_2$ lies in the fact that the instability of the ^{14}C nucleus produces the phenomenon of **radioactivity,** which is a process of atomic decomposition in which subatomic particles and radiation are released.

Radioactivity is one nuclear phenomenon among several that are enormously important and interesting. Even before getting into the substance of this chapter, the vast pool of information that is in the public domain suggests the application of nuclear chemistry to medicine and industry. And in the process, it makes us mindful of the hazards of ignorance or abuse. One of the most dramatic ways in which nuclear reactions differ from "ordinary" chemical reactions is that energy changes in nuclear reactions are usually much greater than in ordinary chemical changes. In fact, a difference of several orders of magnitude is not uncommon.

Another unique characteristic that serves to distinguish chemical from nuclear changes is the **transmutation** of one element into another. Nuclear reactions have enabled scientists to realize the Alchemist's dream—except that the possibility of changing lead into gold tells us nothing about technological and economic feasibility. At this point, we have raised topics for exploration, and we have used terms that need to be defined and explained systematically.

14.1 A HISTORICAL REVIEW

The Atomic Nucleus

Quite early in the text we learned that **protons** and **neutrons** are the principal particles that compose the nucleus of an atom. Collectively, we call them **nucleons.** The various combinations of neutrons and protons that can exist, even for fleetingly short periods of time, are called **nuclides.** We can describe **isotopes** of an element as being composed of atoms that have different nuclides. For example, the isotopes of hydrogen: **protium,** (normal hydrogen atom) $_1^1H$; **deuterium,** $_1^2H$ (or, $_1^2D$); and **tritium,** $_1^3H$ (or, $_1^3T$), are composed of three different nuclides—all of which have the same atomic number (i.e., number of protons) and the same nuclear charge. While the three nuclides of hydrogen exhibit marked similarity in their ordinary chemical behavior, they have distinctly different mass numbers (total number of nucleons). Furthermore, the obviously different ratios of neutrons to protons, such as we note in the hydrogen isotopes, seem to be related to the stability of many nuclides in a manner we shall study very shortly.

Radioactivity

Radioactivity is the term that refers to the spontaneous decomposition of unstable atomic nuclei. The French physicist Henri Becquerel was the one who asked the questions and tried the experiments that led to its discovery. Becquerel, having noticed that fluorescence was associated with X-rays, wondered if naturally fluorescing substances produced X-rays. At first, Becquerel used crystals of a uranium salt that fluoresced in sunlight, and he found that a photographic negative of some object could be made. This seemed to support his hypothesis that the fluorescence caused by sunlight produced X-rays. But, during a cloudy period when he left some of the covered uranium salts in a drawer with some keys that rested on unexposed photo plates, he discovered a vivid negative image of the keys. Becquerel had to conclude that there was some kind of radiation, quite distinct from the fluorescence in sunlight, being emitted by the uranium, and in this way radioactivity was discovered. The encouragement that Becquerel gave to a student, the now famous Marie Curie, to look for other radioactive elements led her and her husband Pierre to discover the elements radium and polonium.

The exciting discovery of radioactivity intrigued a number of scientists during the early 1900's. Ernest Rutherford was one of those who studied the ionizing effect of this phenomenon. Becquerel discovered that the penetrating radiations of radioactive materials had the power to cause ionization. However, it was Rutherford who pioneered the work that has led to the discovery of three principal types of radiation:

1. **Alpha (α)** rays were found to have a particle nature. These alpha particles have a mass equal to a helium ion. Furthermore, the alpha particles suggested the theoretical existence of an electrically neutral particle approximately equal in mass to the proton. This predic-

tion of the existence of a yet undetected neutron ($n°$) derived from the fact that *the alpha particle* (helium ion) *has a charge equivalent to two protons, but the mass of the helium ion is equivalent to four protons!* Rutherford found that the relatively massive alpha particles had high ionizing power but rather weak penetrating ability. In fact, alpha particles can be stopped by paper.

2. **Beta (β) rays** were also found to have a particle nature. The beta particles have greater penetrating power than alpha particles (as evidenced by their ability to penetrate thin aluminum foil), but their ionizing effect is not as great. When beta particles were demonstrated to have a charge and an e/m ratio identical to an electron, it was soon concluded the electrons and the particles of radiation are one and the same.

3. **Gamma (γ) rays** comprise a range of the highest energy electromagnetic radiation. They do not have a particle nature and, like the visible light and X-rays that we have already noted as examples of radiant energy, gamma radiation has neither mass nor charge. However, the extraordinary penetrating power of gamma radiation causes it to be the deadliest form of radiation despite the fact that its ionizing power is negligible.

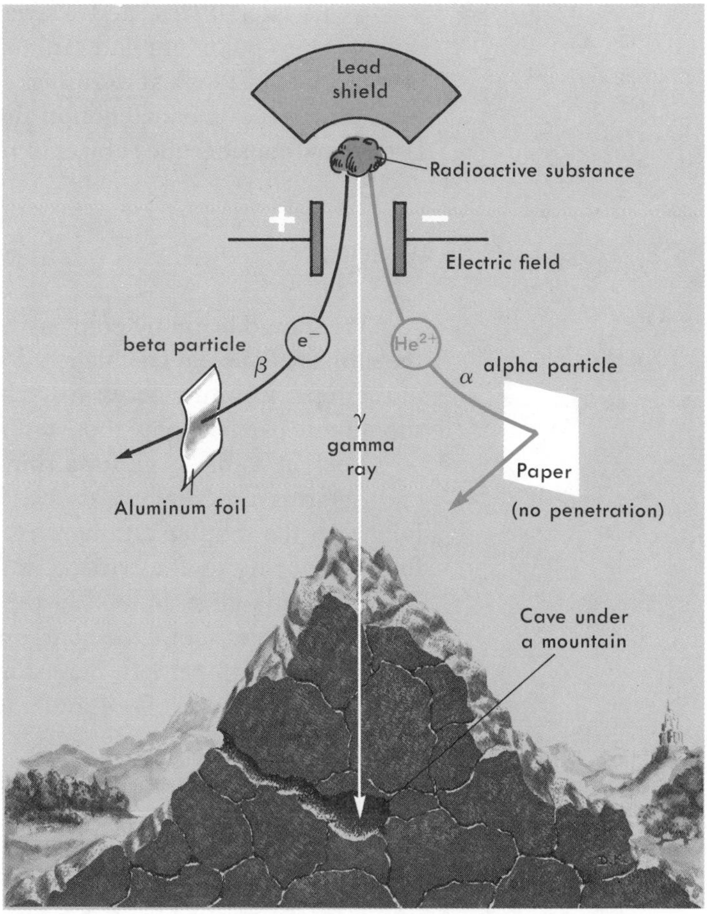

Figure 14–1 The electrical interactions and relative penetrating powers of alpha, beta, and gamma radiation. Note how the positively charged α particle is deflected toward the cathode, the negatively charged β particle toward the anode, and the neutral γ radiation is not affected.

TABLE 14–1 **Characteristics of the Three Principal Types of Radiation**

Common Symbol	α Alpha	β Beta	γ Gamma
Mass	6.64×10^{-24} g	9.11×10^{-28} g	0
Charge	$+2$	-1	0
Nature	Helium ion	Electron	Radiant energy
Ionizing effect	Significant	Slight	Very slight
Penetrating power	Poor	Slight	Powerful

See Figure 14–1 for an illustration of the relative penetrating powers and electrical interactions of the three principal radioactive particles. Their characteristics are summarized in Table 14–1.

In this chapter we shall also make a distinction between radioactive nuclides that occur in nature and those that result from the "bombardment" of atomic nuclei. The bombardment of nuclei is accomplished with various sub-atomic particles that can be accelerated to speeds such that they have remarkably high kinetic energies. The former mode of radioactivity is called **natural radioactivity** while the latter is called **induced radioactivity.** In was natural radioactivity that Henri Becquerel detected in 1896 emanating from a uranium salt. Two years later, Pierre and Marie Curie succeeded in isolating the **radioisotopes** polonium and radium from a uranium ore called pitchblende. We shall look at examples of induced radioactivity in Section 14–3 where the phenomenon of nuclear transmutations is taken up. Let us now consider the subject of nuclear reactions.

14.2 NUCLEAR REACTIONS

We should begin by pointing out that no single chapter can cover the subject of nuclear chemistry adequately. Hence, in this brief survey of the topic we must necessarily choose from among a host of exotic sub-atomic particles only those few needed for a measure of continuity.

Look at Table 14–2 for a summary of the primary particles and energy forms most directly associated with nuclear reactions. We shall follow up the tabular arrangement with a few related comments before proceeding to the writing and balancing of nuclear equations. The conventional style used in the writing of the formal symbols for nuclear particles shows the charge at the lower left of the symbol and the mass at the upper left. For example, a hypothetical particle called "X" having a charge of -1 and a mass of 3 amu should be written:

$$_{-1}^{3}X$$

In addition to alpha, beta, and gamma radiation, **positron** emission is an additional particle associated with radioactivity.

TABLE 14–2 Primary Particles and Radiation Associated with Nuclear Reactions

Name of Particle or Radiation	Common Symbol	Formal Symbol	Charge	Effective Mass
alpha	α	^4_2He	$+2$	4 amu
beta	β^-	$^0_{-1}\text{e}$	-1	0 amu
gamma	γ	$^0_0\gamma$ or $h\nu$	0	0 amu
neutron	n^0	^1_0n	0	1 amu
positron	β^+	^0_1e	$+1$	0 amu
proton	p^+	^1_1H	$+1$	1 amu
deuteron	D	^2_1H	$+1$	2 amu

Positron Emission

The positron, symbolized ^0_1e, is identical to the electron, except for charge. The positron was discovered in 1933 by the American scientist Carl Anderson, while he was doing research on cosmic radiation. Since then, we have learned that positron emission results only from artificially produced radioactive nuclides and never from natural nuclear decay. In any case, the positron is short-lived because it combines very quickly with an electron in a reaction that sees both units converted into a photon of high energy gamma radiation.

$$^0_{-1}\text{e} + {}^0_1\text{e} \rightarrow {}^0_0\gamma$$

Apparently, what happens is that when radioactive "light-weight" isotopes have unstable nuclei because of an insufficient number of neutrons, positron emission occurs. Positron emission is a consequence of a nuclear proton "breaking down" into a neutron and positron.

$$^1_1\text{H} \rightarrow {}^1_0\text{n} + {}^0_1\text{e}$$

Now let us pause for a moment and consider the balancing of nuclear equations. *The rules, known as the Rutherford-Soddy rules, are very simple: (1) the total mass and, (2) the total charge (atomic number) on both sides of the nuclear equation must be equal.* In other words, the sum of the atomic numbers on each side of the equation must be equal and the sum of the mass numbers on each side of the equation must also be equal.

Example 14.1

The "light-weight" radioisotope $^{11}_6\text{C}$ decomposes by positron emission while the relatively "heavy-weight" $^{14}_6\text{C}$ isotope (i.e., too many neutrons for stability) decays by beta emission. Write complete nuclear equations for both reactions.
Solution:
1. Write the symbols for the unstable nuclides and the emitted particles. Leave spaces for the resulting nuclides.

$$^{11}_6\text{C} \rightarrow \underline{\quad ? \quad} + {}^0_1\text{e} \qquad\qquad (1)$$

$$^{14}_6\text{C} \rightarrow \underline{\quad ? \quad} + {}^0_{-1}\text{e} \qquad\qquad (2)$$

---

2. Apply the laws of conservation so that the masses and charges are equal for both sides of each equation. Hence, in equation (1), the produced nuclide must have a nuclear charge of $+5$ and a mass number of 11.

$$^{11}_{6}C \rightarrow ^{11}_{5}? + ^{0}_{1}e \qquad (3)$$

Note: Both sides of the equation have mass numbers adding up to 11 and nuclear charges adding up to $+6$. The isotope having an atomic number of 5 is boron. Therefore the completed equation is:

$$^{11}_{6}C \rightarrow ^{11}_{5}B + ^{0}_{1}e \qquad (4)$$

3. As we have just done in equations (3) and (4), the missing nuclide in the $^{14}_{6}C$ nuclear reaction is:

$$^{11}_{6}C \rightarrow ^{14}_{7}? + ^{0}_{-1}e \qquad (5)$$

Since the addition of $+7$ is needed to accommodate the -1 charge of the emitted electron (i.e., the beta particle), the isotope having an atomic number of 7 is nitrogen. Hence, the equation is:

$$^{14}_{6}C \rightarrow ^{14}_{7}N + ^{0}_{-1}e \qquad (6)$$

Nuclear reactions usually omit the symbolic expression of gamma radiation. As noted earlier, we may symbolize the photon of gamma radiation as $^{0}_{0}\gamma$, or simply $h\nu$, but the fact that any photons of electromagnetic radiation are without mass and charge makes symbolic inclusion pointless. The common practice among scientists is to assume, on the basis of abundant evidence, that most nuclear reactions are accompanied by gamma radiation as nucleons achieve more stable configurations.

In concluding this section, let us look at a few more examples of natural radioactivity involving alpha and beta emission, in addition to **orbital electron capture.** Orbital electron capture, as the term "capture" would seem to suggest, means that some atomic nuclei have a tendency to entrap their own orbital electrons in the decay process. Since orbital electron capture results in no effective change in mass, we presume that the captured electron couples with a proton in the formation of a neutron:

$$^{1}_{1}H + ^{0}_{-1}e \rightarrow ^{1}_{0}n$$

Example 14.2

The $^{18}_{9}F$ isotope decays by either: (1) positron emission or, (2) orbital electron capture. Write equations for both reactions.
Solution:
1. Set up the equation with a space for the decay product:

$$^{18}_{9}F \rightarrow \underline{\quad ? \quad} + ^{0}_{1}e$$

NUCLEAR CHEMISTRY

The decay product must then have a nuclear charge of $+8$ and a mass number that remains unchanged. The final equation is:

$$^{18}_{9}F \rightarrow {}^{18}_{8}O + {}^{0}_{1}e$$

2. In the same manner, we indicate a capture of an orbital electron, and then make sure that the equation demonstrates conservation of charge and mass.

$$^{18}_{9}F + {}^{0}_{-1}e \rightarrow \underline{\quad ? \quad}$$

Interestingly, the decay product is the same although the mode of decay is different:

$$^{18}_{9}F + {}^{0}_{-1}e \rightarrow {}^{18}_{8}O$$

In describing the nuclear decay process, scientists often call an unstable nuclide the **parent,** and the nuclide that results from the nuclear decay is called the **daughter.**

Example 14.3

Write balanced nuclear equations for (1) the decay of $^{14}_{6}C$ by beta emission, and (2) the decay of $^{248}_{100}Fm$ by alpha emission.
Solution:
1. Set up the equations with a blank space for the daughter nuclides.

$$^{14}_{6}C \rightarrow \underline{\quad ? \quad} + {}^{0}_{-1}e$$
$$^{248}_{100}Fm \rightarrow \underline{\quad ? \quad} + {}^{4}_{2}He$$

2. Note that in the ^{14}C decay the daughter nuclide must have a mass number of 14 and an atomic number of 7 in order to conform to the Rutherford-Soddy rule. Hence,

$$^{14}_{6}C \rightarrow {}^{14}_{7}\underline{\quad ? \quad} + {}^{0}_{-1}e$$

By the same rule, the daughter nuclide in the ^{248}Fm decay must have a mass number of 244 and an atomic number of 98. Hence,

$$^{228}_{100}Fm \rightarrow {}^{244}_{98}\underline{\quad ? \quad} + {}^{4}_{2}He$$

3. With reference to the periodic table, we find the actual products:

$$^{14}_{6}C \rightarrow {}^{14}_{7}N + {}^{0}_{-1}e$$
$$^{248}_{100}Fm \rightarrow {}^{244}_{98}Cf + {}^{4}_{2}He$$

We can also apply the Rutherford-Soddy rule toward the identification of the type of radiation absorbed or emitted if the daughter nuclides are known. We shall investigate this type of nuclear change more closely when we consider nuclear fission and fusion reactions. Nevertheless, let us look at Example 14.4 for a preview.

Example 14.4

Write balanced nuclear equations for:

1. $^{8}_{5}B$ decaying to form $^{8}_{4}Be$ with particle emission

2. $^{125}_{54}Xe$ decaying to form $^{125}_{53}I$ with *no* particle emission

3. $^{144}_{58}Ce$ decaying to form $^{144}_{59}Pr$ with radiation

4. $^{230}_{92}U$ decaying to form $^{226}_{90}Th$ with particle emission

Solution:

Write the parent-daughter nuclides for each reaction. Leave a blank space in which we can assign the proper mass number and charge so that the emitted (or absorbed) particle will permit the nuclear equation to balance.

1. $^{8}_{5}B \rightarrow {}^{8}_{4}Be + {}^{0}_{+1}?$

The particle having a charge of $+1$ and mass number of zero is a *positron*. Hence,

$$^{8}_{5}B \rightarrow {}^{8}_{4}Be + {}^{0}_{+1}e$$

2. The most feasible explanation of $^{125}_{54}Xe \rightarrow {}^{125}_{53}I$ without radiation would be to assume *orbital electron capture:*

$$^{125}_{54}Xe + {}^{0}_{-1}e \rightarrow {}^{125}_{53}I$$

3. $^{144}_{58}Ce \rightarrow {}^{144}_{59}Pr + {}^{0}_{-1}?$

The answer has to be *beta* emission:

$$^{144}_{58}Ce \rightarrow {}^{144}_{59}Pr + {}^{0}_{-1}e$$

4. $^{230}_{92}U \rightarrow {}^{226}_{90}Th + {}^{4}_{2}?$

The emitted particles are obviously *alpha:*

$$^{230}_{92}U \rightarrow {}^{226}_{90}Th + {}^{4}_{2}He$$

Exercise 14.1

1. Write the equation for $^{238}_{92}U$ emitting an alpha particle.
2. What product is formed as a result of positron emission from $^{121}_{53}I$ decay?
3. What happens when $^{15}_{7}N$ gains a proton in the formation of $^{12}_{6}C$?
4. When a $^{235}_{92}U$ nucleus is split by a high energy neutron, $^{143}_{56}Ba$ and $^{90}_{36}Kr$ are formed. What else must be involved in the nuclear reaction?

14.3 NUCLEAR STABILITY

Up to this point we have looked at a number of examples of nuclear decay. The rates of decay vary from particles per microsecond to particles per millennia. And, as we have learned, the modes of decay include alpha, beta, gamma, positron emission, and electron capture.

There are questions that must certainly intrigue us now. Why are some nuclei stable and others not? Are there some guidelines that might allow us to predict the modes of nuclear decay? And finally, we wonder if there might be some relationship between the numbers of neutrons and protons in the nucleus and the phenomenon of radioactivity.

Although our understanding of nucleon interaction is imperfect, it is generally believed that there is some property of neutrons that effectively holds the atomic nucleus together. Otherwise we would be at quite a loss to explain how an extraordinarily dense mass of nuclear protons succeeds in staying together when their "like" charges would lead us to expect vigorous forces of repulsion.

Regardless of our inability to explain how the neutrons bind the nucleus together, we can still build on the hypothesis that suggests strongly that they do. Then, looking at the numbers of neutrons and protons in various nuclides, we find some interesting correlations.

Odd and Even Numbers

Nuclear stability seems to be related to odd and even numbers of neutrons and protons according to the following guidelines:

1. The largest number of stable isotopes appears to have *even* numbers of protons and *even* numbers of neutrons.
2. An unusually high degree of stability seems to be associated with nuclei that have neutrons or protons adding up to the even numbers of 2, 8, 20, 28, 50, and 82. It would seem that even numbers of protons relate to stability in a manner similar to electron pairing outside the nucleus.

Now, before we attempt to make predictions regarding which nuclides are most likely to be radioactive, there remains at least one more important correlation between numbers of nucleons and stability.

Neutron-to-Proton Ratio (n/p)

It is common for stable low atomic number isotopes to have a $n/p = 1$. Examples are $^{4}_{2}He$, $^{12}_{6}C$, $^{16}_{8}O$, $^{20}_{10}Ne$, $^{24}_{12}Mg$, and so on. However, as the number of protons increases with our progression up the atomic number scale, it seems as though more neutrons are needed for stability. A reasoned hypothesis suggests that a stronger application of neutron "cement" is required to hold the "like-charged" protons together. Above lead, the concept of n/p ratio as it relates to nuclear stability becomes irrelevant for the most part since all elements having an atomic number of 84 or higher are naturally radioactive. Now, let us look at Figure 14–2 for a sampling of nuclides in which stability and instability are related to n/p ratios. The samples listed in Table 14–3 will be shown to be either within or outside of a **band of stability**.

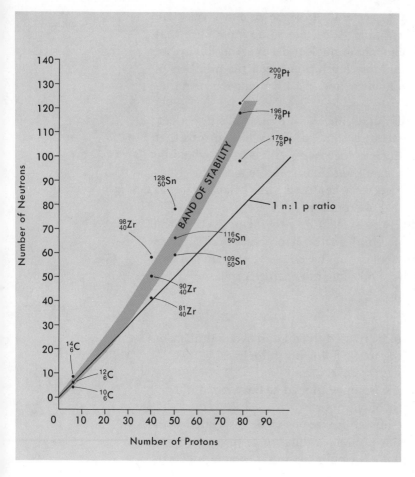

Figure 14–2 Graph showing n/p ratio as it relates to nuclide stability. Selected examples from Table 14-2 are used to illustrate nuclides that fall within and outside the diagonal stripe, called the Band of Stability.

TABLE 14–3 Relating the Stability of Isotopes to the n/p Ratio

Low n/p ratio and mode of decay	Stable isotope— ideal n/p ratio	High n/p ratio and mode of decay
$^{10}_{6}C$ decays by $^{0}_{+1}e$ emission	$^{12}_{6}C$	$^{14}_{6}C$ decays by $^{0}_{-1}e$ emission
$^{81}_{40}Zr$ decays by $^{0}_{+1}e$ emission or electron capture	$^{90}_{40}Zr$	$^{98}_{40}Zr$ decays by $^{0}_{-1}e$ emission
$^{109}_{50}Sn$ decays by $^{0}_{+1}e$ emission or electron capture	$^{116}_{50}Sn$	$^{120}_{50}Sn$ decays by $^{0}_{-1}e$ emission
$^{176}_{78}Pt$ decays by $^{4}_{2}He$ or $^{0}_{+1}e$ emission or electron capture	$^{196}_{78}Pt$	$^{200}_{78}Pt$ decays by $^{0}_{-1}e$ emission

An interesting phenomenon called the **radioactive decay series** involves all of the elements from atomic number 84 (polonium) and above. The nuclide having the highest mass number and atomic number possible for a stable nucleus is $^{209}_{83}\text{Bi}$. Hence, all of the naturally occurring isotopes above $^{209}_{83}\text{Bi}$ belong to one of several radioactive decay series. This means that stability is gained by a stepwise process of successive nuclide decay reactions. The series are described as:

1. **The Uranium Series** in which $^{238}_{92}\text{U}$ decays stepwise until the stable nuclide $^{206}_{82}$ results.
2. **The Thorium Series** where $^{232}_{90}\text{Th}$ evolves finally, by a series of nuclear disintegrations, into $^{208}_{82}\text{Pb}$.
3. **The Actinium Series** where $^{227}_{89}\text{Ac}$ decays in a number of steps to produce finally $^{207}_{82}\text{Pb}$.

Example 14.5

The first four steps in a decay series beginning with the synthetic nuclide $^{235}_{92}\text{U}$ involve the emission of α and β^- particles in the order: $\alpha, \beta^-, \alpha, \beta^-$. Write the balanced nuclear equation for each decay step.

Solution:

1. Let us arrange the steps and then fill in the blanks so that $^{4}_{2}\text{He}$ and $_{-1}^{0}e$ emissions are consistent with Rutherford-Soddy rules.

   $^{235}_{92}\text{U} \rightarrow\ ^{231}_{90}\ \underline{\quad\textbf{?}\quad} +\ ^{4}_{2}\text{He}$

   $^{231}_{90} \rightarrow\ ^{231}_{92}\ \underline{\quad\textbf{?}\quad} +\ _{-1}^{0}e$

   $^{231}_{91}\ \underline{\quad\textbf{?}\quad} \rightarrow\ ^{227}_{79}\ \underline{\quad\textbf{?}\quad} +\ ^{4}_{2}\text{He}$

   $^{227}_{89}\ \underline{\quad\textbf{?}\quad} \rightarrow\ ^{227}_{90}\ \underline{\quad\textbf{?}\quad} +\ _{-1}^{0}e$

2. Identifying the daughter nuclides by their atomic numbers, we can complete the equations.

   $^{235}_{92}\text{U} \rightarrow\ ^{231}_{90}\text{Th} +\ ^{4}_{2}\text{He}$

   $^{231}_{90}\text{Th} \rightarrow\ ^{231}_{91}\text{Pa} +\ _{-1}^{0}e$

   $^{231}_{91}\text{Pa} \rightarrow\ ^{227}_{89}\text{Ac} +\ ^{4}_{2}\text{He}$

   $^{227}_{89}\text{Ac} \rightarrow\ ^{227}_{90}\text{Th} +\ _{-1}^{0}e$

Let us look at Figure 14–3 in order to see a graphical representation of a naturally occurring radioactive decay series, the thorium series. The successive decay steps in the thorium series are listed in Table 14–4, which we shall use as a data base for the graph.

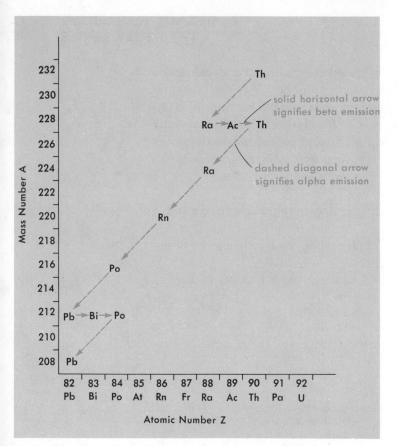

Figure 14–3 The nuclear decay series for thorium.

Up to this point we have concerned ourselves with the phenomenon of natural radioactivity. But all of us, thanks to a life-time exposure to the mass media, are surely aware of scientific and technological advances in the field of nonnatural (artificial) radioactivity. Let us now investigate how the collisions of nucleons lead sometimes to new and strange creations.

TABLE 14–4 The Thorium Series

Radioactive Nuclide (Parent)	Type of Decay	Reaction Product (Daughter)
$^{232}_{90}\text{Th}$	$^{4}_{2}\text{He}$	$^{228}_{88}\text{Ra}$
$^{228}_{88}\text{Ra}$	$^{0}_{-1}\text{e}$	$^{228}_{89}\text{Ac}$
$^{228}_{89}\text{Ac}$	$^{0}_{-1}\text{e}$	$^{228}_{90}\text{Th}$
$^{228}_{90}\text{Th}$	$^{4}_{2}\text{He}$	$^{224}_{88}\text{Ra}$
$^{224}_{88}\text{Ra}$	$^{4}_{2}\text{He}$	$^{220}_{86}\text{Rn}$
$^{220}_{86}\text{Rn}$	$^{4}_{2}\text{He}$	$^{216}_{84}\text{Po}$
$^{216}_{84}\text{Po}$	$^{4}_{2}\text{He}$	$^{212}_{82}\text{Pb}$
$^{212}_{82}\text{Pb}$	$^{0}_{-1}\text{e}$	$^{212}_{83}\text{Bi}$
$^{212}_{83}\text{Bi}$	$^{0}_{-1}\text{e}$	$^{212}_{84}\text{Po}$
$^{212}_{84}\text{Po}$	$^{4}_{2}\text{He}$	$^{208}_{82}\text{Po (stable)}$

In a sense, modern science has succeeded where the alchemist failed. Instead of discovering the legendary Philosopher's Stone so that common metals can be transmuted into gold, we have devised remarkable "shooting-galleries." We call our shooting-galleries **particle accelerators** (or "atom-smashers") and in them we bombard target atoms with neutrons, alpha particles, deuterons, and protons—all of which are listed in Table 14–2. When these particles in addition to other possible positive ions such as $^{12}_{6}C$ nuclei are caused to move at very high speeds, they gain sufficient kinetic energies so that nuclei of target atoms may be permanently altered.

One type of particle accelerator is called a **cyclotron.** The cyclotron was developed in 1932 by the American scientists Ernest Lawrence and Milton Livingston. In the cyclotron, selected positive ions are made to move at extremely high velocities in circular paths within hollow **"D"** shaped electrodes, called **dees.** Above and below the *dees* large magnets establish a strong magnetic field perpendicular to the path of particles (i.e., the positive ions). Alternating the polarity of the two *dees* each time a particle crosses the gap between them causes the particles to accelerate. Hence, the magnets cause the particles to move in a circular path that describes a wider and wider radius because of

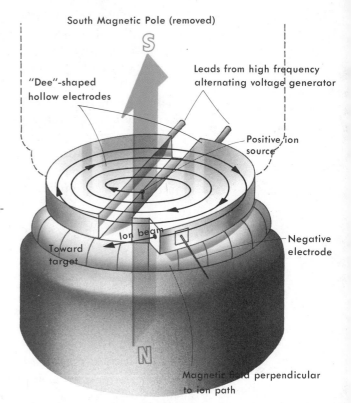

Figure 14–4   Diagrammatic representation of the cyclotron particle accelerator.

the constant acceleration. Finally, the particle exists through an opening at the outer edge of one of the *dees* and hits a target. See Figure 14–4 for an idealized diagram of a cyclotron.

Modern cyclotrons can accelerate protons, deuterons, and alpha particles until they have kinetic energies ranging from about 10 to 50 million electron volts (MeV). This represents a huge improvement over natural sources of positive ion emission or primitive ion accelerating devices. For example, in 1932, the British team of John Cockcroft and Ernst Walton used a proton accelerator to produce the first successful nuclear reaction. They were able to transform lithium into beryllium by bombarding the lithium with protons having less than 1 MeV energy.

$$\ce{^7_3Li} + \ce{^1_1H} \rightarrow \ce{^8_4Be}$$

The $\ce{^8_4Be}$ nuclide proved to be very unstable and it immediately disintegrated into 2 alpha particles.

$$\ce{^8_4Be} \rightarrow \ce{^4_2He} + \ce{^4_2He}$$

The primary fascination with the Cockcroft-Walton experiment was that their measurements, related to the conservation of matter and energy in a nuclear disintegration, proved the accuracy of Einstein's law of mass-energy equivalence. We shall deal with that topic shortly.

A more modern development, spurred by a need for higher energy particles capable of penetrating the nuclei of heavy element atoms, is an accelerator having a magnetic field that can be synchronized to increase with the increasing speed of the particle "bullet." This machine, appropriately called a **synchrotron,** permits a particle to reach speeds approaching the speed of light because of the synchronization of the variable magnetic field with the increasing energy of the particle. The synchrotron allows the positive ions to acquire energies in the range of billions of electron volts **(BeV).** In fact, the name **Bevatron** has been applied to machines having such a capability.

Artificial Radioactivity

Let us now try to give some measure of historical perspective to the whole topic of artificial radioactivity, which is another way of saying "induced" radioactivity. The term "induced" suggests more clearly that we are speaking of causing even nonradioactive nuclides to become radioactive as a result of particle bombardment.

The first recorded induced nuclear transformation came from Ernest Rutherford's bombardment of $\ce{^{14}N}$ by alpha particles. He used radium as the alpha emitter and found that the exposure of $\ce{^{14}N}$ to the alpha particles resulted in the formation of $\ce{^{17}O}$ and the emission of a proton.

The equation is:

$$^{14}_{7}N + ^{4}_{2}He \rightarrow ^{17}_{8}O + ^{1}_{1}H$$

Later, in 1934, the French team of Irene Curie-Joliot and her husband, Frederic Joliot, used polonium as an alpha source in order to bombard aluminum. In this nuclear event they discovered a more complex two-step disintegration series:

(first step) $\quad ^{27}_{13}Al + ^{4}_{2}He \rightarrow ^{30}_{15}P + ^{1}_{0}n$

(second step) $^{30}_{15}P \rightarrow ^{30}_{14}Si + ^{0}_{+1}e$

Hence, with the pioneering work of Rutherford, the Joliots, and the development of particle accelerators, we have the foundation for transmuting elements.

14.6 DETECTION AND MEASUREMENT OF RADIOACTIVITY

For a number of very important reasons, such as basic scientific research, practical applications in the allied health fields and industry, and for the safe-guarding of life against the hazards of radiation, we must have ways of detecting and measuring radiation due to nuclear reactions.

Laboratory personnel who work regularly with radiation or radioactive isotopes need to observe certain safety precautions in order to avoid an accumulating dosage. Laboratories and personnel can be monitored to assure safe levels of radioactivity. Photographic film badges, Geiger-Müller counters, and the use of standard labels can serve this purpose (Fig. 14–5).

The Geiger-Müller Counter

We pointed out earlier that alpha and beta particles have the ability to cause ionization when they collide with electrons of atoms. The transfer of kinetic energy to the atom's electron enables the electron to move free of the atom in the process of ionization. There are several instruments designed to measure radioactive decay rates. One of the most common of these is the Geiger-Müller counter (Fig. 14–6). A typical instrument is composed of a detecting tube and an amplifier

Figure 14–5   Standard label for radiation safety.

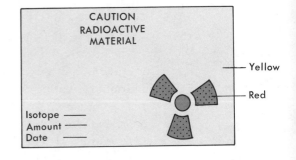

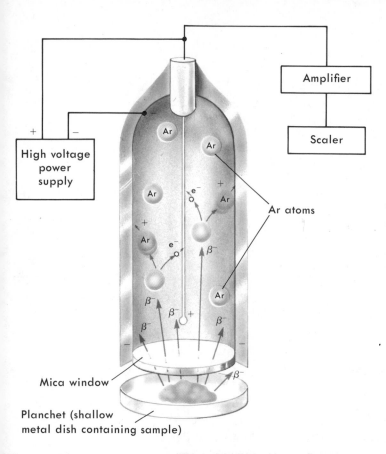

Figure 14–6   Geiger-Müller tube. The entering high-energy particle causes the argon atom to ionize. The liberated electrons are then attracted strongly to the central anode and amplified proportionately.

for the production of audible "clicks," or recording on an electronic counter, called a **scaler.** The Geiger-Müller counter is usually a metal tube containing a mica window. Since beta particles, and especially alpha particles, have very limited penetrating power, the window is necessary. The tube contains argon gas. The cylinder wall serves as the cathode in which a centrally located wire is the anode. A high voltage is established between the central wire and the cylinder wall so that any ions produced will be strongly and quickly attracted to the electrodes. The electron flow from the anode wire is amplified for counting. The amount of radioactive material in a sample is proportional to the current produced by the ionization of the gas.

Scintillation Counter

Another type of detector suitable for nonionizing gamma rays and weak beta particles is the **scintillation counter.** There are two varieties of the scintillation counter—liquid and solid. Scintillation means the production of a flash of light. There are some substances, called **phosphors,** that produce this spark of light when a nuclear particle or a photon of gamma radiation transfers its energy to a phosphor molecule. As the excited phosphor returns to its ground state, the flash of light is emitted. The intensity of the light flash is proportional to the

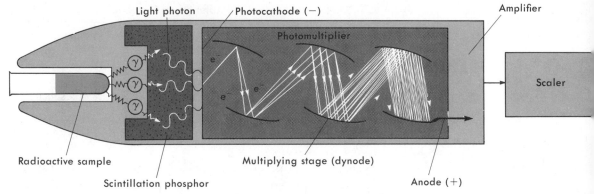

Light photon   Photocathode (−)   Amplifier

                        Photomultiplier

                                                            Scaler

Radioactive sample   Multiplying stage (dynode)

Scintillation phosphor   Anode (+)

Figure 14—7   A diagrammatic representation of a scintillation counter.

energy of the radiation. The light flash is detected by a **photomultiplier tube** (Fig. 14–7) and amplified for registration on the scaler.

Some phosphors are NaI crystals for gamma ray detection, special organic crystals for beta counting, and a variety of organic compounds dissolved in organic solvents for liquid scintillation. Dissolving some sample in the scintillator solution is best for low energy beta particles. A diagrammatic scintillation counter is shown in Figure 14–7. In this diagram, the gamma radiation causes light emission from the phosphor. The light photon causes an electron to be ejected from the first dynode. This photoelectric effect is multiplied by the dynodes to a growing cascade of electrons that are finally collected by the final dynode where the current has been effectively multiplied thousands of times. There may be as many as fifteen dynodes in a photomultiplier. The final current is amplified for the scaler.

Biological Hazards of Radiation

As our attention turns to the interactions of radiation with living organisms, such as human beings, the ionizing of atoms and the rupturing of molecules take on a new dimension. When radiation levels exceed the normal, ever-present levels (called **background** radiation) owing to electromagnetic radiation (UV, X-ray, gamma ray) and naturally occurring radioisotopes in the Earth's crust, we must be especially concerned because of the effect of ionizing radiation on living organisms, living tissue, and the variety of organic molecules that compose living cells. The effects of radiation depend, apparently, on the amount of exposure to radiation and the energy of the radiating particles or photons. Because of the clearly evident direct link between radiation and such effects as cancer, leukemia, genetic defects, and so on, the Atomic Energy Commission is now very careful about establishing guidelines that describe allowable radiation dosage.

We are warned clearly about the danger of skin cancer due to excessive ultraviolet radiation as we strive to suntan our bodies. The controls on the use of X-ray machines are much more rigid than they

TABLE 14–6 Effect of Radiation on Humans

Dose in Rems	Probable Effect on Humans
0–50	Possible genetic mutation
50–100	Headache, dizziness, listlessness
100–200	Radiation sickness and hair loss
200–500	Severe bleeding and tissue destruction
Over 500*	Death

*This is known as a lethal dose, abbreviated LD.

were a few decades ago. Alpha particles, possibly absorbed in industry in the form of powders or aerosol sprays, present a serious form of ionizing radiation. Beta radiation as a hazard is not great except for the possible exposure to a direct beam, in which case the ionization of protein atoms could be serious.

Gamma radiation is very hazardous. The photons have sufficient energy to destroy the molecules of living protoplasm by gross alternation or "burning."

The effects of high energy radiation may be measured in a variety of ways, using several kinds of units of radiation such as *roentgens* (**R**), **rem** (*roentgen equivalent in man*), **RBE** (*relative biological effectiveness*), and **rads.** The most commonly used unit of radiation dosage is the rem. The number of rems, upon which a table of human hazards might be based, is calculated as the product of the amount of radiation absorbed, rad (*radiation absorbed dose*), and the measure of relative biological effectiveness, RBE, which depends on the type of radiation. A rad, more precisely defined, is the amount of radiation corresponding to 1×10^{-5} J of energy affecting a gram of living tissue. Hence, dosage in terms of a number of rems is expressed by the equation:

$$\text{rems} = \text{rads} \times \text{RBE}$$

See Table 14–6 for a summary of some effects of radiation on humans.

When humans are exposed to radiation doses in excess of 100 rems, they may expect to suffer from "radiation sickness." Radiation sickness usually includes such clinical symptoms as nausea, vomiting, fatigue, and diarrhea.

In any case, there is controversy. The resolution of conflicting attitudes about the dangers of radiation may take many more years of case studies—if indeed a uniformity of opinion ever emerges.

14.7 THE KINETICS OF RADIOACTIVITY

An aspect of radioactivity that has a number of rather important practical applications is the rate at which unstable nuclei decay. The most common method of measuring radioactive decay of a substance is in terms of its **specific activity.** Specific activity is the number of curies, **Ci,** per gram of radioactive substance. A curie is defined as 3.7×10^{10} *dps* (disintegrations per second). This rate of decay is equal approximately to the disintegration activity of a gram of radium. Of course we can modify the Ci unit according to various rates of decay. Some substances decay with extraordinary rapidity, in which case the curie would be a most appropriate unit. Other substances may show a range of relatively slow rates of decay so that millicuries (mCi), micro-curies (μCi), or nanocuries (nCi) would be better. One nCi, for exam-ple, would be equivalent to 37 *dps.* In order to get some kind of quanti-tative grasp of what we mean by "fast" or "slow" rates of disintegra-tion, we have a measurement known as the "half-life" of a radioisotope.

Half-Life

The half-life of a radioactive nuclide is a statistical rather than an absolute measure. We may define the half-life of a substance as the length of time required for half an initial quantity to undergo decay. The time, measured in years (**yr**), days (**da**), minutes (**min**), or seconds (**sec**), varies enormously—from billions of years to microseconds. The initial quantity is important only as it relates to the amount remaining as time passes, but it does not affect the rate of decay. In other words, the half-life of a particular radionuclide is constant. Half periods vary greatly from minute fractions of a second to billions of years. Elements having shorter half-life periods are more radioactive than those having longer periods. The infamous ^{90}Sr, produced by atomic bomb reactions, has a half-life of 28 yr. This half-life can be symbolized as $t_{1/2} = 28$ yr. This means that a given mass of ^{90}Sr will be reduced to half its original mass after 28 years. See Table 14–7 for a sample of the enormous range of half-lives within a single decay series for ^{232}Th.

Look now at Figure 14–8 for a presentation of the concept of half-life in the form of a graph. Let us use $^{137}_{55}$Cs as an example,

TABLE 14–7 The Thorium Decay Series

Radioisotope	Half-Life	Type of Decay
$^{232}_{90}$Th	1.39×10^{10} yr	α
$^{228}_{88}$Ra	6.7 yr	β
$^{228}_{89}$Ac	6.13 hr	β
$^{228}_{90}$Th	1.91 yr	α
$^{224}_{88}$Ra	3.64 da	α
$^{220}_{86}$Rn	52 sec	α
$^{216}_{84}$Po	0.16 sec	α
$^{212}_{82}$Pb	10.6 hr	β
$^{212}_{83}$Bi	60.5 min	α, β
$^{212}_{84}$Po	3.0×10^{-7} sec	α
$^{208}_{81}$Tl	3.1 min	β
$^{208}_{82}$Pb		Stable

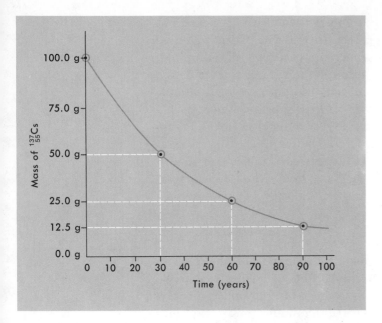

Figure 14—8 Graphical representation of the decay of 100.0 g of $^{137}_{55}Cs$, $t_{1/2} = 30$ yr.

since its half-life of 20 yr lends itself to a relatively uncomplicated diagram. Let us assume additionally that we start with 100.0 grams of the radionuclide.

From the graph we can see that after 30 years, the original mass of $^{137}_{55}Cs$ is reduced to half the initial mass. After the passage of a second half-life (i.e., 60 years) the mass is reduced by half again, and so on.

Example 14.6

Predict the amount of $^{212}_{83}Bi$ remaining if you observe 10.0 g decay over a period of 121 minutes. The $t_{1/2}$ of $^{212}_{83}Bi$ is 60.5 min.
Solution:
The time of 121 minutes is equivalent to 2 half-lives of the radionuclide. Hence, we should expect statistically that 5.0 g will remain after 60.5 min and 2.5 g remains after 121 minutes.

A question that we must surely ask now is how is it possible for scientists to determine half-lives of radioisotopes in the extreme ranges of microseconds or centuries? For reasons that must be obvious, we often cannot enjoy the advantage of direct measurement. Hence, we must turn to firmly established scientific principles from which these data may be derived.

Rate of Radioactive Decay

Let us start our investigation of principles governing radioactive decay with the discovery, by Rutherford and Soddy at the turn of the century, that nuclear decay can be described only in statistical

NUCLEAR CHEMISTRY

terms. Furthermore, the process is beyond our ability to control it. Additionally, we know today that spontaneous nuclear decay is not affected by temperature and the concept of activation energy is irrelevant.

What we do know, empirically, is that *the rate of nuclear decay is a first-order decay phenomenon. In other words, the rate of decay is directly proportional to the amount of radioactive substance.*

Rate of decay $\propto$ N

The first-order decay phenomenon is not unusual in science.

As it often happens in experimentation, from a mass of carefully collected data it is possible to obtain a constant that relates the variables. For example, the density of a given substance at a particular temperature and pressure is a constant that always relates the change in mass to a proportional change in volume.

Similarly, although in a more complex manner, the proportionality constant that relates a decrease in the amount of a radioactive substance to its decreasing rate of nuclear decay can be obtained experimentally. The specific constant for each radionuclide is symbolized, λ, *lambda*. We call λ the **decay constant.** The decay constant has the dimension of *reciprocal time*, which we may describe as *per* second, minute, hour, or whatever is convenient. The rate of decay will be described in a manner consistent with the λ unit so that we can measure nuclide decay in terms of atoms/sec, or atoms/min and so on. An important fact to remember is that *an individual decay constant is related quite specifically to the half-life of an individual radioisotope.*

Although it is beyond the scope of this text to present a rigorous mathematical derivation of the first-order decay equation, we can use the equation on the basis of its empirical origins.

$$\log \frac{N_0}{N} = \frac{\lambda t}{2.30}$$

1. $\log \dfrac{N_0}{N}$ is the logarithm of the ratio of the original amount (N_0) to the final measurement (N) [or initial rate of disintegration such as disintegrations per second (dps), or disintegrations per minute (dpm)].
2. **t** is the time span over which we measure N_0 and N.
3. λ is the decay constant.
4. **2.30** is a conversion factor used to convert logarithms in base e to base 10.

If we arrange the equation in order to solve for the decay constant, we must take the half-life of each radionuclide into account. This is a convenient step because at the half-life, $N = \frac{1}{2} N_0$ (i.e., the amount remaining is half the original).

Hence, at the half-life:

$$\log \frac{N_0}{\frac{1}{2}N_0} = \frac{\lambda t_{1/2}}{2.30}$$

Factoring out the N_0 term, we have $\log 2 = \frac{\lambda t_{1/2}}{2.30}$.

Finding the log of 2 as 0.301, we rearrange the equation

$$t_{1/2} = \frac{0.693}{\lambda}$$

Finally, we have an equation for λ that is most useful:

$$\boxed{\lambda = \frac{0.693}{t_{1/2}}}$$

At this point we should examine a few sample problems so that we can see how the kinetics of radioactivity might be applied usefully.

Dating Organic Remains

One effect of the constant bombardment of our upper atmosphere by cosmic radiation is the production of free neutrons that interact with $^{14}_{7}N$ atoms to form the radionuclide $^{14}_{6}C$ and protons.

$$^{14}_{7}N + ^{1}_{0}n \rightarrow ^{14}_{6}C + ^{1}_{1}H$$

The $^{14}_{6}C$ proceeds to form $^{14}CO_2$, which is taken up by green plants in the process of photosynthesis. Hence, the $^{14}CO_2$ is incorporated eventually into the tissue of the living green plant. Remember, as we describe this process, the $^{14}CO_2$ is chemically identical to non-radioactive CO_2 in which the usual $^{12}_{6}C$ atoms are combined. The essential difference is that while the plant dies and may even be burned as fuel the carbon in the $^{14}CO_2$ continues to emit particles. In fact, $^{14}_{6}C$ is a beta emitter.

$$^{14}_{6}C \rightarrow ^{14}_{7}N + ^{0}_{-1}e$$

Now, if we assume that the ratio of $^{14}_{6}C$ to $^{12}_{6}C$ found in atmospheric CO_2 has remained relatively constant over a period of many centuries, it follows that the age of ancient organic matter might be approximately by comparing the extent of its radioactive decay with the rate of decay in living organic matter. We might use tissue from a living plant or tissue from an animal that eats green plants. There is some reasonable doubt regarding the constant level of ^{14}C production over the centuries, but statistical methods take into account the likelihood of deviations.

Given the information that ^{14}C has a half-life of about 5730 yr, we can date ancient organic artifacts with a fair degree of reliability. We need only to compare the rates of ^{14}C decay in equal amounts of ancient and fresh material. The ancient material, which has long ago ceased metabolizing $^{14}CO_2$, is left only with a fraction of the original amount of ^{14}C. The fresh material, which has been actively incorporating newly formed ^{14}C into its tissues, will be disintegrating at a faster rate.

Example 14.7

Suppose the rate of nuclear decay in a sample of an ancient papyrus scroll found in a cave is compared with the decay rate of an equal mass of freshly cut papyrus plant tissue. The scroll gives an average of 4.20 disintegrations per minute (*dpm*) while the fresh papyrus gives 10.00 *dpm*. Using $T_{1/2}$ for ^{14}C as 5730 yr, estimate the age of the papyrus scroll.

Solution:

1. Write the first-order decay equation as we have modified it.

$$\log \frac{N_o}{N} = \frac{\lambda t}{2.30}$$

2. Organize the data, including the solution for the decay constant

$N_o = 10.00$ *dpm* (new)
$N = 4.20$ *dpm* (ancient)
$t = ?$ $t_{1/2} = 5730$

$$\lambda = \frac{0.693}{t_{1/2}} = \frac{0.693}{5730 \text{ yr}}$$

$$\log \frac{N_o}{N} = \frac{10.00}{4.20} = 0.377$$

3. Solve the equation for t

$$0.377 = \frac{(0.693)(t)}{(5730 \text{ yr})(2.30)}$$

$$\boxed{t = 7170 \text{ year old papyrus}}$$

Finding a Half-Life

Let us now address the question we raised at the beginning of this section. How can we determine the half-life of a radionuclide, especially of it's one like ^{14}C? We certainly cannot sit around 5700 years to get the answer. The method we do use is to gather sufficient data so that the decay constant for a particular radionuclide can be determined—either graphically or by the application of the first-order decay equation. Once we have the decay constant in hand, we can find $t_{1/2}$ easily. The data we need, specifically, can be obtained by use of a radiation detector and counting device (a Geiger counter and scaler, for example).

Example 14.8

Find the half-life of a radioactive isotope if beta emission was measured at 1.42×10^3 *dpm* at 1:00 P.M. and 1.06×3^3 *dpm* at 4:00 P.M. the same afternoon.
Solution
1. Write the modified first-order equation and organize the data

$$\log \frac{N_\circ}{N} = \frac{\lambda t}{2.30}$$

N_0 = original disintegration rate, 1.42×10^3 *dpm*
N = final rate at the end of a fixed time interval, 1.06×10^3 *dpm*
t = 1:00 P.M. to 4:00 P.M. is 3 hr

$$\log \frac{N_\circ}{N} = \frac{1.42 \times 10^3}{1.06 \times 10^3} = 0.127$$

2. Solve for λ

$$\lambda = \frac{(0.127)(2.30)}{3 \text{ hr}} = 0.0974/\text{hr}$$

3. Using the equation $t_{1/2} = \frac{0.693}{\lambda}$

we can find $t_{1/2}$

$$t_{1/2} = \frac{0.693}{0.0974/\text{hr}} = \boxed{7.11 \text{ hr}}$$

Quantitative Transmutation

One more application of the first-order decay phenomenon that illustrates its practicality is that it permits us to determine just about how much of a radionuclide will still be available to us after the passage of time. For example, if a researcher is performing studies on the behavior of a radionuclide in a living organism, it would be essential to know whether or not enough of the material is on hand in order to complete the study. Factors such as amount needed, scarcity, and cost are likely to be related also.

Example 14.9

Suppose a 3.00 mg sample of $^{131}_{53}I$ is available as a beta emitter for studies related to the thyroid gland. The half-life of ^{131}I is 8.05 days. If it is essential that we have at least 0.5 milligram left after 3 weeks, do we have enough $^{131}_{53}I$ on hand to begin the experiment?
Solution:
1. We must find out, as precisely as statistical approximation will allow, just how much of the radioisotope remains after 3 weeks.
2. Let us write the equation and organize the data.

NUCLEAR CHEMISTRY

$$\log \frac{N_o}{N} = \frac{\lambda t}{2.30} = \frac{(0.693)(t)}{(t_{1/2})(2.30)}$$

$N_o = 3.00 \text{ mg}$

$N = ?$

$t_{1/2} = 8.05 \text{ da}$

$t = 3 \text{ weeks} = 21 \text{ da}$

3. Substitute and solve for N

$$\log \frac{3.00 \text{ mg}}{N} = \frac{(0.693)(21 \text{ da})}{(8.05 \text{ da})(2.30)}$$

$$\log \frac{3.00 \text{ mg}}{N} = 0.786$$

Take the *antilog* of both sides of the equation

$$\frac{3.00 \text{ mg}}{N} = 6.11$$

$$N = \frac{3.00 \text{ mg}}{6.11} = \boxed{0.491 \text{ mg}}$$

4. The answer indicates that we will probably not have quite 0.5 milligram. Hence, the experiment ought to be postponed until we obtain more ^{131}I.

Exercise 14.2

1. If the ratio of $^{14}_{6}\text{C}$ in a freshly cut piece of wood to the amount of $^{14}_{6}\text{C}$ in some ashes found in a cave is 1.00:0.76, calculate the age of the cave dwellers. The $t_{1/2}$ for $^{14}_{6}\text{C}$ is 5730 yr.
2. A radioactive sample emitted particles at a rate of 1420 *dpm* at 2:30 P.M. If the rate was 1060 *dpm* at 5:00 P.M., what is the half-life of the isotope?
3. The radioisotope $^{32}_{15}\text{P}$ is a beta emitter that has a half-life of 14.3 da. If a 8.00 g sample of $^{32}_{15}\text{P}$ is used for 24 days, how much will remain?

14.8 MASS—ENERGY RELATIONSHIPS

We have found it convenient to apply the Law of Conservation of Matter to ordinary chemical changes. Indeed, the usefulness of molar mass ratios of reactants and products in stoichiometric reactions depends on the axiom that states that matter can be neither created nor destroyed. We must emphasize, however, that we should add a "for all practical purposes" qualifier to that axiom. Even in an ordinary chemical change it is likely that some very small amount of matter will be converted into an equivalent amount of energy. The popularized Einstein equation:

$$E = mc^2$$

---

expresses that equivalence. However, in the case of ordinary chemical change, the amount of matter converted to energy is so small in comparison to the amount of matter involved that it has no quantitative significance.

But in the realm of the atomic nucleus and in light of the reality of nuclear energy, the interconvertibility of matter and energy takes on new dimensions and great significance.

Nuclear Binding Energy

If we compare the masses of nuclides with the sum of their parts, an interesting discrepancy will be noted. In other words, as far as atomic nuclei are concerned, the mass of a nucleus is not necessarily equal to the sum of its nucleons. Let us look at an example of this curious difference and then look for a possible answer. Consider $^{12}_{6}C$, which has a nuclear mass of 11.9967 amu (this does not include extranuclear electrons whose mass is taken as 0.000549 amu). Now let us consider the sum of the nucleons in the $^{12}_{6}C$ nuclide.

$$6 \text{ protons} = 6 \times 1.00728 \text{ amu} = 6.0437 \text{ amu}$$

$$6 \text{ neutrons} = 6 \times 1.00867 \text{ amu} = 6.0520 \text{ amu}$$

$$\text{The total mass of nucleons} = 12.0957 \text{ amu}$$

We see, indeed, that the whole is not equal to the sum of its parts! The difference between the two, $12.0957 - 11.9967 = 0.0990$ amu, is known as the **mass decrement** or the **mass defect** of the nucleus. This difference in mass, regardless of which label you choose, refers to the loss in matter that occurs when the individual nucleons combine to form a nucleus. Apparently, this "lost" mass is present in the form of energy that holds the nucleus together. In effect, when we calculate the mass decrement of a nucleus we are describing a mass equivalent to what is called the **nuclear binding energy.** The nuclear binding energy more precisely defined is the amount of energy that would have to be absorbed in order to separate an atomic nucleus into its neutron and proton units.

Hence, the most useful understanding we can derive from nuclear binding energies is the relationship between nuclear stability and the binding energy. Clearly, the amount of energy needed to separate an atomic nucleus into its parts is directly proportional to the stability of the nucleus. When binding energies are calculated and plotted graphically as a function of the mass numbers of the elements, we find a pattern. See Figure 14–9 for a graph that illustrates the interrelationships among binding energy, nuclear stability, and mass number. Notice how the binding energies rise sharply until the mass number reaches the 50 to 60 range and then begins a gradual decline. Since

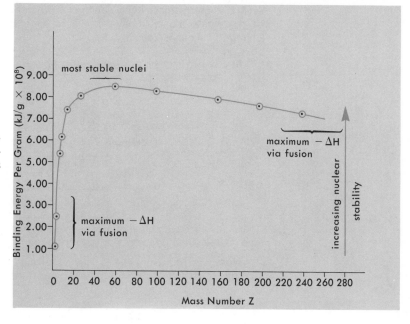

Figure 14–9 Graph illustrating average nuclear binding energy per gram vs. mass number.

the number of nucleons increases as the mass numbers rise, we plot binding energy per gram (rather than binding energy per mole) as a function of mass number in order to present a more uniform basis for comparison.

For example, if the nuclear binding energy of $^{12}_6C$ is 8.91×10^9 kJ/mole, we can calculate readily that this is equivalent to 7.43×10^8 kJ/g.

$$\frac{8.91 \times 10^9 \text{ kJ}}{\cancel{\text{mole C}}} \times \frac{\cancel{\text{mole C}}}{12.00 \text{ g C}} = 7.43 \times 10^8 \text{ kJ/g}$$

The relationships presented in Figure 14–9 will lead us into our next topics of atomic fission and fusion. The maximum energy-producing processes will result from the fusion of light nuclei (relatively low nuclear binding energies) and the fission of heavy nuclei.

Let us illustrate this difference by considering an example in which we shall calculate and compare the nuclear binding energies of ^4_2He and $^{238}_{92}U$ as typical representatives of atoms having light vs. heavy nuclei.

Example 14.10

Earlier we calculated the mass decrement in the $^{12}_6C$ nucleus as 0.0990 amu. Hence, we can describe this as 0.0990 g/mole. In the same manner we can calculate the difference between the nuclear mass and the sum of the nucleons in the case of ^4_2He and $^{238}_{92}U$.
They are:

$$^4_2He = 0.0083 \text{ g/mole}$$

$$^{238}_{92}U = 1.0353 \text{ g/mole}$$

14.8 MASS—ENERGY RELATIONSHIPS

Since the energy-mass equivalence of 1.00 g can be calculated as **8.99 × 10¹⁰ kJ/g,** we have a useful conversion factor for finding nuclear binding energies in kJ/mole. Make those calculations.

Solution:

In order to find the nuclear binding energies of $_2^4$He and $_{92}^{238}$U, we need only apply simple dimensional analysis.

$$_2^4He = \frac{0.0083\ g}{mole} \times \frac{8.99 \times 10^{10}\ kJ}{g}$$

$$_2^4He = \boxed{7.46 \times 10^8\ kJ/mole}$$

$$_{92}^{238}U = \frac{1.0353\ g}{mole} \times \frac{8.99 \times 10^{10}\ kJ}{g}$$

$$_{92}^{238}U = \boxed{9.31 \times 10^{10}\ kJ/mole}$$

In any case we note that nuclear binding energies are vastly more enormous than energies related to ordinary chemical reactions.

Note: You might like to verify the value 8.99 × 10¹⁰ kJ/g. Simply use 1.00 g in the equation $\Delta E = mc^2$, and convert the energy units to joules.

Exercise 14.3

1. Calculate the energy liberated, in kJ, when the mass change in a nuclear reaction is −0.0302 g/mole.
2. Find the nuclear binding energy per mole of $_{14}^{28}$Si if the mass decrement is 0.2722 g.
3. What is the nuclear binding energy per gram of $_{14}^{28}$Si?
4. Calculate the mass decrement in $_{30}^{64}$Zn if the nuclear binding energy is 8.45 × 10⁸ kJ/mole.

14.9 NUCLEAR FISSION

The phenomenon of nuclear fission, i.e., the "splitting" of atomic nuclei, started with the work of the Italian physicist Enrico Fermi in 1934. Fermi suggested that the formation of the transuranium elements might be accomplished by the bombardment of uranium by neutrons. He hypothesized that neutron capture might lead to beta emission and the formation of new elements having a higher mass number. The results, however were astoundingly different.

When the German scientists Otto Hahn and Fritz Strassmann analyzed the products of Fermi's reaction in 1938, they found a small amount of barium. They noted, with some disbelief, that barium has an atomic mass about one-half of uranium! A year later their friends Lise Meitner and Otto Frisch came to the conclusion that neutron bombardment caused the $_{92}^{283}$U nucleus to undergo fission. Meitner and Frisch, working with the knowledge that a trace amount of krypton

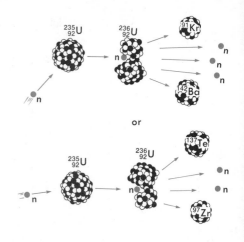

Figure 14–10   Diagrammatic representation of $^{235}_{92}U$ fission chain reaction. The $^{235}_{92}U$ becomes an unstable $^{236}_{92}U$ temporarily. The $^{236}_{92}U$ proceeds to split into $^{142}_{56}Ba$ and $^{91}_{36}Kr$ (or $^{97}_{40}Zr$ and $^{137}_{52}Te$) and release neutrons for further splitting.

also forms in the reaction, were able to explain the fission reaction by the equation:

$$^{238}_{92}U + ^{1}_{0}n \rightarrow \boxed{^{239}_{92}U} \rightarrow ^{139}_{56}Ba + ^{94}_{36}Kr + 3\,^{1}_{0}n$$

unstable
nucleus

Recent history indicates to us quite clearly that the discovery of nuclear fission opened up to scientists the possibility of releasing some of the enormous binding energy via the "splitting" of the nucleus. As the early stages of World War II evolved, a great deal of attention was directed at two heavy isotopes, $^{235}_{92}U$ and $^{239}_{94}Pu$, both of which are so unstable that they can be split by relatively low-energy neutrons. These two isotopes did become the essential components of the atomic bombs that were dropped on Hiroshima and Nagasaki. The high n/p ratios in these unstable isotopes lead to an average of 2 or 3 neutrons released as each nucleus undergoes fission. The released neutrons split other nuclei, which produce more neutrons, and so on—producing an almost instantaneous chain reaction as illustrated in Figure 14–10.

The amount of energy released when a nucleus is split is enormous. Given the information that the average amount of energy released per nucleus split is 3.20×10^{-14} kJ, it doesn't look very impressive. But let us perform an interesting calculation.

Example 14.11

Calculate the energy released by the nuclear fission of 1 kg of $^{235}_{92}U$, presuming only 1 per cent efficiency.
Solution:
1. If we start with the data:
   energy released per nucleus split $= 3.20 \times 10^{-14}$ kJ
   molar mass of $^{235}_{92}U = 235$ g/mole
   mass $= 1$ kg $= 1000$ g
   nuclei per mole $= 6.02 \times 10^{23}$
   We can use dimensional analysis to find the answers.

---

2. Calculate the approximate answer and then determine 1 per cent of the value

$$\Delta E = \frac{3.20 \times 10^{-14} \text{ kJ}}{\text{nucleus}} \times \frac{6.02 \times 10^{23} \text{ nuclei}}{\text{mole}} \times \frac{\text{mole}}{235 \text{ g}} \times \frac{1000 \text{ g}}{1} =$$

$$\Delta E = 8 \times 10^{10} \text{ kJ}$$

$$\text{1 per cent of } 8 \times 10^{10} \text{ kJ} = \boxed{8 \times 10^8 \text{ kJ}}$$

Nearly a billion kJ of energy is impressive indeed!

We can better appreciate the magnitude of the answer in Example 14.11 if we make a simple comparison. For example, a kilogram of dynamite releases about 3×10^3 kJ of energy upon explosion—if we assume 100 per cent efficiency. Compared with a kilogram of $^{235}_{92}\text{U}$ at 1 per cent efficiency, we would need roughly 300 tons of dynamite for the same energy release!

The atomic bomb prototypes were made with the knowledge of the awesome potential of nuclear energy. The problems were primarily technological. The first problem faced by the scientists was to develop a process for obtaining sufficient ^{235}U for a chain reaction and then devising a container (i.e., the bomb itself) in which the chain reaction would be initiated at a predetermined time and place. In effect, distinctions had to be made between a **sub-critical** mass and a **critical** mass. These designations are explained as follows:

1. A sub-critical mass is an amount of fissionable material so small that a chain reaction cannot be sustained. Too many neutrons are lost because the target volume is inadequate.
2. A critical mass is one in which the mass of fissionable material is so large that a relatively efficient chain reaction occurs because few neutrons escape the system without causing additional fission reactions.

The basic condition needed for the construction of an atomic bomb of the fission variety is to separate two sub-critical masses of ^{235}U

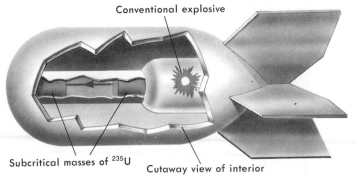

Conventional explosive

Subcritical masses of ^{235}U Cutaway view of interior

Figure 14–11 An imaginary model of an atomic bomb. An ordinary explosive brings two subcritical masses of ^{235}U together so that the critical mass is exceeded.

until a pre-planned conventional explosion bangs them together so that a critical mass is exceeded (Fig. 14–11). The nearly instantaneous chain reaction that follows releases the incredible energy of nuclear fission.

Nuclear Reactors

Nuclear reactors, unlike atomic bombs, are mechanisms so constructed that we can control the rate of nuclear fission. The nuclear reactor is designed so that it is impossible for the fissionable material to exceed a critical mass condition. A typical reactor does this by using a substance such as water (or heavy water, D_2O, deuterium oxide) under a pressure of more than 20 atm/cm^2, which surrounds rods of uranium nuclear fuel. Under this condition, the water (or heavy water) acts to slow down the fission-causing neutrons. Any substance that controls the neutron velocity in this manner is called a **moderator.**

Water is a commonly used moderator because it has a secondary use as a medium for transferring heat. Liquid sodium is also used in this way. In effect, what happens is that water or liquid sodium becomes very hot ($>300°C$) in the nuclear reaction chamber en route to a steam generator. It is a cyclic process, as illustrated in Figure 14–12. The liquid moderator is pumped around and around—absorbing heat from the nuclear fuel and transferring the heat to the relatively cold water in a steam generator. Then the steam is used to drive the turbines that produce electricity.

One other essential part of the nuclear reactor is the battery of metallic **control rods** that allow a finer control of the rate of fission because of their ability to absorb neutrons. By raising or lowering the control rods we can regulate the rate of fission as more or fewer neutrons are available.

However, the nuclear reactor is a mixed blessing. We have problems of nuclear waste disposal, radiation hazards, and related environmental problems that must be weighed in the balance against the benefits of a fuel that, for the time being at least, is cheaper than fossil fuel for the production of electricity. Nuclear reactors produce hardly any air pollution as compared to coal when the coal is burned directly in furnaces. However, modern technology has led to the development of systems in which coal can be used as a fuel such that our air is not polluted annually with tons of gas and ash poisons.

On the other hand, however, the products of nuclear fission must be disposed of because their accumulation affects the efficiency of nuclear reactors. The effect is analogous to any machine or process in which accumulated wastes clutter the system. The nuclear wastes not only contaminate the control rods and reduce their efficiency, but the necessary decontamination procedures result in a disposal problem for contaminants such as ^{90}Sr, ^{239}Pu, and ^{137}Cs—all of which produce deadly gamma radiation. It is no wonder then that knowledgeable

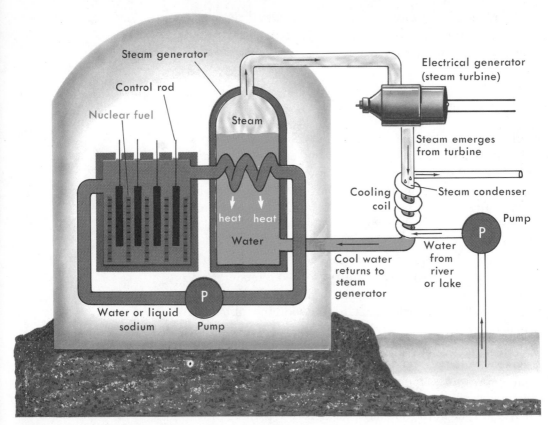

Figure 14–12   A diagrammatic model of a nuclear reactor.

Figure 14–13   Aerial view of the Three Mile Island nuclear power installation near Harrisburg, Pennsylvania.

individuals and concerned environmentalists are fearful of human errors in judgment, mechanical failures, earthquakes and other natural catastrophies, and even sabotage.

A case in point is the near disaster that occurred in April, 1979, at the Three Mile Island reactor near Harrisburg, Pennsylvania (Fig. 14–13). The shocking extent to which inadequate attention to detail, basic incompetence, ignorance, and poor judgment was compounded into what we call "human error" has shifted public opinion on the subject of nuclear energy. Glib assurances that no serious hazards exist have given way to cautious advance and healthy skepticism. The very real possibility of a "China Syndrome" type of "meltdown" or hydrogen gas explosion, both of which have the potential for causing a disaster of enormous proportions, was so real that there is now great concern throughout the world.

While the amounts of nuclear fuel may be held at a sub-critical mass, the leakage of radioactive materials into the environment is a possibility we must take into account. Given the half-life of ^{90}Sr to be about 30 years, and using as a safety guideline the need to wait about 20 half-lives for the nuclear disintegration to result in a "safe" level of radioactivity, we can see that nuclear wastes must be stored under "leak-proof" conditions for 600 years! We wonder how certain we can be about "leak-proof" conditions persisting for generations to come.

Linked to the problem of storage is the need to transport radioactive wastes on railways and by truck. Given the number of derailments and truck mishaps annually, many of which spill corrosive or toxic substances that result in deaths and massive evacuations, the possibility of disaster caused by spilled nuclear wastes adds another component to the moral and political debate about nuclear energy development.

We have some difficult decisions to make, and we have not yet found a way of putting the pro's and con's into the balance in such a way as to resolve opposite opinions into a universally acceptable conclusion.

Breeder Reactors

There is another problem. The reserves of ^{235}U are being exhausted to the extent that we may not have enough to last out this century. Hence, we are looking for alternatives. In addition to the possibilities of solar and geothermal energy, there is the option of using fuels other than ^{235}U—fuels that can be used to produce fissionable materials in nuclear reactors. Such reactors that produce, or "breed," fissionable materials are called **breeder reactors.**

One example of a "breeder" reaction is in the use of relatively abundant ^{238}U, of which geologists estimate an amount sufficient to last about a century. When highly energetic neutrons bombard ^{238}U, it results in the formation of the very unstable radionuclide, ^{239}U,

$$^{238}_{92}U + ^1_0n \rightarrow ^{239}_{92}U$$

The $^{239}_{92}U$ quickly disintegrates by beta emission into $^{239}_{94}Pu$

$$^{239}_{92}U \rightarrow ^{239}_{94}Pu + 2\,^{\,0}_{-1}e$$

Further bombardment of the $^{239}_{94}Pu$ leads to a typical nuclear fission reaction:

$$^{239}_{94}Pu \rightarrow ^{90}_{38}Sr + ^{147}_{56}Ba + 3\,^1_0n$$

In this stepwise reaction, there are 3 neutrons emitted for every 2 neutrons absorbed. Hence, we have the makings of a productive and controllable chain reaction. The attractiveness of the breeder reactor lies clearly in the abundance of ^{238}U that can be converted into fissionable ^{239}Pu by a small amount of a neutron emitter. Now, we ask, "Where are these breeder reactors?"

Not only do we fail to have operational breeder reactors at present, but it is very questionable if they ever will become a significant reality. The technological obstacles are formidable. Not only are the necessary temperatures of operation so high as to make an efficient heat transfer system difficult, but the extraordinarily lethal property of radioactive plutonium introduces an extreme radiation hazard. These and other technological problems, in addition to soaring construction costs, cast a considerable shadow of doubt on the future of breeder reactors.

14.10 NUCLEAR FUSION

The production of energy due to the fusion of light nuclei, such as hydrogen and helium isotopes, has enormous potential. Recent technological advances, principally at the Plasma Physics Center at Princeton, give us hope that nuclear fusion energy may become economically feasible sometime in the 1990's. Our hope is that the incredibly effective fusion reactions that fire our sun and that produced the hydrogen bomb might be harnessed to serve us.

Fusion reactions liberate energy for the same reasons as fission reactions. The mass of the products is less than that of the reactants and the mass difference appears as energy. There are several advantages to fusion reactions as an energy source when compared with fission reactions:

1. The energy produced per mole of nuclide is greater in a fusion reaction. For example, in the fusion of deuterium nuclei,

$$^2_1\text{H} + {}^2_1\text{H} \rightarrow {}^4_2\text{He}$$

the calculated mass decrement is 0.0256 g/mole. Since the mass-energy equivalence is 8.99×10^{10} kJ/mole (see nuclear binding energy in Section 14.8), the total energy released is 3.51×10^{12} kJ. Comparing this with a typical fission reaction involving $^{235}_{92}\text{U}$, for example, we could calculate similarly that the energy released would be 1.93×10^{10} kJ/mole. In other words, we can see in this mole-to-mole comparison that our energy yield from the fusion reaction is 100 times greater than the fission reaction.

2. A second advantage of the fusion reaction is the self-sustaining character of the reaction. The energy released initially as the fusion reaction begins can provide the activation energy for additional fusion reactions.

3. Another advantage is the relative abundance of light nuclei fuel. Naturally occurring deuterium in ocean water (D_2O, heavy water) is the principal source. Although only about 0.015 per cent of ordinary water is composed of heavy water molecules, the vastness of the oceans provides a practically unlimited supply. Also, the production of tritium by the neutron bombardment of lithium provides us with another light and fusable nuclide,

$$^7_3\text{Li} + {}^1_0\text{n} \rightarrow {}^3_1\text{H} + {}^4_2\text{He} + {}^1_0\text{n}$$

tritium

4. Finally, an extremely appealing advantage of fusion is that the production of dangerous nuclear waste material is reduced to insignificant proportions.

Now, we may well ask, "What are we waiting for?" The answers bring us back to the reality of technological problems that must be solved. At the heart of these problems is the necessity for extraordinarily high temperatures—somewhere in the region of 40 to 70 million degrees Celsius! Now you can see why fusion reactions are termed *thermonuclear reactions*.

The need for such high temperatures is so that nuclei capable of undergoing fusion can acquire energies sufficient to overcome the natural force of repulsion due to their like-charges. Atomic nuclei in this kind of super-hot condition must be separated necessarily from their electrons to the extent that a gaseous mixture of ions and electrons results. Such a high-temperature gaseous state of matter is called a **plasma.** The specific technological problem is how to restrict the controlled fusion reaction to the confines of an apparatus that can be used to our advantage. The most likely method of controlling the hot plasma seems to be by use of powerful electromagnets that surround a round cylindrical chamber. Ideally, then, extremely fast moving nuclei can be directed along a circular path by the magnetic field, and the extraordinary heat related to the fusion reaction can be forced to the center of the doughnut-shaped device so that the metal parts are not melted or vaporized.

14.11 PRACTICAL APPLICATION OF NUCLEAR CHEMISTRY

We have already discussed the use of radioisotopes in the dating of artifacts from ancient cultures. But, in addition, there are other applications that we should mention by way of concluding this chapter.

Radioactive isotopes can be used to analyze quantitatively molar fractions of ions in solution that are too small to be measured by other methods. Very small molar fractions (in the microgram range) are described as **trace amounts.** For example, the concentration of a trace amount of chloride ion can be determined by adding silver nitrate solution in which the silver ion has been "tagged" or "labeled" by using ^{110}Ag, which is radioactive. The silver ion causes the precipitation of the chloride ion.

$$^{110}Ag^+ + Cl^- \rightarrow {}^{110}AgCl(s)$$

(radioactive precipitate)

By measuring the radioactivity of the precipitate formed, it is possible to calculate the chloride ion concentration despite the possibility that it may be in the micromolar range.

Another analytical method involves the exposure of trace amounts of material to neutron bombardment. This converts elements to radioactive forms. The method is known as **activation analysis.** Microgram amounts of elements can be detected in this way. The value of this type of analysis is illustrated by the possible measurement of trace amounts of poisonous elements in food material. It may also be used in industry to check on contaminants, uniform dimensions of metal sheets, and corrosion.

Radioisotopes are used extensively in the medical and biochemical fields. Nuclear medicine uses radioisotopes both for diagnosis and for therapy. The use of ^{60}Co in cancer treatment and ^{131}I in the case of thyroid disease is fairly well known. In both examples, the radiation is localized so that very specific tissue (cancerous or diseased) is destroyed. Small amounts of radioactive material may be used to detect circulatory disorders as the flow of blood is followed by a counting device such as a Geiger-Müller tube.

The important field of environmental control explores the use of radiation and radioactive isotopes. The increased yield of better crops has emerged from radiation experiments performed on ungerminated

seeds. Food spoilage has been sharply retarded by irradiation. Neutron and gamma radiation, for example, are effective for the preservation (through sterilization) of food.

Animal gonads are particularly sensitive to all forms of high energy radiation, from X-rays to gamma rays. The effect of this radiation has been to cause genetic mutations by altering the nucleic acid geometry. While the results of some mutations may be hideous or worthless, others lead to the development of superior stock animals. The dangers of DDT for insect pest control are so well-established that an alternative must be found. A great possibility is the use of radiation to sterilize the insects or alter the reproductive cycle so that no live offspring result. Mankind may yet be free of disease-bearing pests without poisoning himself in the process. The Atomic Energy Commission has made, and is making, significant strides in development of useful applications of radioactivity.

Biochemical research laboratories use radioactive elements as **tracers** in an attempt to understand the complexity of biochemical reactions. The use of labeled carbon dioxide in which ^{14}C in used has increased man's understanding of the photosynthetic process of green plants. The growth of plants and the formations of nucleic acids and amino acids can be studied with tritium labeling. In this way, the separation of chromosomes in living cell nuclei can be observed, and protein formation can be measured.

Experiments designed to learn about the rates and complexity of chemical reactions occurring during the process of respiration make use of ^{32}P. Phosphorus is a necessary part of the energy-storing capability of living cells.

When ^{32}P is used as a tracer element, a complex chain of chemical reactions can be followed. When biochemical activity is stopped abruptly at controlled time intervals, the amount of ^{32}P appearing at various points in the chain of reactions can be assayed. This biochemical application of reaction kinetics has illuminated our understanding of the production energy needed to sustain life.

14.1   Explain or define the following terms:
(a) alpha particle
(b) positron
(c) rad
(d) specific activity
(e) nuclear fusion
(f) half-life
(g) cyclotron
(h) nuclide

14.2   Explain or define the following terms:
(a) beta (β^-) particle
(b) gamma radiation
(c) curie unit (Ci)
(d) lethal dose
(e) nucleon
(f) radioisotope
(g) nuclear transmutation
(h) mass decrement

14.3   Explain or define the following terms:
(a) radioactivity
(b) breeder reactor
(c) daughter nuclide
(d) tracer
(e) decay constant
(f) critical mass
(g) relative biological effectiveness (RBE)
(h) rem

14.4   Balance the following equations:
(a) $^{60}_{28}Ni + ^{1}_{0}n \rightarrow ^{60}_{27}Co + ?$
(b) $^{235}_{92}U + ^{1}_{0}n \rightarrow ^{102}_{42}Mo + ^{131}_{50}Sn + ?$
(c) $^{14}_{7}N + ? \rightarrow ^{17}_{8}O + ^{1}_{1}H$

14.5   Balance the following equations:
(a) $^{2}_{1}H + ^{3}_{1}H \rightarrow ^{4}_{2}He + ?$
(b) $^{35}_{16}S \rightarrow ? + ^{0}_{-1}e$
(c) $^{63}_{29}Cu + ^{2}_{1}H \rightarrow ^{64}_{29}Cu + ?$

14.6   Complete and balance the following equations:
(a) $^{9}_{4}Be$ gains an alpha particle in the formation of $^{12}_{6}C$.
(b) $^{14}_{6}C$ is transformed to $^{14}_{7}N$.
(c) When $^{35}_{17}Cl$ gains a proton, it proceeds to emit an alpha particle.

14.7   Complete and balance the following equations:
(a) The alpha emission of $^{17}_{8}O$.
(b) $^{10}_{6}C$ loses a positron.
(c) $^{40}_{19}K$ gains a neutron in the transformation to $^{37}_{17}Cl$.

14.8   Write the name associated with the following symbols:
(a) $^{4}_{2}He$
(b) $^{0}_{-1}e$
(c) $^{0}_{0}\gamma$
(d) λ

14.9   Write the symbol for the following particles:
(a) neutron
(b) half-life
(c) deuteron
(d) positron

14.10   Qualitatively compare the energies liberated in chemical reactions to nuclear transformations

14.11   What is the relationship of neutron-to-proton ratio and isotope stability?

14.12   What property would an element be likely to have if its atoms have a high n/p ratio?

14.13   Calculate the n/p ratio for the following:
(a) ^{52}V
(b) ^{97}Zr
(c) ^{229}Th

14.14   Compare the effects of orbital electron capture and alpha emission on the n/p ratio.

14.15   How would the n/p ratio be affected in the following cases? Explain your answer.
(a) beta emission of $^{14}_{6}C$
(b) positron emission of $^{10}_{6}C$

14.16   Predict the mode of decay for the following and explain your answer.

(a) $^{16}_{7}N$
(b) $^{108}_{49}In$
(c) $^{216}_{85}At$

14.17   Four sequential steps in a decay series starting after the alpha decay of $^{218}_{85}At$ are, in order, β, α, α, β. Write the balanced nuclear equation for each step.

14.18   Use a diagram to explain how a Geiger-Müller counter works.

14.19   Why would you expect $^{35}_{17}Cl$ and $^{37}_{17}Cl$ to have the same chemical properties?

14.20 ^{214}Po has a half-life of 164 microseconds. How useful is this isotope? Explain.

14.21   Convert 0.00350 mCi to nCi. How many disintegrations per second is this?

14.22   If $^{14}_{6}C$ in living plants has a specific activity of 15 counts/min/g, is the $^{14}_{6}C$ method reliable for dating objects 23,000 years old? Why?

14.23   A safety rule dictates that a radioactive substance should be stored for seven half-lives before discarding. How many days should ^{131}I be stored?

14.24   Calculate the age of an ancient pair of fiber sandals if the ^{14}C content of a live plant compared with the sandals is a ratio of 1.14:1.00

14.25   The decay rate of an isotope was 2440 *cpm* on July 3rd. If the rate was 2160 *cpm* on July 7th, what is the half-life?

14.26   Calculate the decay constant of ^{131}I, which has a half-life of 8.07 days.

14.27   How much of a 20.00 g sample of ^{225}Ac will be left after 3.00 days if the half-life is 10.00 days?

14.28   If a radioactive isotope emits 1555 *cpm* at 4 P.M. and 960 *cpm* at 4:30 P.M., what is the half-life?

14.29 ^{35}S has a half-life of 87.1 days. How much of a 10.00 mg sample will remain after 6 months?

14.30   Calculate the decay constant for ^{59}Fe, which has a half-life of 45.00 days.

14.31   What is the essential difference between rads and rems?

14.32   The decay constant for ^{197}Pt is 3.85×10^{-2} hr^{-1}. How much of a 20.00 mg sample remains after 9 hours?

14.33   What effect would a dose of 300 rems be likely to have on a human?

14.34   Calculate the energy released, in kJ, when a nuclear fission reaction results in the complete conversion of 100.0 g of $^{235}_{92}U$ into energy.

14.35   Find the nuclear binding energy of a mole of atoms in which the mass decrement is 0.0184 g.

14.36   Calculate the energy released, in kJ/mole, for the fusion reaction:

$$^{1}_{1}H + ^{3}_{1}H \rightarrow ^{4}_{2}He$$

The nuclear molar masses are:

$^{1}_{1}H = 1.00728$ g/mole
$^{3}_{1}H = 3.01550$ g/mole
$^{4}_{2}He = 4.00150$ g/mole

14.37   What mass decrement would be necessary in order to provide a molar nuclear binding energy of 1.93×10^{10} kJ?

14.38   What is the special danger encountered when alpha emitters are powders or gases?

14.39   Describe three practical applications of radioactivity.

ANSWERS TO NUMERICAL EXERCISES AND PROBLEMS

CHAPTER 1

Exercise 1.1

1. (a) 7.62×10^2
   (b) 5.38×10^5
   (c) 4.26×10^7
   (d) 3.8×10^{-2}
   (e) 1.12×10^{-5}

2. (a) 3.35×10^5
   (b) 7.6×10^7
   (c) 2.8×10^{-5}
   (d) 2.8

3. (a) 6.2

   (b) 7.18×10^7
   (c) 3.08×10^{-7}
   (d) 9.2×10^{-12}

Exercise 1.2

1. (a) $\dfrac{1}{x^{-a}}$ or x^a
   (b) 10^{-7}

   (c) $\dfrac{1}{3^{-1/2}}$ or $3^{1/2}$
   (d) $\dfrac{1}{2^5}$ or 2^{-5}

2. 8.85×10^5
3. 1.67×10^6

Exercise 1.3

1. (a) 10^8
   (b) 2.4×10^9
   (c) 6×10^3
   (d) 3.9×10^4
2. (a) 10^2
   (b) 5×10^{-6}

   (c) 5×10^{-5}
   (d) 5.1×10^{-1}
3. (a) x^{10}
   (b) 2.7×10^{10}
   (c) 6.25×10^{-6}
   (d) 8×10^{12}

4. (a) 6×10^2
   (b) 7×10^{-3}
   (c) 6×10^2
   (d) 2×10^3

Exercise 1.4

1. (a) 4
   (a) 2
   (c) 1
   (d) 3
2. (a) 7.1

   (b) 2.00
   (c) 1.0×10^6
   (d) 69.9
   (e) -7×10^1
3. (a) 3.08

   (b) 0.0764
   (c) 20800
   (d) 14.2

Exercise 1.5

(a) 1.53×10^{-2} m
(b) 2.7×10^{1} m
(c) 1.53×10^{-5} cm
(d) 2.41×10^{-5} cm

(e) 1.2×10^{-6} m
(f) 8.1×10^{4} nm
(g) 2.3×10^{-5} km
(h) 1.5×10^{1} nm

(i) 4.4×10^{-5} μm
(j) 3.5 mm

Exercise 1.6

(a) 2.72×10^{-2} L
(b) 8.4×10^{1} mL
(c) 1.76×10^{-1} mL
(d) 2.0×10^{4} μL

(e) 5.6×10^{1} L
(f) 2×10^{3} λ
(g) 5.4 mL

(h) 2.6×10^{-1} L
(i) 3.1×10^{1} cm^3
(j) 4.9×10^{1} μL

Exercise 1.7

(a) 3.25×10^{-1} g
(b) 1.4×10^{-3} kg
(c) 3.3×10^{3} mg
(d) 3×10^{1} μg

(e) 1.5×10^{-1} μg
(f) 4.280×10^{-3} g
(g) 9×10^{-5} g

(h) 6×10^{-1} g
(i) 2.3×10^{-5} mg
(j) 7.2×10^{4} μg

Exercise 1.8

(a) $-8.3°$C
(b) $392°$F
(c) $-14.0°$C

(d) $-364°$F
(e) $283°$K
(f) $-158°$C

(g) $243°$K
(h) $277°$K

Exercise 1.9

1. 0.643 g/mL
2. 0.19 mL

3. 4.8 cm^3
4. 1.84

5. 1.43 g/L

Questions & Problems

1.1   Smaller: 2.1×10^{-6}
      larger: 2.1×10^{1}
1.5   (a) 2.6 cm
      (b) 0.038 g
      (c) 8300 m
      (d) 7.0 mL
1.6   (a) 4 SF
      (b) 6 SF
      (c) 3 SF
      (d) 4 SF
1.7   (a) 8.4×10^{3}
      (or 8.40×10^{3}, or 8.400×10^{3})

(b) 2.5×10^{-2}
(c) 1.76×10^{8}
    (to 1.76000000×10^{8})
(d) 8.7×10^{-4}
1.8   (a) 2.67617×10^{-1}
      (b) 1.5×10^{-4}
      (c) 2.167×10^{1}
      (d) 8.0161×10^{2}
1.9   (a) 8.30×10^{1}
      (b) 1.822×10^{2}
      (c) 5.41×10^{1}
      (d) 1.36×10^{1}

1.10 (a) 9240
 (b) 7.39
 (c) 2.52×10^2
 (d) 9.12×10^{-3}
 (e) 0.00114
 (f) 88.0
1.11 (a) 4×10^8
 (b) 1.95×10^3
 (c) 8×10^{-5}
 (d) 1.39×10^3
1.12 (a) 2.082×10^{-2}
 (b) 1.206×10^2
 (c) 1.100×10^1
 (d) 2.627×10^3
1.13 (a) 2.5×10^{-3}
 (b) 2.1×10^7
 (c) 3.2×10^2
 (d) 1.00
1.14 (a) 1.3×10^2
 (b) 8.46
 (c) 1.07×10^2
 (d) 1.005×10^3
1.15 (a) 7.527×10^6 g
 (b) 2.120×10^{-3} g
 (c) 7.121×10^{-3} g
 (d) 1.0×10^{-8} g
1.16 (a) 4×10^2 g
 (b) 0.512 g
 (c) 7×10^{-2} g
 (d) 4.5×10^3 μg
 (e) 2 mg
 (f) 6.7×10^2 mg
1.17 (a) 2.32×10^2 mm
 (b) 5.851×10^5 mm
 (c) 1.676×10^{-7} mm
 (d) 9.611×10^{-2} mm
1.18 (a) 4.25×10^3 mm^3
 (b) 18 cm^3

(c) 87 mm^3
(d) 9.2 cm^3
1.19 (a) 3.51×10^{-1} km
 (b) 1.4 cm
 (c) 6.50×10^{-5} cm
 (d) 520 Å
 (e) 2×10^1 μm
 (f) 1.3×10^1 nm
1.20 (a) 1.2×10^2 mL
 (b) 3.7×10^{-2} μL
 (c) 1.85×10^{-1} L
 (d) 28 cm^3
 (e) 8×10^{-5} mL
 (f) 5 mL
1.21 5.53 g/mL
1.22 1.035 g/mL
1.23 0.879
1.24 110 g, 140 mL
1.25 4.2×10^7 erg/cal
1.26 33.4 mL
1.27 0.96 g/cm^3
1.28 924 g
1.29 50.2 mL
1.30 20 g/mL
1.32 (a) 41°F
 (b) −94°F
 (c) 71.1°C
 (d) 193°K
 (e) −43°C
 (f) 264°K
1.33 620 g
1.34 210 g
1.35 20 g
1.36 82 g Al
1.37 55 pellets
1.38 97.2 mL
1.39 6.32×10^{-5} g
1.40 0.7729 g/mL

CHAPTER 2

Exercise 2.1

1. 1.25×10^5 erg
2. 1.25×10^{-2} J
3. 1.0×10^7 cm/sec
4. 639 cal
5. 0.58 cal/g°C

Exercise 2.2

1. 4.8×10^{-4} cal
2. 4.5×10^2 erg
3. 1.4×10^3 erg
4. 7.2 J
5. 2.60×10^3 J
6. 2.1×10^6 cal

Questions & Problems

2.7	2.8×10^3 cm/sec	2.13	2160 cal	2.17	2200 cal
2.8	3×10^2 erg	2.14	38440 cal	2.18	0.209 g
2.9	710 cal	2.15	19 g	2.19	27.4°C
2.10	630 J; 6.30×10^9 erg	2.16	3.2 g	2.20	0.11 cal/g°C
2.12	2860 cal				

CHAPTER 3

Exercise 3.2

1. 28.09 amu
2. 3.4 g
3. 2.22×10^{-3} mole
4. 3.44×10^{-22} g/atom
5. 7.2×10^{21} atoms

Exercise 3.3

1. (a) 199.9 g (d) 342.1 g
   (b) 68.1 g (e) 249.6 g
   (c) 60.0 g 2. 4.5×10^{-2} g
3. 3.53×10^{-3} mole
4. 7.31×10^{-23} g/molecule
5. 2.5×10^{23} molecules

Questions & Problems

3.7	0.25 mole	3.22	3.54×10^{25} atoms	
3.8	6.300×10^{-3} mole	3.23	0.59 g	
3.9	7.3×10^{-2} g	3.24	3.4 g	
3.10	3.27×10^{-22} g	3.25	21.5 moles	
3.11	3.3 g	3.26	(a) 15.3 moles	
3.12	3.9×10^{-3} mole		(b) 5.16×10^{-3} mole	
3.13	2.1 g	3.27	43.6 moles of molecules	
3.14	6.69×10^{24} molecules		87.1 moles of atoms	
3.15	(a) 194.2 g/mole	3.28	136 g Br_2	
	(b) 158.0 g/mole	3.29	(a) 74 g/mole	
	(c) 158.2 g/mole		(b) 921 g/mole	
	(d) 380.5 g/mole	3.30	3.9 moles Ba	
	(e) 82.1 g/mole	3.31	(a) 45.6 g	
	(f) 374 g/mole		(b) 0.187 g	
3.16	2.4×10^{23} ions		(c) 0.972 g	
3.17	2.66×10^{-5} g	3.32	0.100 mole NO_3^-	
3.18	310.2 g/mole	3.33	(a) 2.00 M	
3.19'	18 moles		(b) 1.60×10^{-3} M	
3.20	3.50 moles		(c) 16.2 M	
3.21	4.33×10^{25} atoms			

Exercise 4.7

1. (a) 32.9% K; 67.1% Br
   (b) 54.1% Ca; 43.3% O;
       2.6% H
   (c) 2.2% H; 32.6% S; 65.2% O
   (d) 29.1% Na; 40.5% S; 30.4% O
   (e) 28.0% Fe; 0.8% H; 23.2% P;
       48.0% O
2. 51.20%
3. 52% In; 48% Cl

Exercise 4.8

1. (a) CH_4 (c) $Mg_2P_2O_7$ (e) $C_{12}H_{11}N_4O_2$
   (b) Fe_3O_4 (d) Fe_2O_3 2. Na_3PO_4

Questions & Problems

4.12 1. 48.2% K, 19.7% O, 14.8% C, 17.3% N   4.17 (a) $C_6H_{12}O_6$
      2. 31.0% Fe, 15.6% N, 53.4% O   (b) P_4O_{10}
      3. 72.34% Fe, 27.7% O   (c) $C_6H_3Cl_3$
      4. 17.7% N, 6.3% H, 15.2% C, 60.8% O   4.18 $C_6H_{12}O_3$
4.13 1.41 g   4.19 Cl_2O_7
4.14 As_2O_3 4.20 $C_2H_3O_2$
4.15 C_2H_6O

Exercise 5.3

1. (a) 0.062 mole   (c) 1.7 g   (c) 1.8×10^{-2} g
   (b) 3.6 g   2. (b) 0.20 mole   (d) 6.67×10^{-3} mole

Questions & Problems

5.6   1.6 mole   (b) 80.6 g
5.7   26 g   5.15   (a) 30.6 g
5.8 4.4×10^{-2} g   (b) 327 g
5.9   1.4 g   5.16   35 g
5.10 2.0 g   5.17   74 g
5.11 38.6%   5.18   474 g
5.12 60.0%   5.19   5.75 g
5.13 61.3%   5.20   0.0418 mole
5.14 (a) 16.4 g   5.21   5.12 g

5.22 100 g Cl_2
5.23 39.2 g
5.24 98 g $CaCO_3$
5.25 (a) 5.81 g Ag
      (b) 0.700 mole $AgNO_3$

5.26 (a) 0.735 mole NaOH
      (b) 0.742 g H_2
5.27 0.782 mole NO_2
5.28 2.6 g $KClO_3$/g O_2 vs.
      5.3 g $NaNO_3$/g O_2

CHAPTER 6

Exercise 6.1

1. 0.0134 mole
    (1.34×10^{-2} mole)
2. 7.2×10^{-2} L

3. 5.4×10^{22} molecules
4. 1.96 g/L
5. 44 g/mole

Exercise 6.2

1. 2.15 L
2. 136 mL

3. 3.3 atm = 2.5×10^3 torr
4. 701.4 torr

Exercise 6.3

1. 1.6 L
2. 7×10^3 mL

3. 61 mL
4. 3.0 g/L (@ STP); 67 g/mole

Exercise 6.4

1. 1.3×10^{-2} mole
2. 0.80 atm
3. 51 g/mole

4. 0.79 g
5. 0.46 g/L

Exercise 6.5

1. 0.40 L
2. 0.347 g

3. 0.611 g
4. 0.23 L

Questions & Problems

6.8   231 torr
6.11 252 mL
6.12 33 g/mole
6.13 0.73 L
6.14 65 g/mole
6.15 17 g/L
6.16 897 mL

6.17 73.8 g/mole
6.18 78.4 g/mole
6.19 699 L
6.20 2.60 g; 3.55×10^{22} molecules
6.21 290 g/mole; $C_{21}H_{42}$
6.22 CH_2; 43.2 g/mole; C_3H_6
6.23 C_4H_4

ANSWERS TO NUMERICAL EXERCISES AND PROBLEMS

6.24	486 L	6.28	0.153 L; 0.137 g
6.25	1.00 L; 1.14 L	6.29	110 torr
6.26	0.361 L	6.30	5.14 g/L; 115 amu
6.27	100°K	6.31	469 amu

CHAPTER 7

Questions & Problems

7.6   73.5 kcal/mole

CHAPTER 8

(No numerical problems)

CHAPTER 9

Exercise 9.2

1. (a) 12 g   3. 11 M
    (b) 26 g   4. 6.2×10^{21} molecules
2. 0.890 M

Exercise 9.3

1. Dilute 26.6 ml of stock solution to 250 mL
2. Dilute 8.75 ml of stock solution to 0.650 L
3. Dilute 75.0 ml of stock solution to 3.00 L
4. Dilute 29.2 ml of stock solution to 500 mL
5. Dilute 5.83 ml of stock solution to 350 mL

Exercise 9.4

1. 0.354 torr
2. 0.369 m
3. 5.7 g

Questions & Problems

9.6	1.5 g	9.15	1.0 mL
9.7	0.45 g	9.16	16.2 M
9.8	0.17 M	9.17	2.00 M
9.9	18 M	9.18	0.050 M
9.10	0.83 mL	9.19	13.0 g
9.11	3.8 mL	9.20	0.040 mole
9.12	3.2 mL	9.21	18 mL
9.13	12 g	9.22	100 mL
9.14	0.584 N	9.23	69.9 mL

9.24 572 g
9.25 4.83 L
9.26 1.24 M
9.27 (a) 18.6 M
      (b) 40.3 mL
      (c) 1.1 M
9.28 47.3 g
9.29 (a) 0.76 M, 1.52 N
      (b) 2.8×10^{-2} M, 5.6×10^{-2} N
      (c) 2.34×10^{-3} M, 7.03×10^{-3} N
9.30 0.44 M SO_4^{2-}, 0.88 M Na^+
9.31 0.0726
9.32 19.5 torr
9.33 2.88 g
9.34 10.2 degrees
9.35 $C_{12}H_{15}$
9.36 7.72 g
9.37 76.0 g/mole
9.38 0.480 g
9.39 5.58 degrees
9.40 $C_5H_{10}O_5$
9.41 $C_6H_8N_2$
9.42 0.353 m
9.43 2.07 m
9.44 8.05 g alcohol, 341.95 g H_2O
9.45 576 g
9.46 3.95 g NaCl, 4.47 g KCl
9.47 46.6 g
9.48 79.3 g/mole
9.49 100.60°C
9.50 747 g
9.51 101 g/mole

CHAPTER 10

Exercise 10.4

1. $[OH^-] = 3.1 \times 10^{-12}$ M
2. (a) pH = 0.7
    (b) pH = 1.1
    (c) pH = 2.36
    (d) pH = 8.77
    (e) pH = 10.91

3. (a) pH = 10.3
    (b) pH = 13.0
    (c) pH = 6.57
    (d) pH = 12.72
    (e) pH = 10.14

Exercise 10.5

1. 1.3 g
2. 0.77 mL
3. 0.30 g

4. 1.4 L
5. 240 mL

Exercise 10.6

1. 0.49 N
2. 0.043 N
3. 0.14 N

Questions & Problems

10.15   pH = 1.17
10.16   (a) pH = 1.40
        (b) pH = 3.89
        (c) pH = 2.49
        (d) pH = 7.22
        (e) pH = 9.65
10.17   (a) $[H^+] = 6.3 \times 10^{-5}$ M;
           $[OH^-] = 1.6 \times 10^{-10}$ M
        (b) $[H^+] = 2.0 \times 10^{-2}$ M;
           $[OH^-] = 5.0 \times 10^{-13}$ M
        (c) $[H^+] = 2.5 \times 10^{-12}$ M;
           $[OH^-] = 4.0 \times 10^{-3}$ M
        (d) $[H^+] = 5.0 \times 10^{-9}$ M;
           $[OH^-] = 2.0 \times 10^{-6}$ M
        (e) $[H^+] = 1.6 \times 10^{-1}$ M;
           $[OH^-] = 6.2 \times 10^{-14}$ M
10.18   3.1 g
10.19   6.15 mL
10.20 4.49×10^{-3} L
10.21   2.92 N
10.22   0.223 M; 0.446 N
10.23   0.0922 M

CHAPTER 11

Exercise 11.2

3. 11 M
4. 0.31 M

Exercise 11.3

1. pH = 2.44
2. Ka = 1.66×10^{-10}
3. pH = 1.9

Exercise 11.4

1. 0.75 g per 100 mL
2. $K_{sp} = 1.2 \times 10^{-14}$
3. 4.8×10^{-7} g

Questions & Problems

11.11 $K_c = 30$

11.13 13 M

11.14 17

11.15 0.14

11.16 0.61 M

11.17 Ka $= 5.7 \times 10^{-10}$

11.18 pH $= 4.6$

11.19 Ka $= 1.3 \times 10^{-3}$

11.20 0.63 g/L

11.23 $K_{sp} = 4.4 \times 10^{-27}$

11.24 2.3×10^{-5} g/100 mL

CHAPTER 12

Exercise 12.4

1. 8.18 mg
2. 42.3 mL

Exercise 12.5

1. (c) 0.12 volt
2. (a) 0.48 volt
   (b) 0.46 volt

Questions & Problems

12.21 1.3 g

12.22 0.34 L

12.23 580 mL

12.24 0.31 M

12.25 (b) 2.00 volts

12.26 (a) 1.23 volts
       (b) 1.13 volts

CHAPTER 13

(No numerical problems)

CHAPTER 14

Exercise 14.2

1. 2263 yr
2. 5.93 hr
3. 2.50 g

Exercise 14.3

1. -2.71×10^9 kJ/mole
2. 2.45×10^{10} kJ/mole
3. 8.75×10^8 kJ/g
4. 0.00940 g/mole

Questions & Problems

14.13	(a) 2.26/1	14.28	43.2 min
	(b) 2.43/1	14.29	5.38 mg
	(c) 2.54/1	14.30	0.0154 da^{-1}
14.21	3.50×10^3 nCi, 1.3×10^5 dps	14.32	14.18 mg
14.23	56.5 da	14.34	-8.99×10^{12} kJ
14.24	1074 yr	14.35	-1.65×10^9 kJ/mole
14.25	22.8 da	14.36	-1.91×10^9 kJ/mole
14.26	0.0859 da^{-1}	14.37	2.14×10^{-4} kg
14.27	16.26 g		

TABLES AND CHARTS

A table of modern valences based on commonly used ionic species, alphabetically arranged.

Cations		Anions	
		Acetate	$(CH_3COO^-$ or, $OAc^-)$
Aluminum	Al^{3+}	Arsenate	AsO_4^{3-}
Ammonium	NH_4^+	Arsenite	AsO_3^{3-}
Barium	Ba^{2+}	Bromide	Br^-
Cadmium	Cd^{2+}	Carbonate	CO_3^{2-}
Calcium	Ca^{2+}	Chlorate	ClO_3^-
Chromium (II)	Cr^{2+}	Chloride	Cl^-
Chromium (III)	Cr^{3+}	Chlorite	ClO_2^-
Cobalt (II)	Co^{2+}	Chromate	CrO_4^{2-}
Cobalt (III)		Cyanate	OCN^-
Copper (I)	Cu^+	Cyanide	CN^-
Copper (II)	Cu^{2+}	Dichromate	$Cr_2O_7^{2-}$
Hydrogen	H^+	Dihydrogen phosphate	$H_2PO_4^-$
Iron (II)	Fe^{2+}	Fluoride	F^-
Iron (III)	Fe^{3+}	Hydride	H^-
Lead (II)	Pb^{2+}	Hydrogen carbonate	HCO_3^-
Lithium	Li^+	Hydrogen phosphate	HPO_4^{2-}
Magnesium	Mg^{2+}	Hydrogen sulfate	HSO_4^-
Manganese (II)	Mn^{2+}	Hydrogen sulfite	HSO_3^-
*Mercury (I)	Hg_2^{2+}	Hypochlorite	ClO^-
Mercury (II)	Hg^{2+}	Hydroxide	OH^-
Nickel (II)	Ni^{2+}	Iodide	I^-
Nickel (III)	Ni^{3+}	Nitrate	NO_3^-
Potassium	K^+	Nitrite	NO_2^-
Silver	Ag^+	Oxalate	$C_2O_4^{2-}$
Sodium	Na^+	Oxide	O^{2-}
Strontium	Sr^{2+}	Perchlorate	ClO_4^-
Tin (II)	Sn^{2+}	Permanganate	MnO_4^-
Zinc	Zn^{2+}	†Peroxide	O_2^{2-}
		Phosphate	PO_4^{3-}
		Sulfate	SO_3^{2-}
		Sulfide	S^{2-}
		Sulfite	SO_3^{2-}
		Thiosulfate	$S_2O_3^{2-}$

*The mercury (I) ion normally occurs as a diatomic ion. Each ion of the diatomic unit has the valence, $1+$. Imagine the structural model ($Hg^+ \cdot Hg^+$).

†The peroxide ion is analogous to the mercury (I) diatomic configuration. Imagine the structural model, ($O^- \cdot O^-$).

Names and Formulas for Commonly Used Ionic Species*

Metal Ions				Nonmetal Ions					
1+		**2+**		**3+**		**2−**		**3−**	**1−**
Hydrogen	H^+	Magnesium	Mg^{2+}	Aluminum	Al^{3+}	Oxide O^{2-}	Nitride N^{3-}	Hydride H^-	
Lithium	Li^+	Calcium	Ca^{2+}			Sulfide S^{2-}	Phosphide P^{3-}	Fluoride F^-	
Sodium	Na^+	Strontium	Sr^{2+}					Chloride Cl^-	
Potassium	K^+	Barium	Ba^{2+}					Bromide Br^-	
Rubidium	Rb^+							Iodide I^-	
Cesium	Cs^+								

Transition Metal Ions and Polyatomic Ions

1+	2+	3+	2−	3−	1−
Silver Ag^+	Cadmium Cd^{2+}	Iron(III) (ferric) Fe^{3+}	Peroxide O_2^{2-}		Cyanate OCN^-
Ammonium NH_4^+	Zinc Zn^{2+}		Sulfate SO_4^{2-}	Phosphate PO_4^{3-}	Thiocyanate SCN^-
Copper(I) (cuprous) Cu^+	Mercury(I) (mercurous) Hg_2^{2+}		Carbonate CO_3^{2-}	Arsenite AsO_3^{3-}	Hydroxide OH^-
	Mercury(II) (mercuric) Hg^{2+}		Thiosulfate $S_2O_3^{2-}$	Arsenate AsO_4^{3-}	Nitrate NO_3^-
	Copper(II) (cupric) Cu^{2+}				Cyanide CN^-
					Dihydrogen phosphate $H_2PO_4^-$
	Iron(II) (ferrous) Fe^{2+}				Hydrogen carbonate or Bicarbonate HCO_3^-
	Tin(II) (stannous) Sn^{2+}		Sulfite SO_3^{2-}		
	Lead(II) (plumbous) Pb^{2+}		Oxalate $C_2O_4^{2-}$		Hydrogen sulfate or Bisulfate HSO_4^-
					Hypochlorite ClO^-
	Cobalt(II) (cobaltous) Co^{2+}		Chromate CrO_4^{2-}		Chlorite ClO_2^-
	Manganese(II) (manganous) Mn^{2+}		Dichromate $Cr_2O_7^{2-}$		Chlorate ClO_3^-
	Chromium(II) Cr^{2+}		Monohydrogen phosphate HPO_4^{2-}		Perchlorate ClO_4^-
					Permanganate MnO_4^-
					Acetate $C_2H_3O_2^-$ or OAc^-

*Note: The digit 1 is usually omitted for 1+ and 1− ions.

Vapor Pressure of Water

Temp. °C	Torr	Temp. °C	Torr
0	4.6	28	28.3
5	6.5	29	30.0
10	9.2	30	31.8
15	12.8	31	33.7
16	13.6	32	35.7
17	14.5	33	37.7
18	15.5	34	39.9
19	16.5	35	42.2
20	17.5	40	55.3
21	18.6	50	92.5
22	19.8	60	149.3
23	21.0	70	233.7
24	22.4	80	355.1
25	23.8	90	525.8
26	25.2	100	760.0
27	26.7		

Brønsted-Lowry Acid-Base Conjugate Pairs

Acid	Base
$HClO_4$	ClO_4^-
HI	I^-
H_2SO_4	HSO_4^-
HNO_3	NO_3^-
HCl	Cl^-
H_3O^+	H_2O
H_2SO_3	HSO_3^-
HSO_4^-	SO_4^{2-}
H_3PO_4	$H_2PO_4^-$
HF	F^-
HNO_2	NO_2^-
$HOAc$	OAc^-
$Al(H_2O)_6^{3+}$	$Al(H_2O)_5(OH)^{2+}$
H_2S	HS^-
HSO_3^-	SO_3^{2-}
NH_4^+	NH_3
HCN	CN^-
HCO_3^-	CO_3^{2-}
HPO_4^{2-}	PO_4^{3-}
HS^-	S^{2-}
H_2O	OH^-
CH_3OH	CH_3O^-
NH_3	NH_2^-
OH^-	O^{2-}
H_2	H^-

INCREASING STRENGTH (acid, downward)

INCREASING STRENGTH (base, downward)

Equilibrium Constants of Weak Acids at 25°C

	Name	First Ionization Reaction	K_a
INCREASINGLY WEAK ACIDS	oxalic acid	$H_2C_2O_4 \rightleftharpoons H^+ + HC_2O_4^-$	5.6×10^{-2}
	sulfurous acid	$H_2SO_3 \rightleftharpoons H^+ + HSO_3^-$	1.7×10^{-2}
	phosphoric acid	$H_3PO_4 \rightleftharpoons H^+ + H_2PO_4^-$	5.9×10^{-3}
	hydrofluoric acid	$HF \rightleftharpoons H^+ + F^-$	6.7×10^{-4}
	nitrous acid	$HNO_2 \rightleftharpoons H^+ + NO_2^-$	5.1×10^{-4}
	acetic acid	$HOAc \rightleftharpoons H^+ + OAc^-$	1.8×10^{-5}
	hydrocyanic acid	$HCN \rightleftharpoons H^+ + CN^-$	4.8×10^{-10}

Simplified Table of the Solubility of Common Salts in Water at 20°C

Anion	Cation	Solubility
acetate, chlorate, nitrate	nearly all	soluble
chloride, bromide, iodide	lead(II), silver, mercury(I)	insoluble
	all others	soluble
hydroxide	group I metals, barium, strontium	soluble
	all others*	insoluble
sulfate	mercury(I) and (II), calcium, barium, strontium, silver, lead(II)	insoluble
	all others	soluble
carbonate, phosphate, chromate	group I metals,** ammonium	soluble
	all others	insoluble
sulfide	group I metals, ammonium, magnesium, calcium, barium	soluble
	all others	insoluble

*$Ca(OH)_2$ slightly soluble.
**Li_3PO_4 insoluble.

INTERNATIONAL ATOMIC WEIGHTS

Based on the assigned relative atomic mass of $^{12}C = 12$

The following values apply to elements as they exist in materials of terrestrial origin and to certain artificial elements. When used with footnotes, they are reliable to ±1 in the last digit, or ±3 if that digit is in small type. (A value in parentheses denotes mass number of most stable known isotope. These have been added to the IUPAC table.)

	SYMBOL	ATOMIC NUMBER	ATOMIC WEIGHT		SYMBOL	ATOMIC NUMBER	ATOMIC WEIGHT		SYMBOL	ATOMIC NUMBER	ATOMIC WEIGHT
actinium	Ac	89	(227)	hahnium[h]	Ha	105	(260)	promethium	Pm	61	(147)
aluminum	Al	13	26.9815[a]	helium	He	2	4.00260[b,c]	protactinium	Pa	91	231.0359[a]
americium	Am	95	(243)	holmium	Ho	67	164.9303[a]	radium	Ra	88	226.0254[a,f,g]
antimony	Sb	51	121.75	hydrogen	H	1	1.0080[b,d]	radon	Rn	86	(222)
argon	Ar	18	39.948[b,c,d,g]	indium	In	49	114.82	rhenium	Re	75	186.2
arsenic	As	33	74.9216[a]	iodine	I	53	126.9045[a]	rhodium	Rh	45	102.9055[a]
astatine	At	85	(210)	iridium	Ir	77	192.22	rubidium	Rb	37	85.4678[c]
barium	Ba	56	137.34	iron	Fe	26	55.847	ruthenium	Ru	44	101.07
berkelium	Bk	97	(247)	krypton	Kr	36	83.80	rutherfordium[h]	Rf	104	(261)
beryllium	Be	4	9.01218[a]	kurchatovium[h]	Ku	104	(261)	samarium	Sm	62	150.4
bismuth	Bi	83	208.9806[a]	lanthanum	La	57	138.9055[b]	scandium	Sc	21	44.9559[a]
boron	B	5	10.81[c,d,e]	lawrencium	Lr	103	(257)	selenium	Se	34	78.96
bromine	Br	35	79.904[c]	lead	Pb	82	207.2[d,g]	silicon	Si	14	28.086[d]
cadmium	Cd	48	112.40	lithium	Li	3	6.941[c,d,e]	silver	Ag	47	107.868[c]
calcium	Ca	20	40.08	lutetium	Lu	71	174.97	sodium	Na	11	22.9898[a]
californium	Cf	98	(249)	magnesium	Mg	12	24.305[c]	strontium	Sr	38	87.62[g]
carbon	C	6	12.011[b,d]	manganese	Mn	25	54.9380[a]	sulfur	S	16	32.06[d]
cerium	Ce	58	140.12	mendelevium	Md	101	(256)	tantalum	Ta	73	180.9479[b]
cesium	Cs	55	132.9055[a]	mercury	Hg	80	200.59	technetium	Tc	43	98.9062[f]
chlorine	Cl	17	35.453[c]	molybdenum	Mo	42	95.94	tellurium	Te	52	127.60
chromium	Cr	24	51.996[c]	neodymium	Nd	60	144.24	terbium	Tb	65	158.9254[a]
cobalt	Co	27	58.9332[a]	neon	Ne	10	20.179[c]	thallium	Tl	81	204.37
copper	Cu	29	63.546[c,d]	neptunium	Np	93	237.0482[b]	thorium	Th	90	232.0381[a]
curium	Cm	96	(245)	nickel	Ni	28	58.71	thulium	Tm	69	168.9342[a]
dysprosium	Dy	66	162.50	niobium	Nb	41	92.9064[a]	tin	Sn	50	118.69
einsteinium	Es	99	(254)	nitrogen	N	7	14.0067[b,c]	titanium	Ti	22	47.90
erbium	Er	68	167.26	nobelium	No	102	(254)	tungsten	W	74	183.85
europium	Eu	63	151.96	osmium	Os	76	190.2	uranium	U	92	238.029[b,c,e]
fermium	Fm	100	(255)	oxygen	O	8	15.9994[b,c,d]	vanadium	V	23	50.9414[b,c]
fluorine	F	9	18.9984[a]	palladium	Pd	46	106.4	wolfram	W	74	183.85
francium	Fr	87	(223)	phosphorus	P	15	30.9738[a]	xenon	Xe	54	131.30
gadolinium	Gd	64	157.25	platinum	Pt	78	195.09	ytterbium	Yb	70	173.04
gallium	Ga	31	69.72	plutonium	Pu	94	(244)	yttrium	Y	39	88.9059[a]
germanium	Ge	32	72.59	polonium	Po	84	(210)	zinc	Zn	30	65.37
gold	Au	79	196.9665[a]	potassium	K	19	39.102	zirconium	Zr	40	91.22
hafnium	Hf	72	178.49	praseodymium	Pr	59	140.9077[a]				

Reproduced by permission of the International Union of Pure and Applied Chemistry. From *Pure and Applied Chemistry*. **21**(1), (1970).

[a] Mononuclidic element.
[b] Element with one predominant isotope (about 99 to 100% abundance).
[c] Element for which the atomic weight is based on calibrated measurements.
[d] Element for which variation in isotopic abundance in terrestrial samples limits the precision of the atomic weight given.
[e] Element for which users are cautioned against the possibility of large variations in atomic weight due to inadvertent or undisclosed artificial isotopic separation in commercially available materials.
[f] Most commonly available long-lived isotope.
[g] In some geological specimens this element has a highly anomalous isotopic composition, corresponding to an atomic weight significantly different from that given.
[h] Name and symbol not officially accepted.

TABLE OF ATOMIC WEIGHTS

(Based on Carbon-12)

	Symbol	Atomic No.	Atomic Weight		Symbol	Atomic No.	Atomic Weight
Actinium	Ac	89	227	Mendelevium	Md	101	[256]
Aluminum	Al	13	26.9815	Mercury	Hg	80	200.59
Americium	Am	95	[243]*	Molybdenum	Mo	42	95.94
Antimony	Sb	51	121.75	Neodymium	Nd	60	144.24
Argon	Ar	18	39.948	Neon	Ne	10	20.183
Arsenic	As	33	74.9216	Neptunium	Np	93	[237]
Astatine	At	85	[210]	Nickel	Ni	28	58.71
Barium	Ba	56	137.34	Niobium	Nb	41	92.906
Berkelium	Bk	97	[249]	Nitrogen	N	7	14.0067
Beryllium	Be	4	9.0122	Nobelium	No	102	[253]
Bismuth	Bi	83	208.980	Osmium	Os	76	190.2
Boron	B	5	10.811	Oxygen	O	8	15.9994
Bromine	Br	35	79.909	Palladium	Pd	46	106.4
Cadmium	Cd	48	112.40	Phosphorus	P	15	30.9738
Calcium	Ca	20	40.08	Platinum	Pt	78	195.09
Californium	Cf	98	[251]	Plutonium	Pu	94	[242]
Carbon	C	6	12.01115	Polonium	Po	84	210
Cerium	Ce	58	140.12	Potassium	K	19	39.102
Cesium	Cs	55	132.905	Praseodymium	Pr	59	140.907
Chlorine	Cl	17	35.453	Promethium	Pm	61	[145]
Chromium	Cr	24	51.996	Protactinium	Pa	91	231
Cobalt	Co	27	58.9332	Radium	Ra	88	226.05
Copper	Cu	29	63.54	Radon	Rn	86	222
Curium	Cm	96	[247]	Rhenium	Re	75	186.2
Dysprosium	Dy	66	162.50	Rhodium	Rh	45	102.905
Einsteinium	Es	99	[254]	Rubidium	Rb	37	85.47
Erbium	Er	68	167.26	Ruthenium	Ru	44	101.07
Europium	Eu	63	151.96	Samarium	Sm	62	150.35
Fermium	Fm	100	[253]	Scandium	Sc	21	44.956
Fluorine	F	9	18.9984	Selenium	Se	34	78.96
Francium	Fr	87	[223]	Silicon	Si	14	28.086
Gadolinium	Gd	64	157.25	Silver	Ag	47	107.870
Gallium	Ga	31	69.72	Sodium	Na	11	22.9898
Germanium	Ge	32	72.59	Strontium	Sr	38	87.62
Gold	Au	79	196.967	Sulfur	S	16	32.064
Hafnium	Hf	72	178.49	Tantalum	Ta	73	180.948
Hahnium	Ha	105	[260]	Technetium	Tc	43	[99]
Helium	He	2	4.0026	Tellurium	Te	52	127.60
Holmium	Ho	67	164.930	Terbium	Tb	65	158.924
Hydrogen	H	1	1.00797	Thallium	Tl	81	204.37
Indium	In	49	114.82	Thorium	Th	90	232.038
Iodine	I	53	126.9044	Thulium	Tm	69	168.934
Iridium	Ir	77	192.2	Tin	Sn	50	118.69
Iron	Fe	26	55,847	Titanium	Ti	22	47.90
Krypton	Kr	36	83.80	Tungsten	W	74	183.85
Kurchatovium	Ku	104	[257]	Uranium	U	92	238.03
Lanthanum	La	57	138.91	Vanadium	V	23	50.942
Lawrencium	Lw	103	[257]	Xenon	Xe	54	131.30
Lead	Pb	82	207.19	Ytterbium	Yb	70	173.04
Lithium	Li	3	6.939	Yttrium	Y	39	88.905
Lutetium	Lu	71	174.97	Zinc	Zn	30	65.37
Magnesium	Mg	12	24.312	Zirconium	Zr	40	91.22
Manganese	Mn	25	54.9380				

*A value given in brackets denotes the mass number of the longest-lived or best-known isotope.

PERIODIC TABLE OF THE ELEMENTS

I A

I A	II A	III B	IV B	V B	VI B	VII B	VIII B			I B	II B	III A	IV A	V A	VI A	VII A	VIII A

1 $1s^1$
H 1.0080

3 (He) $2s^1$
Li 6.941
4 (He) $2s^2$
Be 9.012

2 $1s^2$
He 4.0026

5 $2s^2\,2p^1$
B 10.81
6 (He) $2s^2\,2p^2$
C 12.011
7 (He) $2s^2\,2p^3$
N 14.007
8 (He) $2s^2\,2p^4$
O 15.999
9 (He) $2s^2\,2p^5$
F 18.998
10 (He) $2s^2\,2p^6$
Ne 20.179

11 (Ne) $3s^1$
Na 22.99
12 (Ne) $3s^2$
Mg 24.31

13 (Ne) $3s^2\,3p^1$
Al 26.98
14 (Ne) $3s^2\,3p^2$
Si 28.09
15 (Ne) $3s^2\,3p^3$
P 30.974
16 (Ne) $3s^2\,3p^4$
S 32.06
17 (Ne) $3s^2\,3p^5$
Cl 35.453
18 (Ne) $3s^2\,3p^6$
Ar 39.948

19 (Ar) $4s^1$
K 39.102
20 (Ar) $4s^2$
Ca 40.08
21 (Ar) $3d^1\,4s^2$
Sc 44.96
22 (Ar) $3d^2\,4s^2$
Ti 47.90
23 (Ar) $3d^3\,4s^2$
V 50.94
24 (Ar) $3d^5\,4s^1$
Cr 52.00
25 (Ar) $3d^5\,4s^2$
Mn 54.94
26 (Ar) $3d^6\,4s^2$
Fe 55.85
27 (Ar) $3d^7\,4s^2$
Co 58.93
28 (Ar) $3d^8\,4s^2$
Ni 58.71
29 (Ar) $3d^{10}\,4s^1$
Cu 63.55
30 (Ar) $3d^{10}\,4s^2$
Zn 65.37
31 (Ar) $3d^{10}\,4s^2\,4p^1$
Ga 69.72
32 (Ar) $3d^{10}\,4s^2\,4p^2$
Ge 72.59
33 (Ar) $3d^{10}\,4s^2\,4p^3$
As 74.92
34 (Ar) $3d^{10}\,4s^2\,4p^4$
Se 78.96
35 (Ar) $3d^{10}\,4s^2\,4p^5$
Br 79.904
36 (Ar) $3d^{10}\,4s^2\,4p^6$
Kr 83.80

37 (Kr) $5s^1$
Rb 85.47
38 (Kr) $5s^2$
Sr 87.62
39 (Kr) $4d^1\,5s^2$
Y 88.91
40 (Kr) $4d^2\,5s^2$
Zr 91.22
41 (Kr) $4d^4\,5s^1$
Nb 92.91
42 (Kr) $4d^5\,5s^1$
Mo 95.94
43 (Kr) $4d^5\,5s^2$
Tc (99)[a]
44 (Kr) $4d^7\,5s^1$
Ru 101.07
45 (Kr) $4d^8\,5s^1$
Rh 102.91
46 (Kr) $4d^{10}$
Pd 106.4
47 (Kr) $4d^{10}\,5s^1$
Ag 107.87
48 (Kr) $4d^{10}\,5s^2$
Cd 112.40
49 (Kr) $4d^{10}\,5s^2\,5p^1$
In 114.82
50 (Kr) $4d^{10}\,5s^2\,5p^2$
Sn 118.69
51 (Kr) $4d^{10}\,5s^2\,5p^3$
Sb 121.75
52 (Kr) $4d^{10}\,5s^2\,5p^4$
Te 127.60
53 (Kr) $4d^{10}\,5s^2\,5p^5$
I 126.90
54 (Kr) $4d^{10}\,5s^2\,5p^6$
Xe 131.30

55 (Xe) $6s^1$
Cs 132.91
56 (Xe) $6s^2$
Ba 137.34
57 (Xe) $5d^1\,6s^2$
La* 138.91
72 (Xe) $4f^{14}\,5d^2\,6s^2$
Hf 178.49
73 (Xe) $4f^{14}\,5d^3\,6s^2$
Ta 180.95
74 (Xe) $4f^{14}\,5d^4\,6s^2$
W 183.85
75 (Xe) $4f^{14}\,5d^5\,6s^2$
Re 186.2
76 (Xe) $4f^{14}\,5d^6\,6s^2$
Os 190.2
77 (Xe) $4f^{14}\,5d^7\,6s^2$
Ir 192.2
78 (Xe) $4f^{14}\,5d^9\,6s^1$
Pt 195.09
79 (Xe) $4f^{14}\,5d^{10}\,6s^1$
Au 196.97
80 (Xe) $4f^{14}\,5d^{10}\,6s^2$
Hg 200.59
81 (Xe) $4f^{14}\,5d^{10}\,6s^2\,6p^1$
Tl 204.37
82 (Xe) $4f^{14}\,5d^{10}\,6s^2\,6p^2$
Pb 207.2
83 (Xe) $4f^{14}\,5d^{10}\,6s^2\,6p^3$
Bi 208.98
84 (Xe) $4f^{14}\,5d^{10}\,6s^2\,6p^4$
Po (210)
85 (Xe) $4f^{14}\,5d^{10}\,6s^2\,6p^5$
At (210)
86 (Xe) $4f^{14}\,5d^{10}\,6s^2\,6p^6$
Rn (222)

87 (Rn) $7s^1$
Fr (223)
88 (Rn) $7s^2$
Ra (226)
89 (Rn) $6d^1\,7s^2$
Ac† (227)
104 (Rn) $5f^{14}\,6d^2\,7s^2$
Ku ?[b] (260)
105 (Rn) $5f^{14}\,6d^3\,7s^2$
Hab (260)

* Lanthanoid Series

58 (Xe) $4f^2\,6s^2$
Ce 140.12
59 (Xe) $4f^3\,6s^2$
Pr 140.91
60 (Xe) $4f^4\,6s^2$
Nd 144.24
61 (Xe) $4f^5\,6s^2$
Pm (147)
62 (Xe) $4f^6\,6s^2$
Sm 150.4
63 (Xe) $4f^7\,6s^2$
Eu 151.96
64 (Xe) $4f^7\,5d^1\,6s^2$
Gd 157.25
65 (Xe) $4f^7\,5d^1\,6s^2$
Tb 151.93
66 (Xe) $4f^9\,6s^2$
Dy 162.50
67 (Xe) $4f^{10}\,6s^2$
Ho 164.93
68 (Xe) $4f^{11}\,6s^2$
Er 167.26
69 (Xe) $4f^{12}\,6s^2$
Tm 168.93
70 (Xe) $4f^{13}\,6s^2$
Yb 173.04
71 (Xe) $4f^{14}\,5d^1\,6s^2$
Lu 174.97

† Actinoid Series

90 (Rn) $6d^2\,7s^2$
Th 232.04
91 (Rn) $5f^2\,6d^1\,7s^2$
Pa (231)
92 (Rn) $5f^3\,6d^1\,7s^2$
U 238.03
93 (Rn) $5f^4\,6d^1\,7s^2$
Np (237)
94 (Rn) $5f^6\,7s^2$
Pu (244)
95 (Rn) $5f^7\,7s^2$
Am (243)
96 (Rn) $5f^7\,6d^1\,7s^2$
Cm (245)
97 (Rn) $5f^8\,6d^1\,7s^2$
Bk (247)
98 (Rn) $5f^9\,7s^2$
Cf (249)
99 (Rn) $5f^{10}\,7s^2$
Es (254)
100 (Rn) $5f^{11}\,7s^2$
Fm (255)
101 (Rn) $5f^{13}\,7s^2$
Md (256)
102 (Rn) $5f^{14}\,7s^2$
No (254)
103 (Rn) $6d^1\,7s^2$
Lr (257)

In several cases, atomic weight is rounded off to four or five significant figures. See inside back cover for listing of 1969 International Atomic Weights. Electron configurations taken from *Theoretical Inorganic Chemistry* by M. Clyde Day and Joel Selbin. Reinhold Publishing Corporation, except numbers 104 and 105, which are predicted by analogy.

[a] Value in parentheses denotes mass number of most stable known isotope.

[b] Name and symbol are not officially accepted. Kurchatovium, Ku, has been proposed by Russian investigators and rutherfordium, Rf, by American investigators for element 104.

Color key: Gray symbol denotes gaseous element; red, liquid element; and black, solid element.

RELATIVE SIZES FOR ATOMS AND IONS

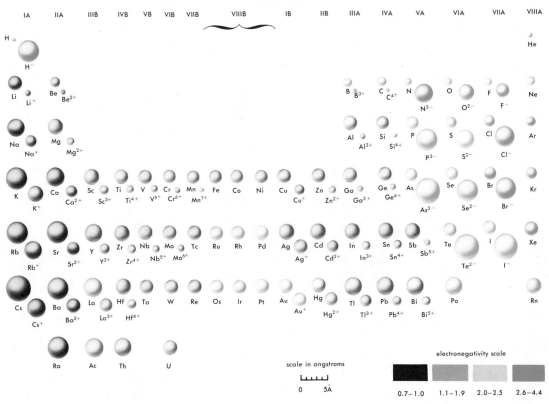

| IA | IIA | IIIB | IVB | VB | VIB | VIIB | VIIIB | IB | IIB | IIIA | IVA | VA | VIA | VIIA | VIIIA |

scale in angstroms

0   5Å

electronegativity scale

| 0.7–1.0 | 1.1–1.9 | 2.0–2.5 | 2.6–4.4 |

Table of Logarithms

	0	1	2	3	4	5	6	7	8	9
1.0	.0000	.0043	.0086	.0128	.0170	.0212	.0253	.0294	.0334	.0374
1.1	.0414	.0453	.0492	.0531	.0569	.0607	.0645	.0682	.0719	.0755
1.2	.0792	.0828	.0864	.0899	.0934	.0969	.1004	.1038	.1072	.1106
1.3	.1139	.1173	.1206	.1239	.1271	.1303	.1335	.1367	.1399	.1430
1.4	.1461	.1492	.1523	.1553	.1584	.1614	.1644	.1673	.1703	.1732
1.5	.1761	.1790	.1818	.1847	.1875	.1903	.1931	.1959	.1987	.2014
1.6	.2041	.2068	.2095	.2122	.2148	.2175	.2201	.2227	.2253	.2279
1.7	.2304	.2330	.2355	.2380	.2405	.2430	.2455	.2480	.2504	.2529
1.8	.2553	.2577	.2601	.2625	.2648	.2672	.2695	.2718	.2742	.2765
1.9	.2788	.2810	.2833	.2856	.2878	.2900	.2923	.2945	.2967	.2989
2.0	.3010	.3032	.3054	.3075	.3096	.3118	.3139	.3160	.3181	.3201
2.1	.3222	.3243	.3263	.3284	.3304	.3324	.3345	.3365	.3385	.3404
2.2	.3424	.3444	.3464	.3483	.3502	.3522	.3541	.3560	.3579	.3598
2.3	.3617	.3636	.3655	.3674	.3692	.3711	.3729	.3747	.3766	.3784
2.4	.3802	.3820	.3838	.3856	.3874	.3892	.3909	.3927	.3945	.3962
2.5	.3979	.3997	.4014	.4031	.4048	.4065	.4082	.4099	.4116	.4133
2.6	.4150	.4166	.4183	.4200	.4216	.4232	.4249	.4265	.4281	.4298
2.7	.4314	.4330	.4346	.4362	.4378	.4393	.4409	.4425	.4440	.4456
2.8	.4472	.4487	.4502	.4518	.4533	.4548	.4564	.4579	.4594	.4609
2.9	.4624	.4639	.4654	.4669	.4683	.4698	.4713	.4728	.4742	.4757
3.0	.4771	.4786	.4800	.4814	.4829	.4843	.4857	.4871	.4886	.4900
3.1	.4914	.4928	.4942	.4955	.4969	.4983	.4997	.5011	.5024	.5038
3.2	.5051	.5065	.5079	.5092	.5105	.5119	.5132	.5145	.5159	.5172
3.3	.5185	.5198	.5211	.5224	.5237	.5250	.5263	.5276	.5289	.5302
3.4	.5315	.5328	.5340	.5353	.5366	.5378	.5391	.5403	.5416	.5428
3.5	.5441	.5453	.5465	.5478	.5490	.5502	.5514	.5527	.5539	.5551
3.6	.5563	.5575	.5587	.5599	.5611	.5623	.5635	.5647	.5658	.5670
3.7	.5682	.5694	.5705	.5717	.5729	.5740	.5752	.5763	.5775	.5786
3.8	.5798	.5809	.5821	.5832	.5843	.5855	.5866	.5877	.5888	.5899
3.9	.5911	.5922	.5933	.5944	.5955	.5966	.5977	.5988	.5999	.6010
4.0	.6021	.6031	.6042	.6053	.6064	.6075	.6085	.6096	.6107	.6117
4.1	.6128	.6138	.6149	.6160	.6170	.6180	.6191	.6201	.6212	.6222
4.2	.6232	.6243	.6253	.6263	.6274	.6284	.6294	.6304	.6314	.6325
4.3	.6335	.6345	.6355	.6365	.6375	.6385	.6395	.6405	.6415	.6425
4.4	.6435	.6444	.6454	.6464	.6474	.6484	.6493	.6503	.6513	.6522
4.5	.6532	.6542	.6551	.6561	.6571	.6580	.6590	.6599	.6609	.6618
4.6	.6628	.6637	.6646	.6656	.6665	.6675	.6684	.6693	.6702	.6712
4.7	.6721	.6730	.6739	.6749	.6758	.6767	.6776	.6785	.6794	.6803
4.8	.6812	.6821	.6830	.6839	.6848	.6857	.6866	.6875	.6884	.6893
4.9	.6902	.6911	.6920	.6928	.6937	.6946	.6955	.6964	.6972	.6981
5.0	.6990	.6998	.7007	.7016	.7024	.7033	.7042	.7050	.7059	.7067
5.1	.7076	.7084	.7093	.7101	.7110	.7118	.7126	.7135	.7143	.7152
5.2	.7160	.7168	.7177	.7185	.7193	.7202	.7210	.7218	.7226	.7235
5.3	.7243	.7251	.7259	.7267	.7275	.7284	.7292	.7300	.7308	.7316
5.4	.7324	.7332	.7340	.7348	.7356	.7364	.7372	.7380	.7388	.7396
5.5	.7404	.7412	.7419	.7427	.7435	.7443	.7451	.7459	.7466	.7474
5.6	.7482	.7490	.7497	.7505	.7513	.7520	.7528	.7536	.7543	.7551
5.7	.7559	.7566	.7574	.7582	.7589	.7597	.7604	.7612	.7619	.7627
5.8	.7634	.7642	.7649	.7657	.7664	.7672	.7679	.7686	.7694	.7701
5.9	.7709	.7716	.7723	.7731	.7738	.7745	.7752	.7760	.7767	.7774

	0	1	2	3	4	5	6	7	8	9
6.0	.7782	.7789	.7796	.7803	.7810	.7818	.7825	.7832	.7839	.7846
6.1	.7853	.7860	.7868	.7875	.7882	.7889	.7896	.7903	.7910	.7917
6.2	.7924	.7931	.7938	.7945	.7952	.7959	.7966	.7973	.7980	.7987
6.3	.7993	.8000	.8007	.8014	.8021	.8028	.8035	.8041	.8048	.8055
6.4	.8062	.8069	.8075	.8082	.8089	.8096	.8102	.8109	.8116	.8122
6.5	.8129	.8136	.8142	.8149	.8156	.8162	.8169	.8176	.8182	.8189
6.6	.8195	.8202	.8209	.8215	.8222	.8228	.8235	.8241	.8248	.8254
6.7	.8261	.8267	.8274	.8280	.8287	.8293	.8299	.8306	.8312	.8319
6.8	.8325	.8331	.8338	.8344	.8351	.8357	.8363	.8370	.8376	.8382
6.9	.8388	.8395	.8401	.8407	.8414	.8420	.8426	.8432	.8439	.8445
7.0	.8451	.8457	.8463	.8470	.8476	.8482	.8488	.8494	.8500	.8506
7.1	.8513	.8519	.8525	.8531	.8537	.8543	.8549	.8555	.8561	.8567
7.2	.8573	.8579	.8585	.8591	.8597	.8603	.8609	.8615	.8621	.8627
7.3	.8633	.8639	.8645	.8651	.8657	.8663	.8669	.8675	.8681	.8686
7.4	.8692	.8698	.8704	.8710	.8716	.8722	.8727	.8733	.8739	.8745
7.5	.8751	.8756	.8762	.8768	.8774	.8779	.8785	.8791	.8797	.8802
7.6	.8808	.8814	.8820	.8825	.8831	.8837	.8842	.8848	.8854	.8859
7.7	.8865	.8871	.8876	.8882	.8887	.8893	.8899	.8904	.8910	.8915
7.8	.8921	.8927	.8932	.8938	.8943	.8949	.8954	.8960	.8965	.8971
7.9	.8976	.8982	.8987	.8993	.8998	.9004	.9009	.9015	.9020	.9026
8.0	.9031	.9036	.9042	.9047	.9053	.9058	.9063	.9069	.9074	.9079
8.1	.9085	.9090	.9096	.9101	.9106	.9112	.9117	.9122	.9128	.9133
8.2	.9138	.9143	.9149	.9154	.9159	.9165	.9170	.9175	.9180	.9186
8.3	.9191	.9196	.9201	.9206	.9212	.9217	.9222	.9227	.9232	.9238
8.4	.9243	.9248	.9253	.9258	.9263	.9269	.9274	.9279	.9284	.9289
8.5	.9294	.9299	.9304	.9309	.9315	.9320	.9325	.9330	.9335	.9340
8.6	.9345	.9350	.9355	.9360	.9365	.9370	.9375	.9380	.9385	.9390
8.7	.9395	.9400	.9405	.9410	.9415	.9420	.9425	.9430	.9435	.9440
8.8	.9445	.9450	.9455	.9460	.9465	.9469	.9474	.9479	.9484	.9489
8.9	.9494	.9499	.9504	.9509	.9513	.9518	.9523	.9528	.9533	.9538
9.0	.9542	.9547	.9552	.9557	.9562	.9566	.9571	.9576	.9581	.9586
9.1	.9590	.9595	.9600	.9605	.9609	.9614	.9619	.9624	.9628	.9633
9.2	.9638	.9643	.9647	.9652	.9657	.9661	.9666	.9671	.9675	.9680
9.3	.9685	.9689	.9694	.9699	.9703	.9708	.9713	.9717	.9722	.9727
9.4	.9731	.9736	.9741	.9745	.9750	.9754	.9759	.9763	.9768	.9773
9.5	.9777	.9782	.9786	.9791	.9795	.9800	.9805	.9809	.9814	.9818
9.6	.9823	.9827	.9832	.9836	.9841	.9845	.9850	.9854	.9859	.9863
9.7	.9868	.9872	.9877	.9881	.9886	.9890	.9894	.9899	.9903	.9908
9.8	.9912	.9917	.9921	.9926	.9930	.9934	.9939	.9943	.9948	.9952
9.9	.9956	.9961	.9965	.9969	.9974	.9978	.9983	.9987	.9991	.9996

The Electron Configurations of the Atoms of the Elements

Element	Atomic Number	1s	2s	2p	3s	3p	3d	4s	4p	4d	4f	5s
H	1	1										
He	2	2										
Li	3	2	1									
Be	4	2	2									
B	5	2	2	1								
C	6	2	2	2								
N	7	2	2	3								
O	8	2	2	4								
F	9	2	2	5								
Ne	10	2	2	6								
Na	11	Neon core			1							
Mg	12				2							
Al	13				2	1						
Si	14				2	2						
P	15				2	3						
S	16				2	4						
Cl	17				2	5						
Ar	18	2	2	6	2	6						
K	19	Argon core						1				
Ca	20							2				
Sc	21						1	2				
Ti	22						2	2				
V	23						3	2				
Cr	24						5	1				
Mn	25						5	2				
Fe	26						6	2				
Co	27						7	2				
Ni	28						8	2				
Cu	29						10	1				
Zn	30						10	2				
Ga	31						10	2	1			
Ge	32						10	2	2			
As	33						10	2	3			
Se	34						10	2	4			
Br	35						10	2	5			
Kr	36	2	2	6	2	6	10	2	6			
Rb	37	Krypton core										1
Sr	38											2
Y	39									1		2
Zr	40									2		2
Nb	41									4		1
Mo	42									5		1
Tc	43									6		1
Ru	44									7		1
Rh	45									8		1
Pd	46									10		
Ag	47									10		1
Cd	48									10		2

Element	Atomic Number	4d	4f	5s	5p	5d	5f	6s	6p	6d	7s
In	49	10		2	1						
Sn	50	10		2	2						
Sb	51	10		2	3						
Te	52	10		2	4						
I	53	10		2	5						
Xe	54	10		2	6						
Cs	55	10		2	6			1			
Ba	56	10		2	6			2			
La	57	10		2	6	1		2			
Ce	58	10	2	2	6			2			
Pr	59	10	3	2	6			2			
Nd	60	10	4	2	6			2			
Pm	61	10	5	2	6			2			
Sm	62	10	6	2	6			2			
Eu	63	10	7	2	6			2			
Gd	64	10	7	2	6	1		2			
Tb	65	10	9	2	6			2			
Dy	66	10	10	2	6			2			
Ho	67	10	11	2	6			2			
Er	68	10	12	2	6			2			
Tm	69	10	13	2	6			2			
Yb	70	10	14	2	6			2			
Lu	71	10	14	2	6	1		2			
Hf	72	10	14	2	6	2		2			
Ta	73	10	14	2	6	3		2			
W	74	10	14	2	6	4		2			
Re	75	10	14	2	6	5		2			
Os	76	10	14	2	6	6		2			
Ir	77	10	14	2	6	9					
Pt	78	10	14	2	6	9		1			
Au	79	10	14	2	6	10		1			
Hg	80	10	14	2	6	10		2			
Tl	81	10	14	2	6	10		2	1		
Pb	82	10	14	2	6	10		2	2		
Bi	83	10	14	2	6	10		2	3		
Po	84	10	14	2	6	10		2	4		
At	85	10	14	2	6	10		2	5		
Rn	86	10	14	2	6	10		2	6		
Fr	87	10	14	2	6	10		2	6		1
Ra	88	10	14	2	6	10		2	6		2
Ac	89	10	14	2	6	10		2	6	1	2
Th	90	10	14	2	6	10		2	6	2	2
Pa	91	10	14	2	6	10	2	2	6	1	2
U	92	10	14	2	6	10	3	2	6	1	2
Np	93	10	14	2	6	10	5	2	6		2
Pu	94	10	14	2	6	10	6	2	6		2
Am	95	10	14	2	6	10	7	2	6		2
Cm	96	10	14	2	6	10	7	2	6	1	2
Bk	97	10	14	2	6	10	9	2	6		2
Cf	98	10	14	2	6	10	10	2	6		2
Es	99	10	14	2	6	10	11	2	6		2
Fm	100	10	14	2	6	10	12	2	6		2
Md	101	10	14	2	6	10	13	2	6		2

(Krypton core underlies all rows.)

INDEX

References to figures are in *italic* type.
References to tables are followed by t.

Ammonia, covalent bonding of, 225
  sp^3 hybridization of, 240–241, *241*
Amphiprotic ions, 287
Amphiprotism, 286
Amphoterism, 286–288
Amplitude, of electromagnetic radiation, 173, *173*
AMU. See *Atomic mass unit.*
Analytical balance, 26, *26*
Analytical chemistry, 4
Anderson, Carl, 413
Angstrom, defined, 18
Anhydride(s), acid, 124
  basic, 123
Anion, 211
Anion nomenclature, 97t
Anode, 351, *353*
Anodic oxidation, 351
Antifreeze, 392
Antiparallel electron spin, *193,* 194
(aq), 247
Aqueous solution, 43, *43,* 78
Area, metric units of, 16–18
Aromatic hydrocarbons, 368, 369t, 384–386, 384t
Arrhenius, Svante, 280
Arrhenius theory, 280
  hydrolysis in, 288
  neutralization in, 279
Artificial radioactivity, 422
Asymmetry, bond, polarity and, 232
atm. See *Atmosphere.*
Atmosphere, 147
  force of, 137
Atmospheric pressure, standard, 147
Atom(s), Bohr, 180–185
  Dalton's solid sphere concept of, 61
  gaseous, ionization energies of, 214t
  new model of, 188–192
  nuclear model of, 64
  relative sizes of, 467t
  Rutherford model of, 171–172, *171*
  wave model of, 186–187
Atomic bomb, 437, *438*
Atomic mass, 68–69, *68*
Atomic mass unit, 65
Atomic number, 68, *68*
Atomic self-destruction, spiral of, 171–172, *172*
Atomic size, and electronegativity, 229
Atomic weights, of elements, 464t, 465t
Attometer, defined, 18
Aufbau principle, 194, 194t, *195*
Avogadro, Amadeo, 69
Avogadro's hypothesis, and gases, 142–145
Avogadro's number, 68
Azimuthal quantum number. See *Orbital quantum number.*

Background radiation, 425
Balance, 25, *26*
Balanced equations, meaning of, 116–119
Balmer, Johann, 180
Band of stability, 417, *418*
Barometer, 146
  mercury, 137
Base(s), 278, 301
  and litmus paper, 279
  and phenolphthalein, 279

Base(s) (*continued*)
  Arrhenius theory of, 280
  as proton acceptor, 281
  Brønsted-Lowry theory of, 281
  defined, 279–280
  Lewis concept of, 290
  strong, 283
  weak, 283
Basic anhydrides, 123
Becquerel, Henri, 410
Benzene, 384
Benzene ring, 384–385
Beta rays, 411, *411,* 412t, 413t
Bevatron, 422
Binary compounds, formula writing for, 91
  nomenclature of, 95–97
    summary of, 96
Biochemistry, 4
Bohr, Niels, 180
Bohr atom, 180–185
Bohr frequency rule, 182
Boiling point(s), 40
  of solvents, 274t
Boltzmann, Ludwig, 141
Bond(s), 211
  carbon, 366
  covalent, 211, 222–226, *222*
    in Lewis concept, 290
    multiple, See *Network bond.*
  hydrogen, 236–237, *236, 237*
  ionic, 218–221, 221t
  metallic, 233–235, *234*
  network, 235–236, *235, 236*
  polar vs. nonpolar, 226–227
  symmetrical vs. asymmetrical, polarity and, 232
Born, Max, 188
Boron trifluoride, sp^2 hybridization of, 241–242, *243*
Boyle, Robert, 147
Boyle's law, 146–148
  and Charles' law, 154–157
Branched chain, 366
Brønsted, Johannes, 280
Brønsted-Lowry acid-base conjugate pairs, 462t
Brønsted-Lowry theory, 280–286
  hydrolysis in, 288
Buret, 21, *22*
Butane, structure of, 370
n-Butane, *374*
Butanoic acid, 399
Butanone, 398
Butylene, isomers of, 378–379
1-Butyne, 382
Butyric acid, 399, 405

Calorie, defined, 49
Calorimeter, 50, *51*
Carbon, and organic chemistry, 365–366
  structural characteristics of, 366
Carbon atoms, number of, prefix for, 372
Carbon dioxide, *sp* hybridization of, 243t
Carbon-12, and atomic mass unit, 65
  and mole, 70
Carbon-14 dating, 430
Carbonyl compounds, 395
Carbonyl group, 395, 398

Oxalic acid, 399, 400
Oxidation, 87
  anodic, 351
  cathodic, 351
  defined, 120, 331, *332*
Oxidation half-reaction, 333
Oxidation numbers, 87, 90t, 331, *332*
Oxidation number method, 335, 336–340
  summary of, 340
Oxidation state, 87
Oxidation-reduction equations, 119
Oxidation-reduction reactions. See *Redox reactions.*
Oxidizing agent, 331, 332, 332t
Oxyacid, 98
Oxyanion, 96
Oxyanion nomenclature, 97t
Oxygen, *sp*³ hybridization of, 240–241, *241*

Palmitic acid, 400t, 401, 405t
Paraffins, 370, 377. See also *Alkanes.*
Parameters, 145, 146t
  of gases, 145, 146t
Parallel electron spin, *193,* 194
Parent nuclide, 415
Partial pressure, and Dalton's law, 148–152, *149*
Particle(s), primary, and nuclear reactions, 412, 413t
  subatomic, 63–65
Particle accelerator, 421, *421*
Pauli, Wolfgang, 192
Pauli exclusion principle, 192–194, 200
Pauling, Linus, 229
Pentane, isomers of, 375
3-Pentanone, 398
Per, 19
Percentage composition, 83
Percentage yield, 131
Perfect gas, 141
Period(s), in periodic table, 204
Periodic law, 204
Periodic table, 203–207, *205*
  "geography" of, *203*
  orbital blocks in, 206, *206*
Periodic table of the elements, 466t
Periodicity, chemical, 203–207
pH, 293–295, 293t, *295*
Phase separation, 255, *255*
  density and, 256, *256*
  polarity and, 256, *256*
Phenolphthalein, bases and, 279
  in acid-base titration, 300
Phenols, 388
Phosphors, 424
Photoelectric effect, 180, *180*
Photomultiplier tube, 425, *425*
Physical changes, in matter, 42
Physical chemistry, 4
Picometer, defined, 18
Pictorial representation, of electron distribution, 195
Pipet, 21, *22*
Planck, Max, 179
Planck's constant, 185
Plasma, 443
Polar bonds, 226
Polar covalent solutes, and water, 249–251, *250, 251*
Polarity, and immiscibility, 256, *256*
  degree of, 227–228, *228*

Polarity (*continued*)
  in multi-atom molecules, 232–233, *232, 233*
  induced, 253, *253*
  magnetic, 226, *227*
  of water molecule, 232–233, *233*
Polar molecules, 226, *226*
Polonium, 410
Polyatomic ions, names and formulas of, 461t
Polymerization, addition, 380
Polymers, 380
Polyprotic acid, 286
p orbital, 189, *190, 191,* 192t
Positron, 413, 413t
Positron emission, 412, 413
Potential(s), 350
  standard electrode, 354–357, *355,* 356t
  standard reduction, 355, 356t
Potential difference, 350
Potential energy, 45, *45, 46*
Precipitate, 43
Precision, 11, *12*
Prefix, for carbon atom number, 372
Pressure, and Boyle's law, 146–148
  and reaction rates, 307, 310
  atmospheric, 146, *146*
    standard, 147
  defined, 146
  of gas, temperature effect on, 137, *139*
  partial, and Dalton's law, 148–152, *149*
  vapor, lowering of, Raoult's law and, 270–271
    solute and, 268–269, *269*
    of water, 151, *151,* 151t, 462t
    and lowering of freezing and boiling points,
      271–272, *272*
  vs. volume, 147, *147*
Primary alcohol, 393
Principal quantum number, 183, 198, 198t, 200, 201t
  spectroscopic notation and, 198t
Principle of complementarity, 187
Principle of maximum multiplicity, 196
Prism, *177,* 178
Probability regions, electron, 189, *190,* 192t
Product(s), 42, 113
Proof, 392
Propanal, 397
Propane, structure of, 370, *370*
Propanoic acid, 399
Propanone, 395, 398
Propionaldehyde, 397
Propionic acid, 399
Propyl alcohol, isomers of, 388, 389, *389*
n-Propyl alcohol, 389, *390*
Propyne, 382
Protium, 410
Proton, 65, 409, 413t
  properties of, 67t
Proton acceptor, 281
Proton donor, 281
Proton transfer, 281, *281*

Quantitative transmutation, 432
Quantum (quanta), 179
Quantum jump, 181
Quantum number(s), 197–203
  combinations of, 201t
  magnetic, 198, 201, 201t